Foundations of Genetic Programming

Springer
Berlin
Heidelberg
New York
Barcelona
Hong Kong
London
Milan
Paris
Tokyo

William B. Langdon · Riccardo Poli

Foundations of Genetic Programming

With 117 Figures and 12 Tables

William B. Langdon

Computer Science
University College, London
Gower Street
London, WC1E 6BT
UK
W.Langdon@cs.ucl.ac.uk

Riccardo Poli

Department of Computer Science
The University of Essex
Wivenhoe Park
Colchester, CO4 3SQ
UK
rpoli@essex.ac.uk

Library of Congress Cataloging-in-Publication Data

Langdon, W.B. (William B.)
Foundations of genetic programming/William B. Langdon, Riccardo Poli.
p. cm.
Includes bibliographical references and index.
ISBN 3540424512 (alk. paper)
1. Genetic programming (Computer science) I. Poli, Riccardo, 1961-II. Title

QA 76.623.L35 2001
006.3'1-dc21

2001049394

ACM Subject Classification (1998): F.1.1, D.1.2–3, G.2.1, G.1.6, G.1.2, E.1, G.3, I.2.6, I.2.8, I.1.1–3

ISBN 3-540-42451-2 Springer-Verlag Berlin Heidelberg New York

Springer-Verlag Berlin Heidelberg New York
is a member of BertelsmannSpringer Science+Business Media GmbH
http://www.springer.de

Printed in Germany

Typesetting: Camera-ready by the authors
Cover Design: design & production, Heidelberg
Printed on acid-free paper SPIN 10848206 – 45/3142SR – 5 4 3 2 1 0

To

Caterina and Ludovico R.P.

Preface

Genetic programming (GP) has been highly successful as a technique for getting computers to automatically solve problems without having to tell them explicitly how to do it. Since its inception more than ten years ago genetic programming has been used to solve practical problems but along with this engineering approach there has been interest in how and why it works. This book consolidates this theoretical work.

One of the goals of any theoretical work is to better understand the subject. This is useful in its own right and as an aid to designing improvements. We will describe several new genetic operators that arose naturally from theoretical work and suggest modest changes to the way existing GP systems could be used on specific problems to yield improved performance. No doubt these operators and suggestions will be of direct practical interest, even to those who are not interested in "theory" for its own sake.

Genetic programming is one of a wide range of evolutionary computation techniques, such as evolutionary strategies and evolutionary programming, being itself a descendent of one of the oldest, Genetic Algorithms (GAs). It is nice to be able to report in this book that theoretical results from the "new boy", GP, can be directly applied to GAs. Since GP is more expressive than GAs, it can be viewed as a generalisation of GAs. In the same way, GP theory is a generalisation of GA theory, although, in fact, some recent advances in GP theory came first and the corresponding GA theory was derived by specialising the more general GP theory. In effect we are getting GA theory for free, from the GP theory. In this way the various strands of evolutionary computation theory are themselves coming together (although convergence is some way off).

The title of our book has, itself, a genetic pedigree. Its direct ancestor is a workshop of the same name held at the first Genetic and Evolutionary Computation Conference [Banzhaf *et al.*, 1999], which we organised (together with Una-May O'Reilly, Justinian Rosca and Thomas Haynes) in July 1999, in Orlando. Prior to this (starting in 1990) there has been a long-running series of workshops called Foundations of Genetic Algorithms (FOGA). More generally, the inspiration for "Foundations of Genetic Programming" came from a panel called "The next frontiers of AI: the role of foundations", held at EPIA 1995 [Pinto-Ferreira and Mamede, 1995]. On that occasion Riccardo put forward the view that the foundations of Artificial Intelligence (AI) are fundamental principles which are common to all disciplines within AI, be they artificial neural networks, evolutionary computation, theorem proving, etc. (see figures on the next page). The common feature of these techniques is search (although the representation being used to express solutions and the search used may be radically different). In our opinion search (be it deterministic or stochastic, complete or incomplete, blind, partially sighted, heuristic, etc.), the related representation, operators and objective functions are the foundations of AI. So Foundations of Genetic Programming should not be viewed only as a collection of techniques that one needs to know in order to be able to do GP well but also as a first attempt to chart and explore the mechanisms and fundamental principles behind genetic programming as a search algorithm. In writing this book we hoped to cast a tiny bit of light onto the theoretical foundations of Artificial Intelligence as a whole.

Acknowledgements

We would like to thank Andy Singleton, Trevor Fenner, Tom Westerdale, Paul Vitanyi, Peter Nordin, Wolfgang Banzhaf, Nic McPhee, David Fogel, Tom Haynes, Sidney R. Maxwell III, Peter Angeline, Astro Teller, Rafael Bordini, Lee Spector, Lee Altenberg, Jon Rowe, Julian Miller, Xin Yao, Kevin P. Lucas, Martijn Bot, Robert Burbidge, the people of the Chair of Systems Analysis (University of Dortmund), the Centrum voor Wiskunde en Informatica, Amsterdam, University College, London, and the members of the EEBIC group at the University of Birmingham.

We would also like to thank Axel Grossmann, Aaron Sloman, Stefano Cagnoni, Jun He, John Woodward, Vj Varma, Tim Kovacs, Marcos Quintana-Hernandez and Peter Coxhead for their useful comments on drafts of the book.

Finally, we would like to thank numerous anonymous referees of our work over several years for particularly helpful comments and suggestions.

October 2001

W.B. Langdon
Riccardo Poli

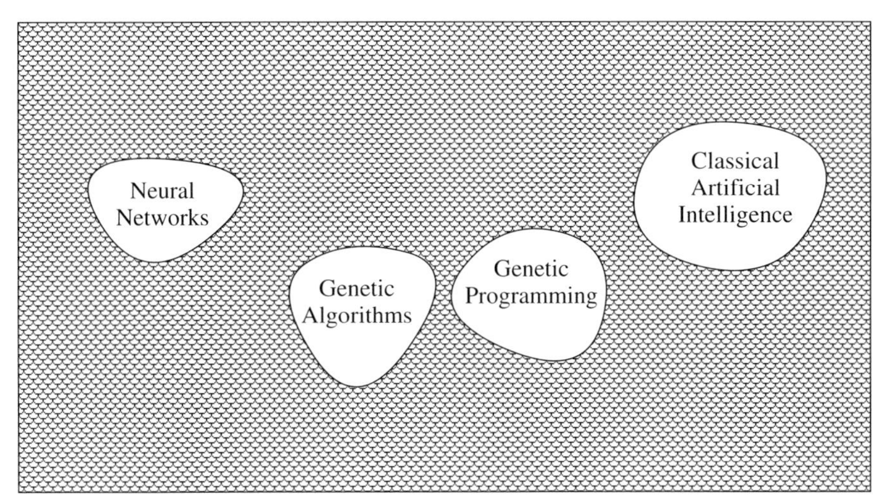

Artificial Intelligence can be seen as a cluster of islands in the sea.

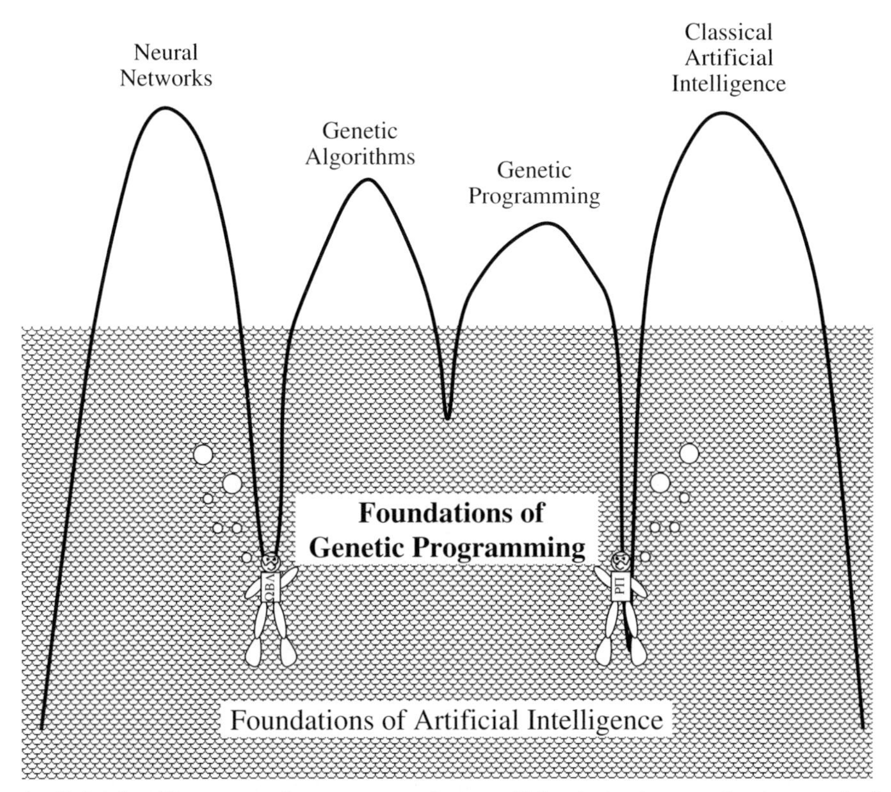

Artificial Intelligence can be seen as a cluster of islands in the sea *sharing a set of common foundations* (cross-sectional view).

Contents

1. Introduction

Living creatures are divided into species. Over time (which may be comparatively short or very long) species change, i.e. they evolve. There are several essential components to natural evolution. Individuals (e.g. animals or plants) within a species population are different. As a result, some live longer and are more likely to have children that survive to adulthood than others (natural selection). These animals (or plants etc.) are said to be fitter (cf. "survival of the fittest"). Sometimes the variation has a genetic component, i.e. the children may *inherit* it from their parents. Consequently after a number of generations the proportion of individuals within the species with this favourable inheritable characteristic tends to increase. That is, over time the species as a whole changes or evolves.

Each animal or plant grows from an egg or seed to a full grown adult. Its development is controlled in part by its genes. These are the genetic blueprints which specify this development process and so the final form of the individual. Of course other factors (often simply called "the environment") or just plain luck have a large impact on the animal or plant. The relative importance of genes, which each of us inherits from our parents, and "the environment" remains deeply controversial.

Even today in many species each adult produces children whose genes come only from that adult. For example bulbs not only have seeds (sex) but also form daughter bulbs which grow directly from the parent's root. They are initially physically part of the same plant but may eventually split from it. Except for random copying errors (mutations) the new plant is genetically identical to the original.

Many of the species we are used to seeing use sex to produce children. Sex means the genes in each child are a combination of genes from its two (or more) parents. The process whereby the genes of both parents are selected, manipulated and joined to produce a new set of genes for the child is called *crossover*. (The term recombination is also occasionally used).

Why Nature invented sex is by no means clear [Ridley, 1993] but notice that it has several important properties. First children are genetically different from both their parents. Second, if individuals in the population contain different sets of genes associated with high-fitness, a child of two high fitness parents may inherit both sets of genes. Thus sex provides a mechanism whereby new beneficial combinations of existing genes can occur in a single individual plant or animal. Since this individual may itself have children, the new combination of genes can in principle spread through the population.

The process of copying genes is noisy. While Nature has evolved mechanisms to limit copy errors they do occur. These are known as *mutations*. If errors occur when copying genes in cells which are used to create children, then the mutations will be passed to the children. It is believed that most mutations are harmful but sometimes changes occur that increase the fitness of the individual. By passing these to the children the improvement may over successive generations spread through the population as a whole.

Notice that although there are no explicitly given commands or even goals, natural evolution has over time produced a great many novel and sophisticated solutions to everyday (in fact life or death) problems. The idea that blind evolution can do this has proved very seductive.

Computer scientists and engineers have been exploiting the idea of natural evolution within their computers (artificial evolution) for many years. Initially the theoretical foundations of automatic problem solving using evolutionary computation were quite weak, but in recent years the theory has advanced considerably. This book concentrates on the theoretical foundations of one such technique, genetic programming (GP) [Koza, 1992, Banzhaf *et al.*, 1998a], which uses artificial evolution within the computer to automatically create programs. However genetic programming is a generalisation of older artificial evolutionary techniques called genetic algorithms (GAs) and so many of our theoretical results actually also cover results for GAs as special cases. Therefore these advances in GP theory can also be applied to GAs. This is part of a trend towards the advancement and coming together of the theory behind the various strands of evolutionary computing.

We will defer more details of genetic programming until Section 1.2, and Section 1.3 will give an outline of the rest of this book, but first we will consider evolution (either in Nature or in the computer) as a search process.

1.1 Problem Solving as Search

Genetic programming has been applied successfully to a large number of difficult problems such as automatic design [Koza *et al.*, 1999], pattern recognition [Tackett, 1993], data mining [Wong and Leung, 2000], robotic control [Banzhaf *et al.*, 1997], synthesis of artificial neural architectures (neural networks) [Gruau, 1994a, Gruau, 1994b], bioinformatics [Handley, 1995], generating models to fit data [Whigham and Crapper, 1999], music [Spector and Alpern, 1994] and picture generation [Gritz and Hahn, 1997]. (See also Appendix A).

Throughout this book we will treat evolution as a search process [Poli and Logan, 1996]. That is, we will consider all possible animals, plants, computer programs, etc. as our search space, and view evolution as trying a few of these possibilities, deciding how "fit" they are and then using this information to decide which others to try next. Evolution keeps on doing this until it finds an individual which solves the problem. I.e. has a high enough "fitness". In practice search spaces are huge or even infinite and evolution (even Natural evolution with a whole planet's resources) cannot try all possibilities.

Any automatic program generation technique can be viewed as searching for any program amongst all possible programs which solve our problem. Naturally there are an enormous number of possible programs. (Given certain restrictions we can calculate how many there are; see page 114). As well as being huge, program search spaces are generally not particularly benign. They are usually assumed to be neither continuous nor differentiable, and so classical numerical optimisation techniques are incapable of solving our problems. Instead heuristic search techniques, principally stochastic search techniques, like GP, have been used. In this and the following chapters we shall cast a little light on the search space that GP inhabits and how GP moves in it.

Often genetic algorithm and GP search is like that shown in Figure 1.1. The dark-gray area represents the search space. Each dot (circular, square or triangular) is an individual. Different symbols are used to represent individuals of different generations in a population made of five individuals: circles represent the initial population (Gen 0), squares represent generation 1 and triangles represent generation 2. The initial population is scattered randomly throughout the search space. Some of its members (such as the circle at the top left of the figure) are better than others and so they are selected to have children (produced using crossover) more often. Other individuals are selected less frequently or not at all (such as the circle at the bottom of the figure). The creation of the generation 1 program at the left of Figure 1.1 (grey square) is caused by the mixing of the genes (by crossover) of the two initial programs at the left of Figure 1.1. This crossover event is represented by the arrows leaving the two parent programs and pointing to their child program. Very often crossover will produce offspring that share some of the characteristics of the parents and can be somehow considered as being "halfway" between the parents in our idealised example. As a result of selecting fitter individuals and of crossover producing children between their parents, the new generation will be confined to a smaller region of the search space. For example, the individuals at generation 1 (squares) are all within the light-gray area, while the individuals at generation 2 (triangles) will be confined to the white area.

One important question that the theory of evolutionary algorithms wants to answer is: "How can we characterise, study and predict this search?" According to the different approaches used to try to answer this question we can identify a number of major strands in the theoretical foundations of evolutionary algorithms.

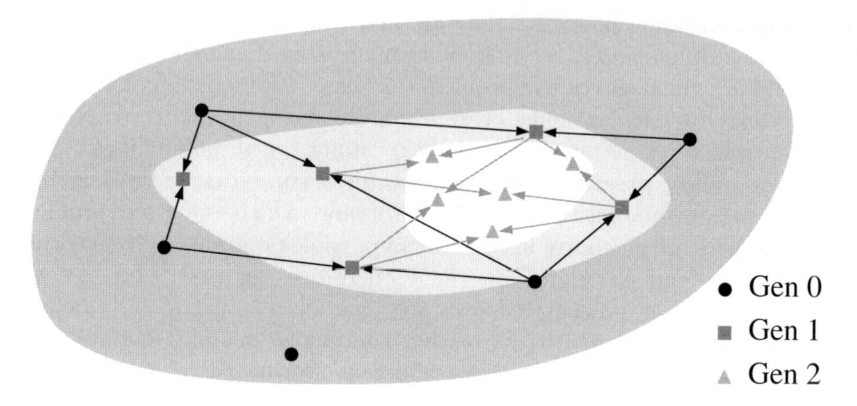

Fig. 1.1. Typical behaviour of evolutionary algorithms involving selection and crossover

1.1.1 Microscopic Dynamical System Models

These approaches are typified by Markov chain analysis which models the fine details of the individuals within the artificial population and how the genetic operations probabilistically change them over time [Nix and Vose, 1992, Davis and Principe, 1993, Rudolph, 1994, Rudolph, 1997c, Rudolph, 1997a, Rudolph, 1997b, Vose, 1999, Kellel *et al.*, 2001]. In these models the whole population is represented as a point in a multidimensional space, and we study the trajectory of this point to determine attractors, orbits, etc. The idea is illustrated in Figure 1.2. Typically these are exact models but with a huge number of parameters. Such models have been studied extensively in the genetic algorithm literature, but it is only very recently that Markov chain models for GP have been proposed [Poli *et al.*, 2001], and so, they are not discussed in this book.

1.1.2 Fitness Landscapes

In its simplest form a fitness landscape can be seen as a plot where each point in the horizontal direction represents all the genes in an individual (known as its genotype) corresponding to that point. The fitness of that individual is plotted as the height. If the genotypes can be visualised in two dimensions, the plot can be seen as a three-dimensional map, which may contain hills and valleys. Large regions of low fitness can be considered as swamps, while large regions of similar high fitness may be thought of as plateaus. Search techniques can be likened to explorers who seek the highest point. This summit corresponds to the highest-fitness genotype and so to the optimal solution (cf. Figure 1.3).

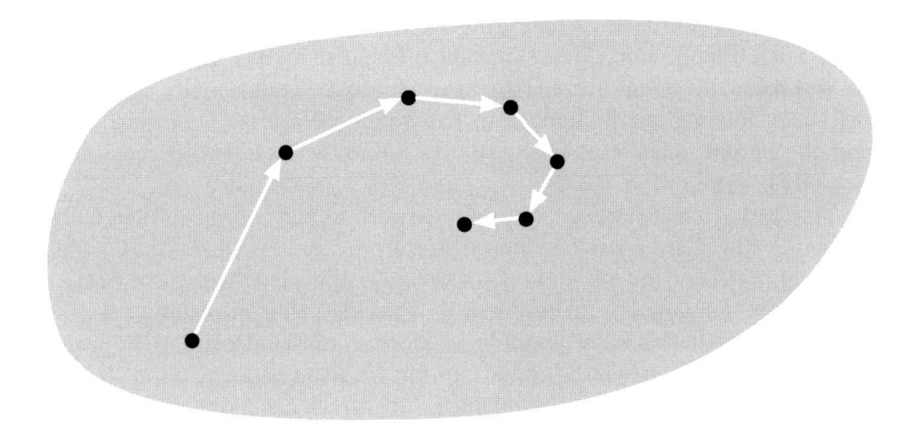

Fig. 1.2. Dynamical system model of genetic algorithms. The grey area represents the set of all possible states the population can be in. For a simple GA there are $\binom{2^N+M-1}{2^N-1}$ states (N, number of binary genes; M, size of population). The population starts in one state (represented by the lower left dot) and each generation moves (cf. white arrows) to the next.

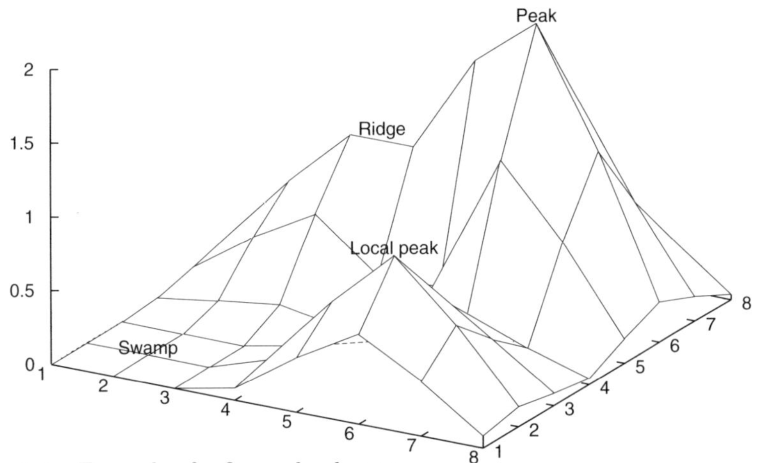

Fig. 1.3. Example of a fitness landscape

Greedy search techniques which only consider the local neighbourhood can be likened to short-sighted explorers, who just climb up the slope they are currently on. This is often a good strategy which leads to the top of the local hill, which is a local optimum. However the problem landscape may be structured so that this is not the global optimum. The local landscape may be flat (so there is no local gradient to guide the explorer) or it may lead the explorer via an unnecessarily long path to the local peak.

We can think of GAs as operating via a population of explorers scattered across such a landscape. Those that have found relatively high fitness points are rewarded by being allowed to have children. Traditionally mutation is seen as producing small changes to the genotype and so exploring a point near the parent, while crossover explores points between two parents and so may be far from either.

Notice that the fitness landscape incorporates the concept of neighbours, i.e. points that can be reached in one step from where we are now. Changing the neighbourhood changes the fitness landscape. This can have profound implications for how easy a problem is. Provided that all fitness values are different, it is theoretically possible to reorder any landscape so a local hill climber will quickly reach the global summit from any starting point. However in general it is not practical to find such a transformation due to the vast size of practical search spaces and consequently even larger number of different transformations.

Chapter 2 describes fitness landscapes in more detail. Chapter 9 maps the fitness landscape of the Santa Fe ant trail problem, while Section 8.3.1 shows the parity problem's fitness landscape is particularly awful.

1.1.3 Component Analysis

Component analysis concentrates on how genes or combinations of genes spread. Such combinations can be thought of as replicating themselves in their own right. These replicants are components of the higher level replicants represented by individual animals (or plants etc.) which reproduce and have children (who are also animals, plants etc.). [Dawkins, 1976] calls this idea the "selfish gene".

The component analysis approach focuses on the propagation of subcomponents of individuals. In genetic algorithms the individuals are represented in the computer by strings of bits. Component analysis considers what happens inside the strings and looks at single bits or groups of bits. For example, suppose the population contains just two bit strings: 0000 and 1111. Figure 1.4 shows the effects of crossing over the individuals 0000 and 1111 at their middle point to produce individuals 0011 and 1100. The component approach looks not at the effect on the individuals but at the effect on bit patterns. For example, before crossover the population contains three instances of the bit pattern 00, as shown in the upper part of Figure 1.4. (The bit string 0000 contains the bit pattern 00 in three different ways, i.e. using bits 1 and 2, 2 and 3 or 3 and 4). After crossover the population contains two instances of bit pattern 00. The deletion of one instance is represented by the cross in the lower part of Figure 1.4. The same crossover operation creates some other components. For example, the bit patterns 10 and 01. These are represented by the white dots in the lower part of Figure 1.4.

In genetic programming subcomponents could be single primitives, entire expressions, or even groups of expressions. For example, in GP we can con-

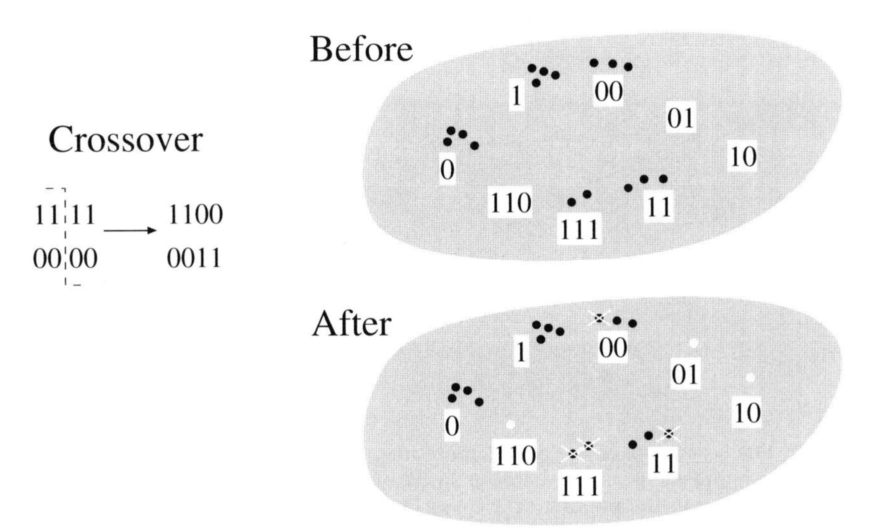

Fig. 1.4. Some of the effects of a crossover on the components of the bit strings 1111 and 0000 (left). The dots represent cases of the corresponding bit string components. After crossover (lower right) the white dots indicate instances of matching strings which did not exist before (i.e. schema creation). The crossed out dots indicate strings which no longer match (i.e. schema disruption) and black dots those that are unchanged.

sider how many copies of particular program components there are in total in the population and how this evolves. For example Chapter 10 considers how many "+" functions there are in the population within certain levels of the programs and analyses if they increase or decrease and even if they become extinct. [Langdon, 1998c, Chapter 8] gives a similar analysis for a different problem.

Chapter 3 describes various component analysis approaches in more detail.

1.1.4 Schema Theories

An alternative approach to understanding how genetic algorithms and genetic programming search, is based on the idea of dividing the search space into subspaces (called *schemata*[1]). The schemata group together all the points in the search space that have certain macroscopic quantities in common. For example all the programs which start with an IF statement could be grouped together in one schema.

Traditionally in genetic algorithms each gene is represented by a single bit. The complete set of genes is represented by a string of such bits. So,

[1] The word "schemata" is the plural for "schema".

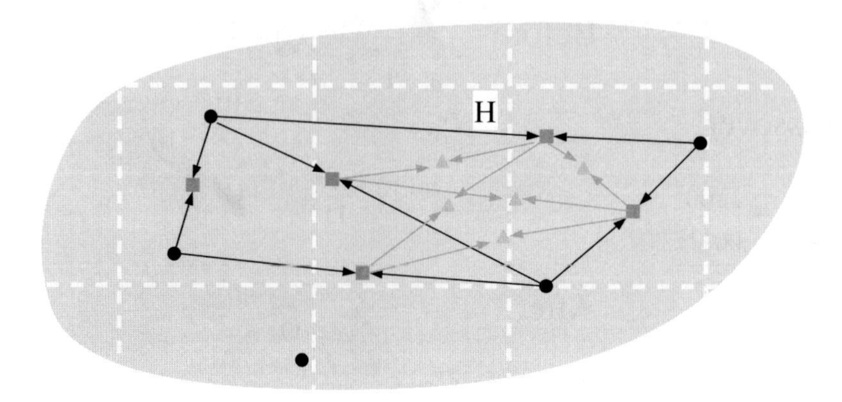

Fig. 1.5. The search space is divided into subspaces (schemata), shown as large squares. The populations are the same as in Figure 1.1 but instead of keeping track of every individual in the population we consider only how much of the population is in each schema.

as another example, we can group together every string with a particular pattern of genes in one schema. For example our schema could be everyone who has the allele 0 in locus 4 and the allele 1 in loci 5 and 6, i.e. all the strings with pattern 011 three bits from the left end. These form a subset of the whole search space (actually one-eighth of it).

Using such subsets it is possible to study how and why the individuals in the population move from one subspace to another (*schema theorems*). This idea is depicted in Figure 1.5. In the figure schemata are represented as squares. If we focus our attention on schema H, a reasonable way of considering the search might be to look at the number of individuals within H at generation t, $m(H,t)$. This number varies over time. In the initial generation $m(H,0)$ is 0. In the first generation $m(H,1)$ is 2, while $m(H,2) = 3$, etc. A *schema theorem* would model mathematically how and why $m(H,t)$ varies from one generation to the next. Chapters 3–6 will discuss the schema theories at length.

The concept of schema is very general and can be used to characterise the search space itself as well as the motion of the population in the search space. For example, as we will show in Chapter 7, one can divide the search space into sets (schemata) of programs with the same number of instructions, programs with the same behaviour or having the same fitness.

1.1.5 No Free Lunch Theorems

The no free lunch theorems (often abbreviated to NFL) are a series of results that prove that no search algorithm is superior to any other algorithm on average across all possible problems. A consequence of this is that if an algorithm, say genetic programming, performs better than random search on a class of problems, that same algorithm will perform worse than random search on a different class of problems [Wolpert and Macready, 1997]. From this one might draw the erroneous conclusion that there is no point in trying to find "better" algorithms. However, since typically we are not interested in all possible problems, this is not the case. A search algorithm can do better than random search provided that the search bias explicitly or implicitly embedded in such an algorithm matches the regularities present in the class of problems of interest.

To date no specialised result for GP has been presented, and this topic is not discussed further in the book.

1.2 What is Genetic Programming?

Figure 1.6 shows the essential iterative nature of evolutionary algorithms. In evolution in the computer, like in Nature, there is a population of individuals. The fortunate or fitter ones survive and produce children. These join the population and older individuals die and are removed from it. As in Nature, there are two ways this can happen. The whole population may be replaced by its children (the generational model). For example, in annual plants or animal species where only the eggs survive the winter, the whole population is renewed once a year. Alternatively new children are produced more or less continuously (the steady state model). In the computer, the population is usually of fixed size and each new child replaces an existing member of the population. The individual that dies may be: chosen from the relatively unfit individuals (e.g. the worst) in the population, chosen at random or selected from the child's parents.

In artificial evolution most of the computer resources are devoted to deciding which individuals will have children. This usually requires assigning a fitness value to every new individual. In genetic programming the individuals in the population are computer programs. Their fitness is usually calculated by running them one or more times with a variety of inputs (known as the test set) and seeing how close the program's output(s) are to some desired output specified by the user. Other means of assigning fitness such as coevolution are also used (e.g. [Juille and Pollack, 1996]).

In genetic programming the individual programs are usually written as syntax trees (see Figures 1.6–1.11) or as corresponding expressions in prefix notation. New programs are produced either by mutation or crossover. There are now many mutation and crossover operators: [Langdon, 1998c, pages 34–36] describes many of the mutation and crossover operators used in GP. Mutation operators make a random change to (a copy of) the parent program. For example, point mutation replaces a node within the tree by another chosen at random which has the same arity (i.e. number of arguments).

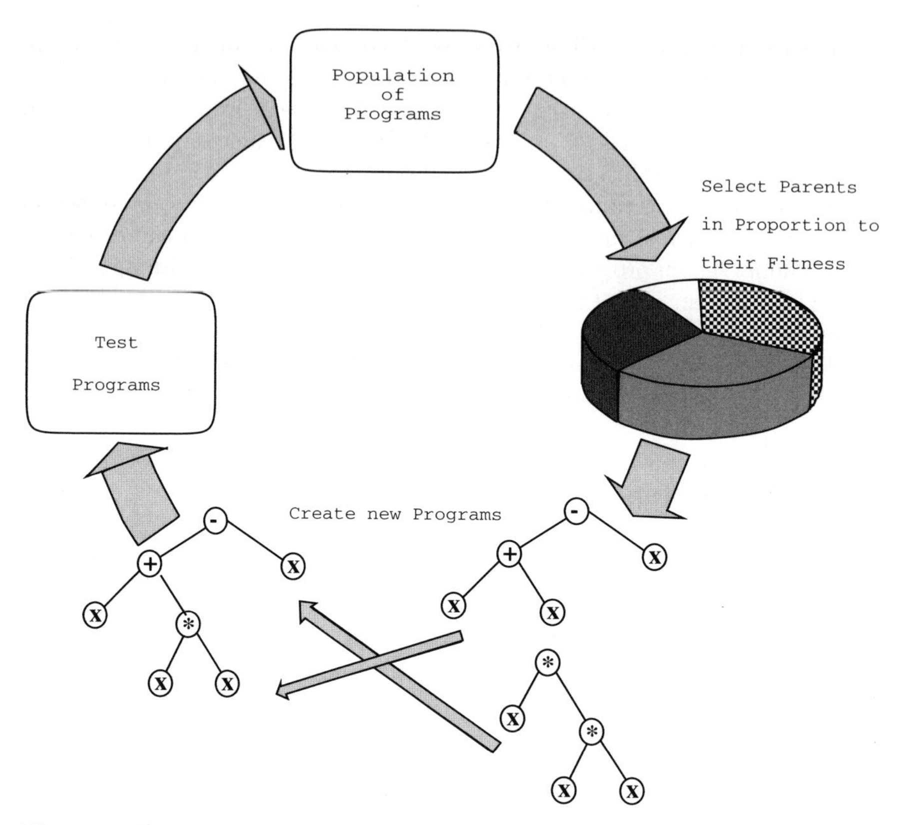

Fig. 1.6. Genetic programming cycle. The GP cycle is like every evolutionary process. New individuals (in GP's case, new programs) are created. They are tested. The fitter ones in the population succeed in creating children of their own. Unfit ones die and are removed from the population.

Thus a leaf (which occurs at the end of a branch, and has arity 0) is replaced by another leaf chosen at random and a binary function (e.g. $+$) is replaced by another binary function (e.g. $\times$).

Crossover works by removing code from (a copy of) one parent and inserting it into (a copy of) the other. In [Koza, 1992] subtree crossover removes one branch from one tree and inserts it into another. This simple process ensures that the new program is syntactically valid (see Figure 1.7).

1.2.1 Tree-based Genetic Programming

As a very simple example of genetic programming, suppose we wish a genetic program to calculate $y = x^2$. Our population of programs might contain a program that calculates $y = 2x - x$ (see Figure 1.8) and another that calculates $y = \frac{x}{\frac{x}{x-x^3}} - x$ (Figure 1.9). Both are selected from the population

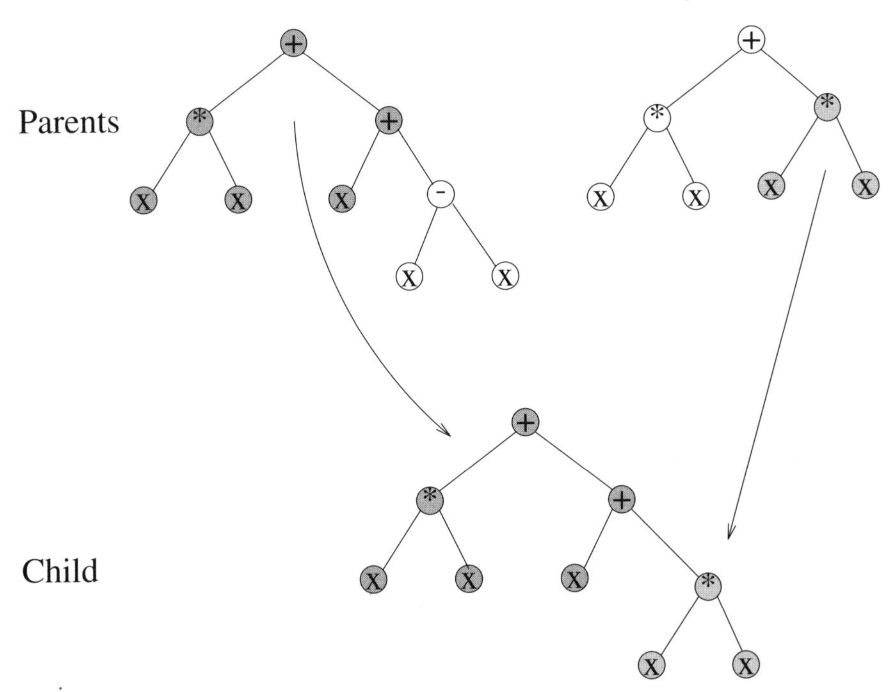

Fig. 1.7. Genetic programming subtree crossover: $x^2 + (x + (x - x))$ crossed with $2x^2$ to produce $2x^2 + x$. The right hand subtree $(\times\, x\, x)$ is copied from the right hand parent and inserted into a copy of the left hand parent, replacing the subtree $(-\, x\, x)$ to yield the child program $(+\,(\times\, x\, x)\,(+\, x\,(\times\, x\, x)))$.

because they produce answers similar to $y = x^2$ (Figure 1.11), i.e. they are of high fitness. When a selected branch (shown shaded) is moved from the father program and inserted into the mother (displacing the existing branch, also shown shaded) a new program is produced which may have even higher fitness. In this case the resulting program (Figure 1.10) actually calculates $y = x^2$ and so this program is the output of our GP. The C code for this example is available via anonymous ftp from `ftp://ftp.cs.bham.ac.uk/pub/authors/W.B.Langdon/gp-code/simple`

1.2.2 Modular and Multiple Tree Genetic Programming

Many variations on Koza's tree based genetic programming have been proposed. Of these perhaps Koza's own Automatically Defined Functions (ADFs) [Koza, 1994] are the most widely used. With ADFs the program is split into a main program tree and one or more separate trees which take arguments and can be called by the main program or each other. Evolution is free to decide how, if at all, the main program will use the ADFs as well as what they do. In more recent work even the number of ADFs and their parameters can be left to GP to decide [Koza *et al.*, 1999].

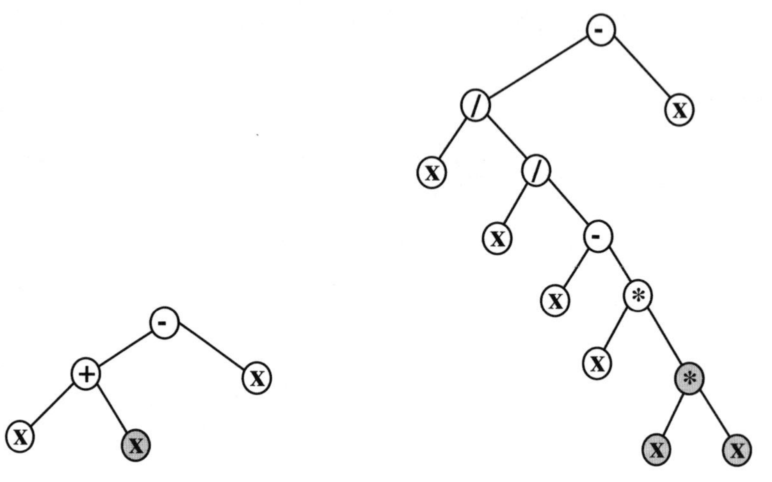

Fig. 1.8. Mum, fitness .64286, $2x - x$ **Fig. 1.9.** Dad, fitness .70588, $\frac{x}{\frac{x}{x-x^3}} - x$

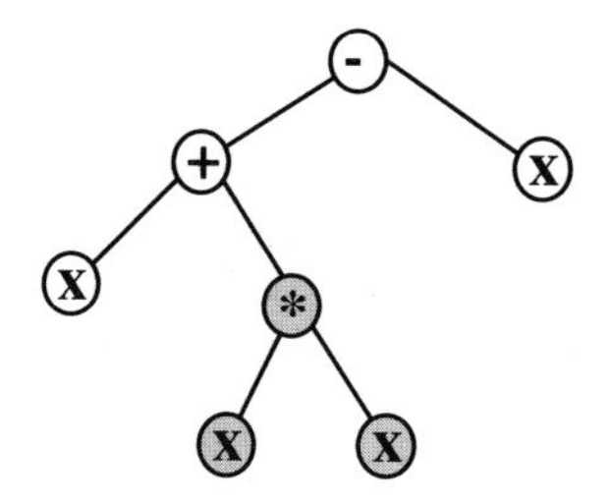

Fig. 1.10. Correct program, fitness 1.0, $x + x^2 - x$

Angeline's genetic library [Angeline, 1994] and Rosca's Adaptive Representation with Learning (ARL) [Rosca and Ballard, 1996] are other extensions to GP. In these, evolved code fragments are automatically extracted from successful program trees and are frozen and held in a library. They are then available to the evolving code via library calls to new primitives in the program tree. Sometimes code can be extracted from the library and returned to the program tree where it can continue to evolve.

Another approach, popular with multi-agent systems, is to have multiple trees, each of which has a defined purpose [Langdon, 1998c]. This can be combined with the ADF approach. So each individual may contain code for a number of agents or purposes. These may call other trees as functions (i.e. as ADFs). The ADFs may be specific to each purpose or shared by the agents or purposes.

Several authors have taken the tree program concept further. Broadly instead of the tree being the program it becomes a program to create another

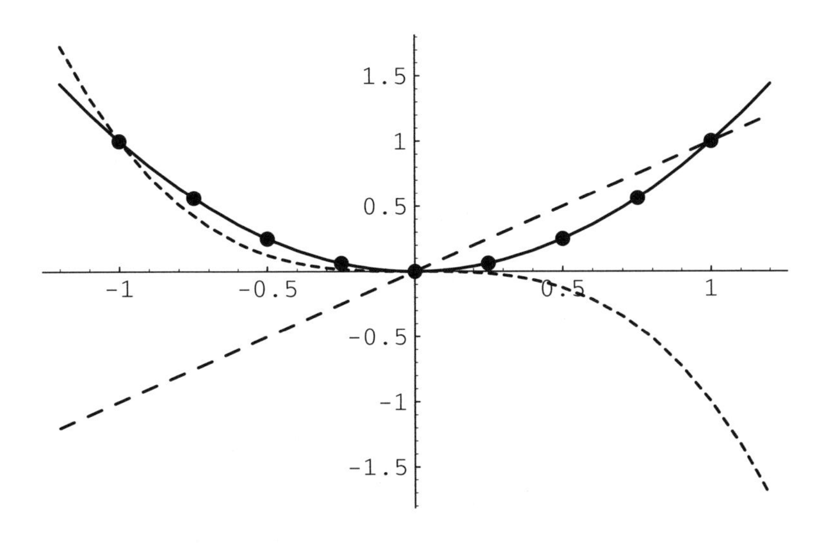

Fig. 1.11. x^2 (solid), test points (dots), values returned by mum ($2x - x$, dashed line) and dad ($\frac{x}{\frac{x}{x-x^3}} - x$, small-dashed line)

program. In [Gruau, 1994b]'s cellular encoding technique the evolved trees become programs to describe artificial neural networks or electrical networks [Koza and Bennett III, 1999] etc. Some draw inspiration from the development of embryos to further separate the phenotype from the genotype. For example [Jacob, 1996]'s GP uses Lindenmayer systems (L-Systems) as intermediates for evolving artificial flowers, while [Whigham and Crapper, 1999] treat the tree as defining a path through a formal grammar. The approach taken in [O'Neill and Ryan, 1999] is similar, but uses an initially linear program to navigate the grammar yielding a computer program written in the user-specified language.

1.2.3 Linear Genetic Programming

Except for Section 8.1 we shall not say much about linear GP systems. Briefly there are several linear GP systems. They all share the characteristic that the tree representation is replaced by a linear chromosome. This has some similarities with conventional genetic algorithms. However, unlike conventional GAs, typically the chromosome length is also allowed to evolve and so the population will generally contain individual programs of different sizes. Typically each chromosome is a list of program instructions which are executed in sequence starting at one end. Linear GP systems can be divided into three groups: stack based, register based and machine code.

In stack-based GP [Perkis, 1994] each program instruction takes its arguments from a stack, performs its calculation and then pushes the result back onto the stack. For example when executing a + instruction, the top two items are both popped from the stack, added together and then the result is pushed back onto the stack. Typically the program's inputs are pushed onto the stack before it is executed and its results are popped from the stack afterwards. To guard against random programs misbehaving, checks may be made for both stack overflow and underflow. The programs may be allowed to continue execution if either happens, e.g. by supplying a default value on underflow and discarding data on overflow [Perkis, 1994].

Register-based and machine-code GP [Nordin, 1997] are essentially similar. In both cases data is available in a small number of registers. Each instruction reads its data from one or more registers and writes its output to a register. The program inputs are written to registers (which the evolved code may be prevented from over writing) before the program is executed and its answer is given by the final contents of one or more registers. In machine-code GP the instructions are real hardware machine instructions and the program is executed directly. In contrast register-based (and all other GP) programs are interpreted, i.e. executed indirectly or compiled before execution. Consequently machine code GP is typically at least ten or twenty times faster than traditional implementations.

1.2.4 Graphical Genetic Programming

While we do not deal explicitly with the theory of other types of GP, the reader may wish to note that programs can be represented in forms other than linear or tree-like. So the chromosomes of GP individuals can also be of these forms and genetic operators can be defined on them. For example PDGP (Parallel Distributed Genetic Programming) represents programs as graphs, see Figure 1.12. PDGP defines a number of graph mutation operators and crossover operations between graphs. The graph edges are directed and are interpreted as data-flow connections between processing nodes. While PDGP defines a fixed layout for the nodes, both the connections between them and the operations they perform are evolved. In a number of benchmark problems PDGP has been shown to produce better performance than conventional tree GP [Poli, 1997, Poli, 1999a].

Another example of graph programs is Teller's PADO (Parallel Architecture Discovery and Orchestration) [Teller and Veloso, 1996, Teller, 1998]. PADO is primarily designed for solving the problem of recognising objects

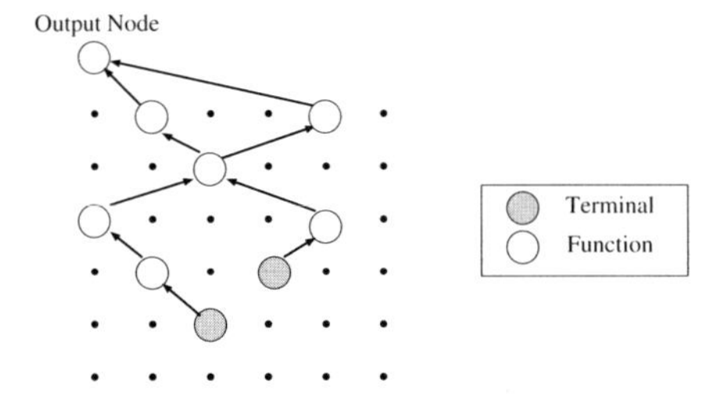

Fig. 1.12. Grid-based representation of graphs representing programs in PDGP (for simplicity inactive nodes and links are not shown).

in computer vision or signal analysis. Again there are a number of mutation operators and crossover operators and the graph edges are directed but they are control-flow connections between processing nodes. While PDGP defines a fixed layout for the nodes, PADO has a looser structure in which the number and location of processing nodes as well as the connections between them and the operations they perform are evolved.

1.3 Outline of the Book

Chapter 2 introduces the notion of fitness landscape and discusses its usefulness in explaining the dynamics of evolutionary search algorithms.

Chapter 3 describes ways of analysing evolution by looking at the propagation of components within the programs rather than the individual programs themselves. This can be likened to [Dawkins, 1976]'s "Selfish Gene" theory of evolution. Early genetic programming schema theorems, such as those due to Koza, Altenberg, O'Reilly and Whigham, fall into this category, which makes them different from the traditional genetic algorithm schema theory [Holland, 1975].

Schema theorems make predictions for the average behaviour of the population at the next generation. Chapter 4 looks at GP schema theorems (such as those due to Poli and Rosca) which provide lower bounds on schema proportions. Like Holland's schema theorem, these do not include the fact that genetic operators can create new instances of schemata as well as deleting (disrupting) them. Therefore we have referred to them as "pessimistic" schema theories. However Chapter 5 describes "exact" schema theories which do make allowance for schema creation. Chapter 6 concludes the presentation of the GP schema theorems by bringing together their implications.

Chapter 7 takes a different view point, by considering the GP search space and how it varies with the size and shape of programs. Some widely applicable results are presented. The formal proofs are deferred to Chapter 8.

We finish in Chapters 9, 10 and 11 with a detailed analysis of two benchmark GP problems and by considering bloat within GP. The analysis uses the component analysis tools of Chapter 3, the schema theorems of Chapters 4–6 and the search space analysis of Chapter 7.

Appendix A gives some pointers to other sources of information on genetic programming, especially those available via the Internet, and is followed by the bibliography, a list of special symbols and a glossary of terms.

Finally, while each chapter finishes with a summary of the results contained within it, Chapter 12 gives our overall conclusions and indications of how the theoretical *foundations of genetic programming* may continue to develop.

2. Fitness Landscapes

The idea of fitness landscapes, firstly proposed in genetics by Sewall Wright [Wright, 1932], is well established in artificial evolution. They are a metaphor used to help visualise problems. In the metaphor landscapes are viewed as being like an area of countryside. Looking vertically down from a great height the countryside appears like a map. This corresponds to the area to be searched. That is, the plan view is the search space.

The height of each point is analogous to the objective or fitness value of the point. Thus the task of finding the best solution to the problem is equivalent to finding the highest mountain. The problem solver is seen as a short-sighted explorer searching for this point.

2.1 Exhaustive Search

In an exhaustive search strategy the explorer simply walks across the landscape in some regular way. For example, he may walk north from the south-western corner of the search space, calculating the height as he goes and noting the location of the highest point he reaches (see Figure 2.1). When the north-western corner is reached, he steps one pace to the east and then repeats the process but heading south. Eventually his straight traversals will lead him to the eastern edge of the search space. In the process he will have exhaustively covered the whole space and will be guaranteed to have walked over the highest point. Such an exhaustive search is, in general, the only certain way to find the best solution in the search space, but it is not feasible in many practical circumstances.

2.2 Hill Climbing

An obvious alternative is for our explorer to start at some point in the countryside and simply walk uphill and until he reaches the top of the mountain. This approach is called hill climbing. If the explorer cannot see the landscape around him (e.g. because of a dense fog) he could still climb a hill by trying one step in each direction and then choosing the steepest direction. How

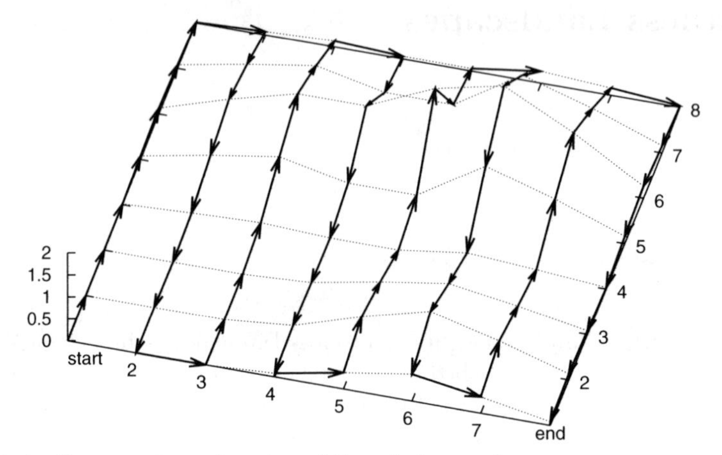

Fig. 2.1. Systematic exploration of the whole search space

successful this will be now depends upon how many peaks there are, and whether or not our starting point leads to the highest peak (see Figure 2.2).

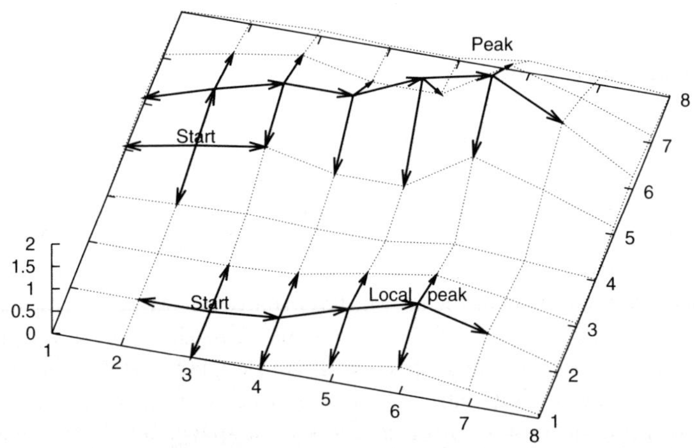

Fig. 2.2. Two hill climbing paths. One reaches the global optimum but the second is trapped at the top of a local peak.

Sometimes people will invert the metaphor at this point. So hills become pits and valleys become mountain ridges. The global optimum is now the deepest point in the landscape and so our explorer walks downhill. That is he follows the path of rivers and streams ever-downwards. The rivers divide the landscape into (possibly) separate drainage basins. If our explorer starts anywhere in the drainage basin of the deepest hole, he is certain to find the best solution (global optimum). This area is known as the solution's "basin of attraction". So if the explorer chooses where he starts at random, his chances of finding the global optimum are directly related to the size of its basin of attraction (see Figure 2.3).

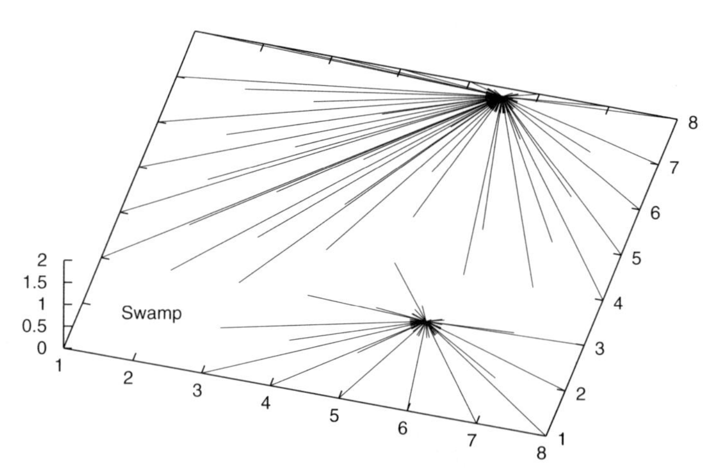

Fig. 2.3. Lines indicate the ultimate location of a hill climber on the landscape shown in Figure 2.2. All the starting points connected to a hill are known as the hill's "basin of attraction". In this example there are two major basins of attraction. The higher peak also has the larger basin. This need not always be the case. A third area is the flat swamp, where the explorer has no gradient to guide him.

2.3 Fitness Landscapes as Models of Problem Difficulty

We can imagine very smooth landscapes where our hill climber moves rapidly to the single hill top as well as very rugged landscapes with many hills which are not as high as the best one. If these hills are isolated from each other by deep valleys, the hill climber will fare worse than if they are interconnected by mountain ridges. But if the ridges are narrow and lead downwards our explorer will still have difficulties (see Figure 2.4).

Where the local gradient leads up to a hill which is not the highest, our explorer might get trapped in a suboptimal place (such hills are variously known as "local optima", "false peaks", etc.).

Some landscapes have been constructed where there is an uphill path to the goal but it is very long (long-path problem). We can imagine our short-sighted hill climber climbing on top of a dry-stone wall which spirals around the hill top and eventually reaches the top. Our explorer remains on top of the wall (since it the highest point locally) and thus walks in a long spiral rather than jumping down the far side of the wall and walking directly to the hill top (see Figure 2.5).

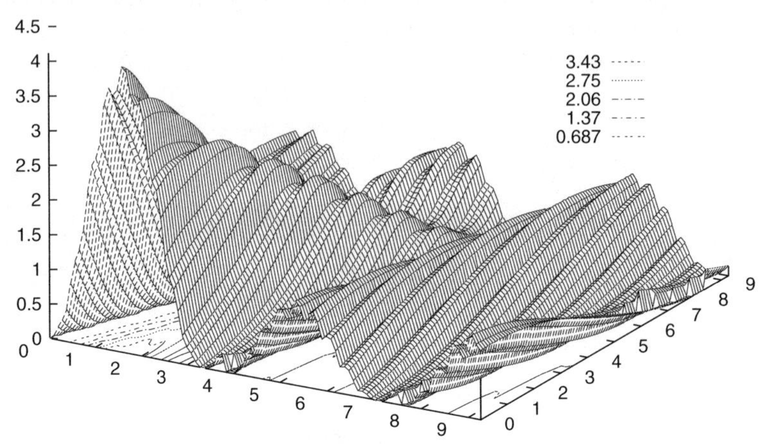

Fig. 2.4. A rugged landscape.

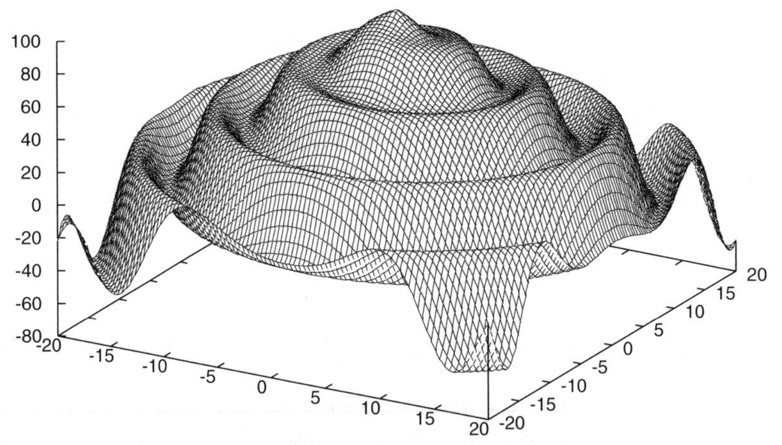

Fig. 2.5. In this landscape the local gradient may eventually lead to the summit but the path is much longer than the direct path.

There might also be totally flat areas. These might be low-lying swamps or high plateaus but in both cases there is no gradient and our hill climbing explorer has no guidance as to where he should go next. There are also problems where the mountain top we seek is isolated by deep moats which render it invisible to hill climbers.

2.4 An Example GP Fitness Landscape

In GP it is harder to use the fitness landscape metaphor, but we can construct a GP example. Consider the classic symbolic regression problem [Koza, 1992].

The target function is $x^4 + x^3 + x^2 + x$ over the range -1 to $+1$ (see the solid line in Figure 2.6). Suppose we are trying to fit it with a tree like that drawn in Figure 2.7. This tree is equivalent to the function $f(x) = \text{alpha} + \text{beta}\ x^2$. The two parameters, alpha and beta, have to be chosen. Figure 2.8 shows the fitness landscape when we vary them both in steps of 1.0.

Although Figure 2.8 gives us a nice picture, there are many aspects that it does not capture. The tree in Figure 2.7, comprising x, addition and multiplication nodes, would appear to be a good starting point to finding a polynomial like $x^4 + x^3 + x^2 + x$. However, Figure 2.8 tells us nothing about how to build on our starting point. Instead, if we use this fitness landscape to optimise alpha and beta we are lead to a local optimum in which both parameters are zero. That is, the constant value of zero is the best fit. In a fixed representation this would be true but genetic programming can do better since the standard genetic operators may alter the structure of the tree. However, it is hard to use fitness landscapes to show this.

2.5 Other Search Strategies

Other search techniques can also be visualised using the fitness landscape metaphor. Smarter explorers may not be as short-sighted, and may be able to see more than a few feet, or might make assumptions about the smoothness of the landscape and make large jumps towards where they calculate the peak "should be" (if their assumptions are correct). There are many enhancements to our short-sighted explorer. In simulated annealing [Kirkpatrick *et al.*, 1983], he is sometimes allowed to climb downhill. The probability of allowing a downward move falls rapidly if the downward move is large or if the explorer has been active for some time. Usually a hill climbing algorithm will have some means of detecting a local peak and may start exploring again from a new randomly chosen start point (hill-climbing with restarts). Tabu search [Glover, 1989] can be thought of as keeping a list of "tabu" places that the explorer should not revisit. In some Artificial Intelligence search techniques, such as $A^\star$, the search can be guided by additional information (known as heuristics) [Russell and Norvig, 1995]. In $A^\star$ once a search is underway it may be possible to exclude large areas of the search space. For example if looking for the cheapest route between two cities, $A^\star$ can stop exploring any partial route which already costs more than a viable route it has found previously.

In evolutionary computation techniques which use a population, such as genetic algorithms and genetic programming, each member of the population is viewed as a separate explorer occupying a fixed location on the search space. Initially these might be scattered at random through the countryside but as time progresses and evolution occurs the individuals should start to cluster in the high-altitude regions.

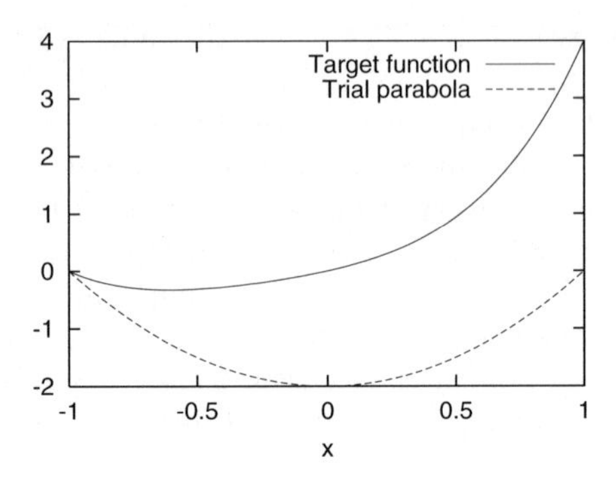

Fig. 2.6. The target function, $x^4 + x^3 + x^2 + x$, and the trial function $f(x) = \text{alpha} + \text{beta } x^2$ (Figure 2.7). $f(x)$ is plotted with alpha = -2 and beta = 2.

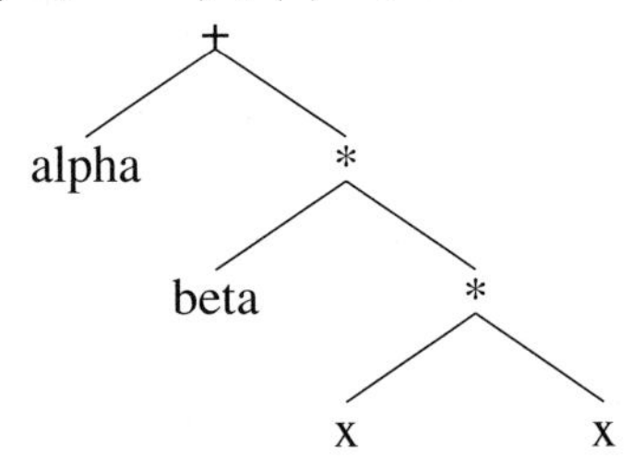

Fig. 2.7. The trial function $f(x) = \text{alpha} + \text{beta } x^2$. The parameters alpha and beta need to be adjusted to minimise the error.

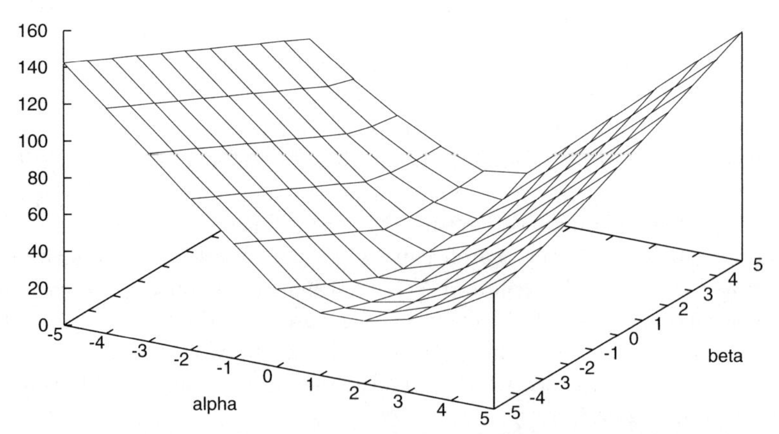

Fig. 2.8. Fitness landscape of alpha and beta. NB the sum of the error across 20 points is plotted; i.e. the global optimum is at the lowest value, rather than the highest.

The metaphor can be used to make clear the global versus local search tradeoff. To be efficient search strategies may need to focus on what they have already discovered about the landscape and look for the optimum in the neighbourhood of good points that they have already discovered. On the other hand by doing so they may miss an even higher hill somewhere where they have not looked yet; i.e. they also have to perform a global exploration.

2.6 Difficulties with the Fitness Landscape Metaphor

The fitness landscape metaphor has considerable explanatory power. However there are severe problems with making it more concrete in practical problems. First large-scale problems are solved by modularisation and reuse. When building a house, we don't initially consider each brick, instead we consider the global properties of brick walls. When we come to make the walls, each brick is not optimally placed for itself but instead bricks are forced to be the same and are processed in a standard fashion. By considering all possibilities, fitness landscapes may conceal useful regularities (i.e. they tend to overlook long-range correlations). One of the goals of genetic algorithms is to discover and exploit regularities. Schema analysis looks at the fitness of patterns and how well a genetic algorithm (such as GP) is exploiting them.

The countryside has only two horizontal dimensions (e.g. latitude and longitude) so it is difficult to imagine a fitness landscape with more than two input parameters. Similarly there are many problems where there are multiple objectives (e.g. design a fast portable computer which also has a long battery life). Here we need a vertical axis for both objectives (e.g. number of spreadsheet cells updated per second and number of minutes the battery remains usable). But how can we visualise a patch of land with two or more altitudes?

Furthermore many problems are discrete in the sense that the parameters of the problem can only take one of a small number of values (often each parameter may only take one of two values: 0 and 1). Thus the countryside metaphor must be stretched to deal with many binary dimensions rather than two continuous ones.

In order to understand what the features of the landscape are, it is not enough to know which points are in the search space and what their fitness is: we also need to know which point is a neighbour of which other point. This information is provided by a neighbourhood relationship. So far, when talking about fitness landscapes, we have assumed that the points in the search space had a natural neighbourhood relationship which we used to plot the landscape, and we assumed that our algorithm (the explorer) could move freely between neighbouring points. For example, in the case of two variables, we implicitly assumed that two points are neighbours if their distance is 1 step. In the case of multiple binary variables, we implicitly assumed that two points are neighbours if their Hamming distance is 1. However, often there is no natural ordering for the structures in the landscape. Also, even if there is one, the action of a search operator may be such that it does not respect the natural neighbourhood relationship available for the domain; i.e. the operator may allow multi-step jumps. In these cases, it may be more informative to consider as neighbours of each point only the points that can be reached with one application of the search operator.

The lack of a natural ordering may prevent us from being able to plot the landscape, even for one- or two-dimensional search spaces. However, we can still visualise the landscape by representing it as a graph where the nodes represent the points in the search space and the links indicate which points are neighbours. In order to visualise the fitness of the points in the search space, the nodes in the graph could be labelled or coloured on the basis of their fitness. Different operators would therefore change the topology of the graph.

In many cases the operators used in search algorithms may not be symmetric. For example, point B might be reachable from point A with one application of a search operator, but point A might not be reachable from B. This would require stretching the fitness landscape metaphor even more. However, it would still be possible to represent the landscape as a graph, but in this case we would need to use oriented links.

Since the operators used in many stochastic search algorithms include a random component, not all moves that an explorer could make from a particular point in the landscape are equally probable. A fitness landscape could not represent this notion. However, the graph-based representation of the search space mentioned above could easily be extended to include this information: it would be sufficient to label each link with the probability that the operator(s) will actually generate that move.

Binary and multi-parent operators also stretch the metaphor. For example, recombination, i.e. the creation of a search point from two known points, complicates things considerably. One could interpret these operators as unary operators by imagining that one of the parents is the point where the explorer currently is, and the others are parameters which modify the behaviour of the operator. Obviously the connectivity of the landscape depends crucially on these "parameters". So, the connectivity of the landscape depends on the composition of the population. Since this varies over time, the landscape topology would also vary over time. This, plus the asymmetries in the operators and the different likelihoods of different moves, would make plotting the landscape quite a challenge. However, it is still possible to imagine the landscape as a graph with time-varying topology and time-varying labels (probabilities) for the links.

2.7 Effect of Representation Changes

In the discussion above, we have ignored the effects of the representation chosen for the points in the search space. For example, we have assumed that if we want to optimise two integers, the parameter encoding used will preserve the natural ordering of integers and so the moves performed by the search operator will correspond to incrementing or decrementing each parameter by one unit. For example, if one used a binary representation for the integers, this could be achieved using a Grey code and a mutation operator which flips exactly one bit. One could have used a different encoding scheme (genotype-to-phenotype mapping), such as the binary code. However, changing the way parameters are encoded can radically change which points are neighbours even if the search operator is kept constant. It is as if the map of the countryside had been folded or sliced up and then put back together so that the countryside was still all there but which bit was next to which had been changed. Naturally remapping the landscape in this way may radically change its characteristics. That is, changing the parameter encoding might transform a smooth landscape into a rugged one (or viceversa), as in Figure 2.9. So, it is really the interplay between representation, operators and fitness function which determines the characteristics of the landscape.

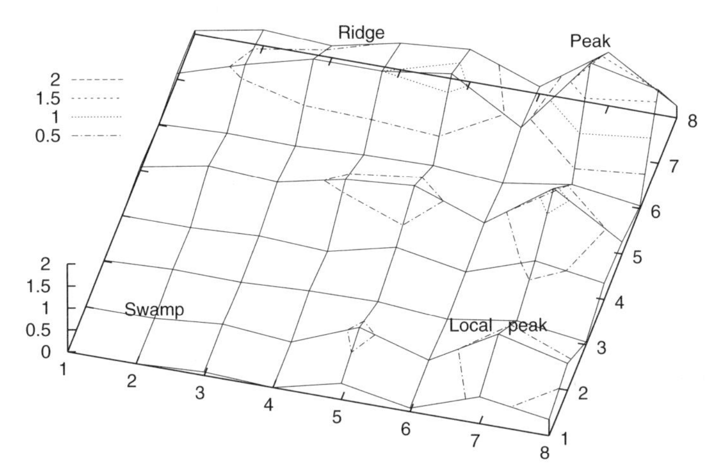

Fig. 2.9. Landscape using binary coding. This is the same landscape as Figures 2.1–2.3 but using a binary coding rather than a Grey coding. The labels ("Peak" etc.) refer to the original figures. Recoding the parameters changes the topology of the landscape and, in this example, introduces more local peaks.

2.8 Summary

The idea of a fitness landscape is appealing. With hill climbing search techniques it provides a useful metaphor allowing us to characterise problems in geometric terms. However, with more complex search techniques apparent landscape complexity is not directly correlated with performance of the search algorithm, and there are variety of problems which make it difficult to use in GP.

3. Program Component Schema Theories

Since John Holland's work in the mid-1970s and his well known schema theorem [Holland, 1975, Goldberg, 1989b], schemata are often used to explain why genetic algorithms work. Schemata are similarity templates representing entire groups of chromosomes. The schema theorem describes how schemata are expected to propagate generation after generation under the effects of selection, crossover and mutation. The usefulness of schemata has been recently criticised (see for example [Altenberg, 1995, Macready and Wolpert, 1996, Kargupta, 1995, Kargupta and Goldberg, 1994]) and many researchers nowadays believe that Holland's schema theorem is nothing more than a trivial tautology of no use whatsoever (see for example [Vose, 1999, preface]). However, as stated in [Radcliffe, 1997], the problem with Holland's schema theorem is not the theorem itself, rather its over-interpretation.

Recently, the attention of genetic algorithm theorists has moved away from schemata to land on Markov chain models [Nix and Vose, 1992, Davis and Principe, 1993, Rudolph, 1997c]. These are very accurate models and have been very useful for obtaining theoretical results on the convergence properties of GAs (for example, on the expected time to hit a solution [De Jong *et al.*, 1995, Rudolph, 1997a] or on GA asymptotic convergence [Nix and Vose, 1992, Rudolph, 1994, Rudolph, 1997b]). However, although results based on Markov chains are very important in principle, very few useful recipes for practical GA users have been drawn from these fine-grain models. One of the reasons for this is that such models lead to computationally intractable equations (e.g. transition matrices including billions of billions of entries) even for tiny populations of very short bit strings.

On the contrary, schema theorems lead to relatively concise descriptions of the way GAs search, which can be easily understood (although such descriptions, being coarsely grained, necessarily model only some of the aspects of GAs). So, a natural way of creating a theory for genetic programming is to define a concept of schema for trees and to extend Holland's schema theorem to GP.

One of the difficulties in obtaining theoretical results using the idea of a schema is that the definition of a schema for GP is much less straightforward than for GAs and a few alternative definitions have been proposed in the literature. All of them define schemata as similarity templates composed of one or multiple trees or fragments of trees. However in some definitions [Koza, 1992, O'Reilly, 1995, O'Reilly and Oppacher, 1995, Whigham, 1995, Whigham, 1996b] components of a schema are *non-rooted* in the sense that the schema can potentially match components anywhere in the tree. This means a schema can be present multiple times within the same program. Therefore, as clarified in Sections 3.3 onwards, these definitions depart significantly from the original idea of interpreting genetic algorithm schemata as subsets of the search space. Instead these definitions (and the related schema theorems) focus on the propagation of program components within the population.

Since component (gene) propagation is also the focus of theoretical genetics, in the Section 3.1 we divert briefly to describe Price's Theorem. Price relates the expected spread in the next generation of any sort of genetic component to its fitness in the current generation. It is very general and can be applied to both genetic algorithms, genetic programming and other evolutionary populations. The remainder of this chapter critically reviews the main results obtained in the theory of GP schema component propagation after briefly recalling Holland's schema theory for binary GAs.

In Chapter 4 we will present and discuss more recent definitions of GP schemata, which are represented at a syntactic level by *rooted* tree fragments. These are schemata whose components are tied to fixed locations (given by the root) within the program tree. At a semantic level these represent sets of programs with certain characteristics, like Holland's schema for genetic algorithms.

3.1 Price's Selection and Covariance Theorem

Price's selection and covariance theorem from population genetics relates the change in frequency of a gene in a population from one generation to the next, to the covariance between the gene's frequency in the original population and the number of offspring produced by individuals in that population:

$$\Delta Q = \frac{\text{cov}(z, q)}{\overline{z}} \tag{3.1}$$

where:

Q = Frequency of a given gene (or linear combinations of genes) in the population.

ΔQ = Change in Q from one generation to the next.

q_i = Frequency of the gene in the individual i (more information is given in Section 3.1.1).

z_i = Number of offspring produced by individual i.

$\overline{z}$ = Mean number of children produced.

cov = Covariance.

The theorem holds: "for a single gene or for any linear combination of genes at any number of loci, holds for any sort of dominance or epistasis (non-linear interaction between genes), for sexual or asexual reproduction, for random or non-random mating, for diploid, haploid or polyploid species, and even for imaginary species with more than two sexes" [Price, 1970][1]. In particular it applies to genetic algorithms [Altenberg, 1994b].

3.1.1 Proof of Price's Theorem

In this section we follow the proof of Price's Theorem [Price, 1970] but recast it and simplify it for use with genetic programming. First we define the additional symbols we shall use:

P_1 = Initial population.
P_2 = Population at next generation (for purposes of the proof generations are assumed to be separated).
M = Size of initial population.
g_i = Number of copies of gene in individual i.
q_i = g_i (We keep the term q_i for compatibility with [Price, 1970]).
$\overline{q}$ = Arithmetic mean of q_i in population P_1.
Q_1 = Frequency of given gene (or linear combinations of genes) in the population; i.e. the number of copies of the gene in population divided by the number of chromosomes it could occupy.
Q_2 = Frequency of gene in population P_2.
z_i = Number of offspring produced by individual i (i.e. the number of individuals in the next population containing code fragments produced from i).
$\overline{z}$ = Mean number of children produced.
g_i' = Number of copies of the gene in all the code fragments in the next population produced by individual i.
q_i' = Frequency of gene in the offspring produced by individual i. Defined by

$$q_i' = \frac{g_i'}{z_i}\ , \text{ if } z_i \neq 0, \text{and}$$
$$q_i' = q_i\ , \text{ otherwise.}$$

$\Delta q_i = q_i' - q_i$.

We shall start with the frequency of the gene in the current population, Q_1, and then find the frequency in the subsequent generation, Q_2. Subtracting

[1] Diploid means the genetic chromosomes are paired, haploid means that they are not (although each cell may have more than one chromosome). In diploid species during sexual reproduction, the chromosome pairs (in general each cell has multiple pairs of chromosomes) separate. When pairs are formed again in the child, one chromosome in each pair comes from the mother and the other from the father. Polyploid means that more than two copies of each chromosome exist.

them yields the change in frequency, which we shall simplify to give Price's Theorem.

$$\begin{aligned} Q_1 &= \frac{\sum g_i}{M} \\ &= \frac{\sum q_i}{M} \\ &= \overline{q} \end{aligned}$$

Each individual in the new population is created by joining one or more code fragments and the number of each gene in the individual is the sum of the number in each of the code fragments from which it was formed. Thus the number of genes in the new population is equal to the number in the code fragments produced by the previous generation.

$$\begin{aligned} Q_2 &= \frac{\sum g_i^{'}}{\sum z_i} \\ &= \frac{\sum z_i q_i^{'}}{\sum z_i} \\ &= \frac{\sum z_i q_i^{'}}{M\overline{z}} \\ &= \frac{\sum z_i q_i}{M\overline{z}} + \frac{\sum z_i \Delta q_i}{M\overline{z}} \\ &= \frac{\sum((z_i - \overline{z})(q_i - \overline{q}) + \overline{z}\, q_i + z_i \overline{q} - \overline{z}\,\overline{q})}{M\overline{z}} + \frac{\sum z_i \Delta q_i}{M\overline{z}} \\ &= \frac{\frac{1}{M}\sum(z_i - \overline{z})(q_i - \overline{q}) + \overline{z}\frac{1}{M}\sum q_i + \overline{q}\frac{1}{M}\sum z_i - \frac{1}{M}\sum \overline{z}\,\overline{q}}{\overline{z}} + \frac{\sum z_i \Delta q_i}{M\overline{z}} \\ &= \frac{\frac{1}{M}\sum(z_i - \overline{z})(q_i - \overline{q}) + \overline{z}\,\overline{q} + \overline{q}\,\overline{z} - \overline{z}\,\overline{q}}{\overline{z}} + \frac{\sum z_i \Delta q_i}{M\overline{z}} \\ &= \frac{\frac{1}{M}\sum(z_i - \overline{z})(q_i - \overline{q}) + \overline{q}\,\overline{z}}{\overline{z}} + \frac{\sum z_i \Delta q_i}{M\overline{z}} \\ &= \frac{\mathrm{Cov}(z, q)}{\overline{z}} + \overline{q} + \frac{\sum z_i \Delta q_i}{M\overline{z}} \\ \Delta Q &= \frac{\mathrm{Cov}(z, q)}{\overline{z}} + \frac{\sum z_i \Delta q_i}{M\overline{z}} \end{aligned}$$

"If meiosis and fertilization are random with respect to the gene, the summation term at the right will be zero except for statistical sampling effects ('random drift'), and these will tend to average out" [Price, 1970] to give Equation 3.1 (page 28); i.e. the expected value of $\sum z_i \Delta q_i$ is zero.

So while survival of an individual and the number of children it has may be related to whether it carries the gene, it is assumed that the production of crossover fragments and their fusing to form offspring is random. That is, selection for reproduction is dependent upon fitness and in general dependent on the presence of specific genes but selection of crossover and mutation points is random and so independent of genes (Section 3.1.4 discusses this further for genetic programming).

3.1.2 Price's Theorem for Genetic Algorithms

Let us assume the population size is unchanged, as is usually the case in genetic algorithms and genetic programming, and two parents are required for each individual created by crossover. The average number of children is given by the proportion of children created by crossover. For example, if all children are created by crossover, i.e. everyone has two parents, then the average number of children, $\overline{z}$, is two. More generally, if we define p_r as the fraction of children that are identical to their (single) parent, p'_m as the fraction that are mutated copies of their (single) parent and p_{xo} as the fraction that are created by crossover between two parents, then $\overline{z} = p_r + p'_m + 2p_{xo}$. Since $p_r + p'_m + p_{xo} = 1$, $\overline{z} = 1 + p_{xo}$ and Equation (3.1) becomes:

$$\Delta Q = \frac{\text{Cov}(z, q)}{1 + p_{xo}} \tag{3.2}$$

3.1.3 Price's Theorem with Tournament Selection

When using tournament selection, each individual's chance of producing children is given only indirectly by its fitness score. The expected number of children for each individual (i.e. ignoring sampling noise) is given directly by its rank r when the population is sorted by score. (The best individual being given rank $r = M$, where M is the population size, and the worst rank being $r = 1$):

$$E(z_r) = M\left((r/M)^T - ((r-1)/M)^T\right) \tag{3.3}$$

where T is the tournament size. With a large population ($M \gg 1$) where individuals do not have identical fitness scores, Equation (3.3) becomes [Blickle, 1996b, Langdon, 1998c]:

$$E(z_r) = T(r/M)^{T-1} \tag{3.4}$$

Thus when using tournament selection in large populations Equations (3.1) and (3.2) can be approximated as [Langdon and Poli, 1998a]:

$$\Delta Q = \frac{T}{\overline{z}}\text{Cov}((r/M)^{T-1}, q) \tag{3.5}$$

$$\Delta Q = \frac{T}{1 + p_{xo}}\text{cov}((r/M)^{T-1}, q)$$

3.1.4 Applicability of Price's Theorem to GAs and GPs

The simplicity and wide scope of Price's Theorem has led Altenberg to suggest that covariance between parental fitness and offspring fitness distribution is fundamental to the power of evolutionary algorithms. Indeed [Altenberg, 1995] shows that Holland's schema theorem [Holland, 1973, Holland, 1975] can be derived from Price's Theorem. This and other analysis leads [Altenberg, 1995, page 43] to conclude: "the Schema Theorem has no implications for how well a GA is performing". Although this might be true for Holland's original result, it is not true for modern schema theories [Stephens and Waelbroeck, 1997, Stephens and Waelbroeck, 1999, Poli, 2000c, Spears, 2000].

While the proof [Price, 1970] assumes discrete generations the result "can be applied to species with overlapping, inter-breeding generations". Thus the theorem can be applied to steady-state genetic algorithms [Syswerda, 1989, Syswerda, 1991].

For the theorem to hold, the genetic operations (crossover and mutation in genetic algorithm terminology) must be independent of the gene. That is, on average there must be no relationship between them and the gene. In large populations random effects will be near zero on average, but in smaller populations their effect may not be negligible. In genetic algorithms selection of crossover and mutation points is usually done independently of the contents of the chromosome and so Price's theorem will hold (except in small populations where random fluctuations may be significant). GP populations are normally bigger (and the number of generations similar) so random effects, "genetic drift", may be expected to be less important.

In standard genetic programming it is intended that the genetic operators should also be independent. However in order to ensure that the resultant offspring are syntactically correct and not too big, genetic operators must consider the chromosome's contents. This is normally limited to just its structure in terms of tree-branching factor (i.e. the average number of arguments that the functions have) and tree depth or size limits. That is, the operators ignore the actual meaning of a node in the tree (e.g. whether it is MUL or ADD) but do consider how many arguments it has. Thus a function with two arguments (e.g. MUL) and a terminal (e.g. x) may be treated differently.

In a large diverse population these factors should have little effect and Price's Theorem should hold. However, when many programs are near the maximum allowed size, a function that has many arguments could be at a disadvantage since the potential offspring containing it have a higher chance of exceeding size limits. Therefore restrictions on program size may on average reduce the number of such functions in the next generation compared to the number predicted by considering only fitness (i.e. by Price's Theorem). [Altenberg, 1994b, page 47] argues that Price's Theorem can be applied to genetic programming and we shall show in Chapter 10 experimental evidence for it based on genes composed of a single GP primitive. However, Section 10.6

contains an example where a depth restriction produces a correlation between genes and genetic operators which, in turn causes $\sum z_i \Delta q_i$ to be non-zero. Similarly correlation between genes and one-point crossover might occur, thus potentially invalidating Equation (3.1).

3.2 Genetic Algorithm Schemata

In the context of genetic algorithms operating on binary strings, a schema (or similarity template) is a string of symbols taken from the set of three symbols {0,1,#}. The character # is interpreted as a "don't care" symbol, so that a schema can represent several bit strings. For example the schema #10#1 represents four strings: 01001, 01011, 11001 and 11011. The number of non-# symbols is called the *order* $\mathcal{O}(H)$ of a schema H. The distance between the furthest two non-# symbols is called the *defining length* $\mathcal{L}(H)$ of the schema. Holland obtained a result (often referred to as "the schema theorem") which predicts how the number of strings in a population matching (or belonging to) a schema is expected to vary from one generation to the next [Holland, 1975] under the effects of selection, bit-flip mutation and one-point crossover.[2] The theorem is as follows:[3]

$$E[m(H,t+1)] \geq m(H,t) \cdot \underbrace{\frac{f(H,t)}{\bar{f}(t)}}_{Selection} \cdot \underbrace{(1-p_m)^{\mathcal{O}(H)}}_{Mutation} \cdot \underbrace{\left[1 - p_{xo} \overbrace{\frac{\mathcal{L}(H)}{N-1}\left(1 - \frac{m(H,t)f(H,t)}{M\,\bar{f}(t)}\right)}^{P_d(H,t)} \right]}_{Crossover} \tag{3.6}$$

where

[2] In one-point crossover, the parent strings are aligned, a crossover point is randomly selected and the offspring are generated by swapping the left-hand sides (w.r.t. the crossover point) of the parents.

[3] This is a slightly different version of Holland's original theorem. It applies when crossover is performed by picking both parents using fitness proportionate selection [Whitley, 1993, Whitley, 1994].

$m(H,t)$ = Number of strings matching the schema H at generation t.
$f(H,t)$ = Mean fitness of the strings matching H.
$\bar{f}(t)$ = Mean fitness of the strings in the population.
p_m = Probability of mutation per bit.
p_{xo} = Probability of crossover.
N = Number of bits in the strings.
M = Number of strings in the population.
$E[m(H,t+1)]$ = Expected number of strings matching the schema H at generation $t+1$.

In genetic algorithms, mutation and crossover points are randomly selected. Also, randomness is present in the selection of the parents. The schema theorem (3.6) avoids getting into the details of these random events by dealing with "expected" (i.e. average) behaviour. It provides a lower bound on the expected number of children in the next generation which match schema H. It is pessimistic in the sense that the lower bound is derived by assuming that whenever disruption of H could occur, it does. The three horizontal curly brackets beneath the equation indicate which operators are responsible for each term. A lower bound for the expected number in the next generation is the same as the number in this generation, $m(H,t)$, multipled by a term for each operator.

The "Selection" term comes from fitness proportionate selection, i.e. if the individuals which match H are of above average fitness they will (on average) have more children than the rest of the population. So, the lower bound for $E[m(H,t+1)]$ increases (or decreases) in proportion to their fitness divided by the average fitness.

The "Mutation" term is the proportion of children that match schema H which still match H after mutation. The chance that an individual bit is not mutated is $(1-p_m)$. This is the same for every bit in the individual. So the chance of two bits not changing is $(1-p_m)^2$, and that for three bits is $(1-p_m)^3$, and so on. Therefore the chance of all defining bits in H remaining the same (i.e. of mutation not disrupting H) is $(1-p_m)^{\mathcal{O}(H)}$. Note inequality (3.6) does not include children who did not match H which now do match H as a result of mutation. (These are known as schema creation events). Equation (3.6) is a lower bound because these (and similar crossover creation events) are neglected.

The "Crossover" term represents a lower bound for the probability that the schema H will survive (i.e. will not be disrupted by) crossover at generation t. This is expressed as one minus an upper bound for the disruption probability. The upper bound is represented by the bracket labelled $P_d(H,t)$ times p_{xo} which represents the probability of performing crossover as opposed to selection followed by cloning. The first term of $P_d(H,t)$, $\frac{\mathcal{L}(H)}{N-1}$, is the chance of choosing (from the $N-1$ possibilities) a crossover point that lies between the defining points of H. ($\frac{\mathcal{L}(H)}{N-1}$ can be thought of as the *fragility* of the schema). For the second term, we use the property of bit string GAs that if both parents match a schema, then so too must their children; i.e. there is no schema disruption. The second term of $P_d(H,t)$ places an upper bound on disruption by calculating the chance that the other parent also matches schema H, and then subtracting it from 1, to give the chance the second parent does not match H.

The theorem can be extended to any selection-with-replacement mechanism, obtaining:

$$E[m(H,t+1)] \geq Mp(H,t) \cdot (1-p_m)^{\mathcal{O}(H)} \cdot \left[1 - p_{xo}\frac{\mathcal{L}(H)}{N-1}(1-p(H,t))\right]$$

where $p(H,t)$ is the probability of selection of the schema H (in fitness proportionate selection $p(H,t) = m(H,t)f(H,t)/(M\bar{f}(t))$).

3.3 From GA Schemata to GP Schemata

In this section we discuss and modify the representation of schemata used in genetic algorithms to make it easier for the reader to understand the genetic programming schema definitions introduced in the following sections.

A genetic algorithm schema is fully determined by the defining bits (the 0's and 1's, the non-# symbols) it contains and by their positions. So, instead of representing a schema as a string of characters, one could equivalently represent it with a set of pairs $[(c_1,i_1),(c_2,i_2),\ldots]$. The terms c_j are strings of characters from the original alphabet (typically 0's and 1's) which represent groups of contiguous defining symbols (bits) which we call the *components* of the schema. The terms i_j are integers giving the positions of the c_j's. For example in a binary genetic algorithms the schema #10#1 would have two components and could be represented as [(10,2),(1,5)], the schema ##11## can be represented as [(11,3)] and 11#111#1 can be represented as [(11,1),(111,4),(1,8)].[4] As an additional example, let us consider a genetic algorithm operating on strings from the alphabet {"(", ")", ".", "-", ":"}. Then the schema ##:-)### represents all the strings with a happy face at position 3, and can be represented as [(:-),3)].

This is an explicit representation for a way of looking at schemata as sets of components that has often been implicitly used in the GA literature. For example, when we say that a schema has been disrupted by crossover, we usually mean that one or more of its components have not been transmitted to the offspring entirely, or have been partly destroyed, rather than thinking that the offspring sample a subspace that is different from the one represented by the schema. (Of course both are true, they are different ways

[4] The notation [(10,2),(1,5)] for the schema #10#1 should not be confused with the set of individuals belonging to the schema {01001,01011,11001,11011} (its semantics). To minimise the risk of confusion we have used square brackets to represent the syntactic representation of schemata, while pairs of curly brackets indicate the set of individuals in a schema.

of thinking about the same effect). Likewise, we explain the *building block hypothesis* [Goldberg, 1989b, Holland, 1975] by saying that GAs work by combining relatively fit, short schemata to form complete solutions. What we mean is that crossover mixes and concatenates the components of low order schemata (schemata with mostly "don't care" symbols) to form higher order ones (schemata with more defined bit positions and less "don't care" symbols). Similarly the GA schema theorem (3.6) is often interpreted as describing the variation of the number of certain groups of bits within the population (the schema components c_j, at their positions i_j) rather than the number of strings sampling the subspace represented by a schema (even though both are true). For example consider $H = \#10\#1$. The theorem could be interpreted as describing the variation in the number of the component 10 at position 2 *and* 1 at position 5 within a population of bit strings, rather than the variations in the proportion of individuals belonging to the set H.

Obviously there is no real distinction between these two ways of interpreting the schema theorem, because counting the number of strings matching (or belonging to) a given schema and counting the number of schema components in the population give the same result. This is because we specify exactly where each component of the schema must be located. So the number of strings in the population sampling H, $m(H, t)$, is the same as the number of times the components $c_1, ..., c_n$ are all simultaneously present in the strings of the population. We term the latter *instantiations* and we will denote them with the symbol $i(H, t)$.

For example consider the following population of bit strings at time t:

```
11001
00011
00001
11011
```

We have

$$\begin{aligned} m(11\#\#\#, t) &= m([(\texttt{11}, 1)], t) &&= i([(\texttt{11}, 1)], t) &&= 2, \\ m(\#\#0\#1, t) &= m([(\texttt{0}, 3), (\texttt{1}, 5)], t) &&= i([(\texttt{0}, 3), (\texttt{1}, 5)], t) &&= 4. \end{aligned}$$

Let us consider again the alphabet {"(", ")", ".", "-", ":"} and the following population:

```
..:-(...
..:-):-)
:-).:-).
:-:-)...
```

One might want to know how many individuals have a happy face at position 3 in the population, i.e. what is the value of $m(\texttt{\#\#:-)\#\#\#}) = m([(\texttt{:-)}, 3)], t)$? (Actually it is 2). Alternatively, one might ask how many happy faces are there in the population at position 3, i.e. what is the value of $i([(\texttt{:-)}, 3)], t)$? (Of course it is again 2).

Now let us try to understand what would happen if we omitted the positional information i_j from all schema pairs. Syntactically, schemata would be sets of components $[c_1, c_2, \ldots]$, such as $H = [\texttt{10}, \texttt{1}]$. Semantically a schema would represent all the strings which include as substrings all c_j's. For example if the strings are six bits long ($N = 6$), $H = [\texttt{11}, \texttt{00}]$ represents the set of all strings in which at least one 11 and one 00 are present, i.e. $H \equiv$ {110000, 011000, ...}. If $N = 3$, $H = [\texttt{10}]$ represents {100, 101, 010, 110}, while $H = [\texttt{10}, \texttt{00}]$ represents {100}. Interestingly, with this definition of schema a string can include multiple copies of the components of a schema. For example, the component $c_1 = \texttt{10}$ is present twice (at positions 1 and 3) in the string 10101. Similarly the components of the schema [10,01] are jointly present four times in the string 10101. We call each of these an *instantiation* of the schema in the string. The four instantiations of [10,01] in 10101 are indicated in Figure 3.1.

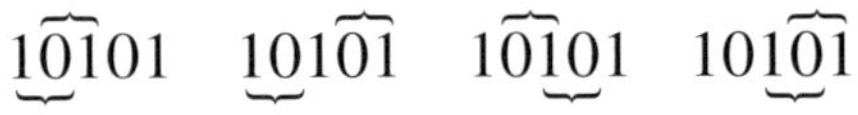

10101   10101   10101   10101

Fig. 3.1. The four joint instantiations of the components of the schema [10,01] in the bit string 10101.

Therefore, if the positional information is omitted from a genetic algorithm schema definition, counting the number of strings belonging to a given schema and counting the number of instantiations of the schema produce different results. So, the number of strings in a given schema *may be different* from the number of joint instantiations of the components of the schema in the population. For example consider the bit string population at time t again:

```
11001
00011
00001
11011
```

We have

$$m([\texttt{11}], t) = 3, \quad i([\texttt{11}], t) = 4,$$
$$m([\texttt{0}, \texttt{1}], t) = 4, \quad i([\texttt{0}, \texttt{1}], t) = 20.$$

Going back to the "happy faces" population at page 36, one might want to know how many individuals have a happy face i.e. what is the value of $m([\texttt{:-)}], t)$ (it is 3). Note this is different from asking how many happy faces there are in the population, i.e. what is the value of $i([\texttt{:-)}], t)$ (it is 5).

These examples illustrate that the two ways of interpreting the schema theorem mentioned in the previous paragraphs are now different. Studying the propagation of components (e.g. the happy faces) is very different from studying the propagation of individuals. When considering individuals one is taking the viewpoint illustrated in Figure 1.5 (page 8), while with components one takes the stance illustrated in Figure 1.4 (page 7).

The previous argument might look academic for binary fixed-length genetic algorithms and most people would think that position-less schema definitions are not very useful for GAs and genetic programming in general. However, position-less schemata might be useful when one uses more complex representations and gene-position-independent fitness functions. For example, a position-less schema-component representation similar to the one presented above was proposed in [Radcliffe, 1991], to represent sets of solutions of travelling salesman problems.

The distinction between positioned and position-less schema components is also very important in genetic programming. In fact, information about the positions of the schema components was omitted in the earlier GP schema definitions reviewed in the rest of this chapter. This has led some authors to concentrate their analysis on the propagation of such components in the population rather than on the way the number of programs sampling a given schema change over time.

3.4 Koza's Genetic Programming Schemata

Koza made the first attempt to explain why genetic programming works, producing an informal argument showing that Holland's schema theorem would also work for GP as well [Koza, 1992, pages 116–119]. The argument was based on the idea of using schemata to define a subspace of all trees. According to Koza's definition, a schema H is represented as a set of program subtrees (S-expressions). For example the schema H=`[(+ 1 x), (* x y)]` represents all programs including at least one occurrence of the expression `(+ 1 x)` and at least one occurrence of `(* x y)`. This definition of schema was probably suggested by the fact that Koza's GP crossover moves subtrees between parents and children. Koza's definition gives only the defining components of a schema and not their position. So the same schema can potentially be matched (instantiated) in different ways in the same program. (Therefore a schema can match a given program multiple times and so $m(H, 5)$ may be different from $i(H, t)$.) For example, the schema H=[x] can be instantiated in two ways in the program `(+ x x)`. One of the two possible instantiations of the more complex schema H=`[(- 2 x), x]` in the program `(+ (- 2 x) x)` is shown in Figure 3.2.

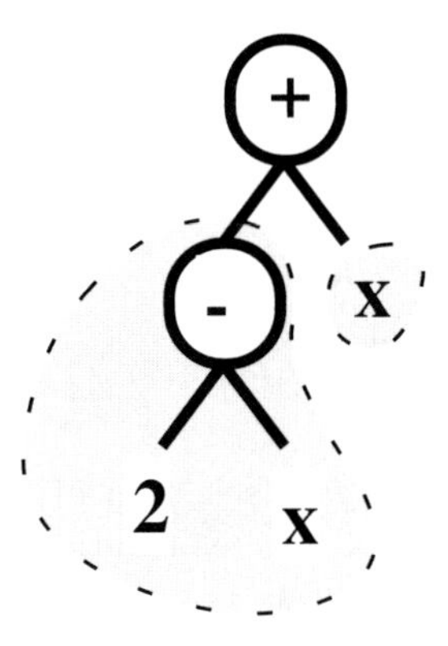

Fig. 3.2. The shaded regions indicate where Koza's schema H=`[(- 2 x), x]` matches the program `(+ (- 2 x) x)`.

3.5 Altenberg's GP Schema Theory

Altenberg produced a probabilistic model of genetic programming which can be considered as the first mathematical formulation of a schema theorem for GP [Altenberg, 1994a].

Assuming that the population is very large, no mutation, and that fitness proportionate selection is used, the frequency of program i in the next generation is:

$$\frac{m(i,t+1)}{M} = (1-p_{xo})\frac{f(i)}{\bar{f}(t)}\frac{m(i,t)}{M} + \quad (3.7)$$

$$p_{xo}\sum_{j,k\in\mathcal{P}}\frac{f(j)f(k)}{\bar{f}^2(t)}\frac{m(j,t)}{M}\frac{m(k,t)}{M}\sum_{s\in\mathcal{S}}P(i\leftarrow j,s)C(s\leftarrow k)$$

where:

$f(i)$ = Fitness of program i.
$\bar{f}(t)$ = Average fitness of the programs in the population.
$\mathcal{P}$ = The population.
$\frac{m(j,t)}{M}$ = Frequency of program j at generation t.
$\mathcal{S}$ = The space of all possible subexpressions extractable from $\mathcal{P}$.
$P(i \leftarrow j, s)$ = Probability that inserting expression s in program j produces program i.
$C(s \leftarrow k)$ = Probability that crossover picks up expression s in program k.

This equation is slightly different from the one reported in [Altenberg, 1994a], where the ratios $\frac{m(i,t+1)}{M}$, $\frac{m(i,t)}{M}$, $\frac{m(j,t)}{M}$ and $\frac{m(k,t)}{M}$ were interpreted as proportions. Formally both equations are correct only in the infinite population limit. The first part of Equation (3.7) has some similarities with Inequality (3.6) when there is no mutation, since both deal with the effect of selection. However Altenberg explicitly considers all ways in which programs can be *created*. This means the second part of (3.7) is more complicated but Altenberg's result is an equation rather than a pessimistic lower bound. Equation (3.7) can be generalised to other selection methods to obtain:

$$E\left[\frac{m(i,t+1)}{M}\right] = (1-p_{xo})p(i,t) + p_{xo} \sum_{j,k\in\mathcal{P}} p(j,t)p(k,t) \sum_{s\in\mathcal{S}} P(i \leftarrow j,s)C(s \leftarrow k) \tag{3.8}$$

where $p(h,t)$ is the probability of selection of program h at generation t. Thanks to the introduction of the expectation operator (E) on its l.h.s., Equation (3.8) applies to both finite populations and infinite populations.

Let us try to understand this equation with an example. Let us consider a population including only two programs: $h_1 =$ `(* (+ 1 x) (+ 1 x))` and $h_2 =$ `(+ 1 (+ 1 x))`:

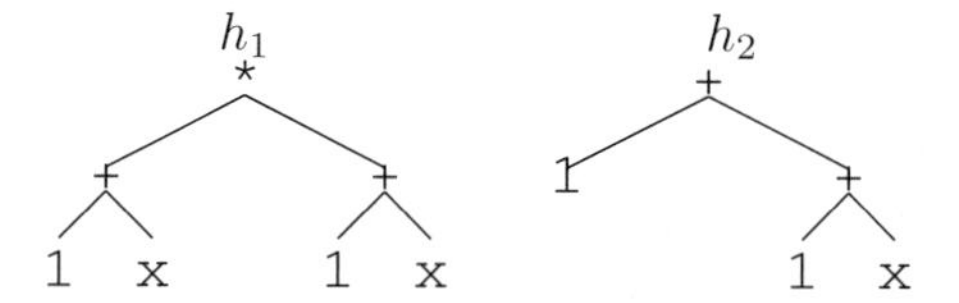

So, $\mathcal{P} = \{h_1, h_2\}$. Then the subexpressions extractable from $\mathcal{P}$ are $s_1 =$`1`, $s_2 =$`x`, $s_3 =$`(+ 1 x)`, $s_4 =$`(* (+ 1 x) (+ 1 x))` and $s_5 =$`(+ 1 (+ 1 x))`:

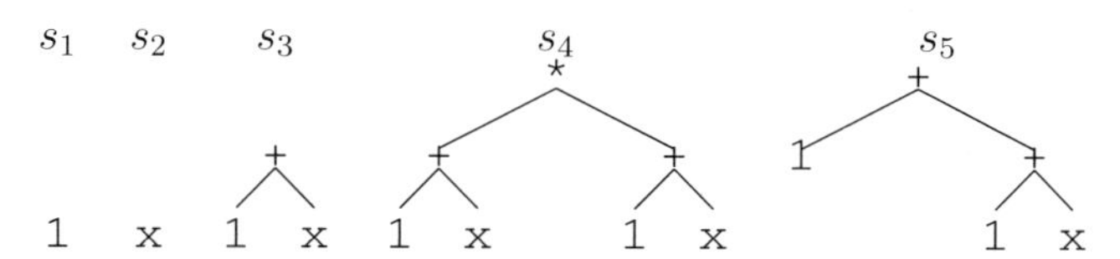

So, $\mathcal{S} = \{s_1, s_2, s_3, s_4, s_5\}$. Let us focus on evaluating $E\left[\frac{m(i,t+1)}{M}\right]$ for program $i = h_1$.

If the crossover points are selected randomly (and we allow the selection of the root node too), the probability that expression s_1 is extracted from program h_1 is $C(s_1 \leftarrow h_1) = C($`1` $\leftarrow$ `(* (+ 1 x) (+ 1 x))`$) = \frac{2}{7}$, since of the seven crossover sites/subtrees in h_1 two lead to extracting the constant `1`. In total:

$$
\begin{array}{ll}
C(s_1 \leftarrow h_1) = C(\texttt{1} \qquad\qquad\qquad\quad \leftarrow \texttt{(* (+ 1 x) (+ 1 x))}) = \frac{2}{7} \\
C(s_2 \leftarrow h_1) = C(\texttt{x} \qquad\qquad\qquad\quad \leftarrow \texttt{(* (+ 1 x) (+ 1 x))}) = \frac{2}{7} \\
C(s_3 \leftarrow h_1) = C(\texttt{(+ 1 x)} \qquad\qquad \leftarrow \texttt{(* (+ 1 x) (+ 1 x))}) = \frac{2}{7} \\
C(s_4 \leftarrow h_1) = C(\texttt{(* (+ 1 x) (+ 1 x))} \leftarrow \texttt{(* (+ 1 x) (+ 1 x))}) = \frac{1}{7} \\
C(s_5 \leftarrow h_1) = C(\texttt{(+ 1 (+ 1 x))} \qquad \leftarrow \texttt{(* (+ 1 x) (+ 1 x))}) = 0
\end{array}
$$

$$
\begin{array}{ll}
C(s_1 \leftarrow h_2) = C(\texttt{1} \qquad\qquad\qquad\quad \leftarrow \texttt{(+ 1 (+ 1 x))}) = \frac{2}{5} \\
C(s_2 \leftarrow h_2) = C(\texttt{x} \qquad\qquad\qquad\quad \leftarrow \texttt{(+ 1 (+ 1 x))}) = \frac{1}{5} \\
C(s_3 \leftarrow h_2) = C(\texttt{(+ 1 x)} \qquad\qquad \leftarrow \texttt{(+ 1 (+ 1 x))}) = \frac{1}{5} \\
C(s_4 \leftarrow h_2) = C(\texttt{(* (+ 1 x) (+ 1 x))} \leftarrow \texttt{(+ 1 (+ 1 x))}) = 0 \\
C(s_5 \leftarrow h_2) = C(\texttt{(+ 1 (+ 1 x))} \qquad \leftarrow \texttt{(+ 1 (+ 1 x))}) = \frac{1}{5}
\end{array}
$$

The next step in applying Equation (3.8) is to calculate the probability, $P(h_1 \leftarrow j, s)$, that inserting expression s in program j produces program h_1, for each possible value of $j \in \mathcal{P}$ and $s \in \mathcal{S}$. Let us start with $j = h_1$ and $s = s_1 = \texttt{1}$. Clearly, the offspring is an instance of h_1 only if the terminal `1` replaces another terminal `1`. Since there are two such terminals in the parent tree (out of the seven subexpressions extractable from it) $P(h_1 \leftarrow h_1, s_1) = \frac{2}{7}$. With similar calculations one finds:

$$
\begin{aligned}
P(h_1 \leftarrow h_1, s_1) &= \tfrac{2}{7} \\
P(h_1 \leftarrow h_1, s_2) &= \tfrac{2}{7} \\
P(h_1 \leftarrow h_1, s_3) &= \tfrac{2}{7} \\
P(h_1 \leftarrow h_1, s_4) &= \tfrac{1}{7} \\
P(h_1 \leftarrow h_1, s_5) &= 0
\end{aligned}
$$

$$
\begin{aligned}
P(h_1 \leftarrow h_2, s_1) &= 0 \\
P(h_1 \leftarrow h_2, s_2) &= 0 \\
P(h_1 \leftarrow h_2, s_3) &= \tfrac{1}{5} \\
P(h_1 \leftarrow h_2, s_4) &= \tfrac{1}{5} \\
P(h_1 \leftarrow h_2, s_5) &= 0
\end{aligned}
$$

With this information it is now possible to calculate the probability, $P(i \leftarrow j, k)$, that crossing over program j with program k leads to program i, $P(i \leftarrow j, k) = \sum_{s \in \mathcal{S}} P(i \leftarrow j, s) C(s \leftarrow k)$. For all four possible values of j and k and for $i = h_1$ we obtain:

$$
\begin{aligned}
P(h_1 \leftarrow h_1, h_1) &= \tfrac{2}{7}\tfrac{2}{7} + \tfrac{2}{7}\tfrac{2}{7} + \tfrac{2}{7}\tfrac{2}{7} + \tfrac{1}{7}\tfrac{1}{7} + 0 = \tfrac{13}{49} \\
P(h_1 \leftarrow h_1, h_2) &= \tfrac{2}{7}\tfrac{2}{5} + \tfrac{2}{7}\tfrac{1}{5} + \tfrac{2}{7}\tfrac{1}{5} + 0 + 0 = \tfrac{8}{35} \\
P(h_1 \leftarrow h_2, h_1) &= 0 + 0 + \tfrac{1}{5}\tfrac{2}{7} + \tfrac{1}{5}\tfrac{1}{7} + 0 = \tfrac{3}{35} \\
P(h_1 \leftarrow h_2, h_2) &= 0 + 0 + \tfrac{1}{5}\tfrac{1}{5} + 0 + 0 = \tfrac{1}{25}
\end{aligned}
$$

This allows us to write Equation (3.8) for program h_1 obtaining:

$$E\left[\frac{m(h_1,t+1)}{M}\right] = (1-p_{xo})p(h_1,t) + \qquad (3.9)$$
$$p_{xo}\left(\frac{13}{49}(p(h_1,t))^2 + \frac{11}{35}p(h_1,t)p(h_2,t) + \frac{1}{25}(p(h_2,t))^2\right)$$

If both programs h_1 and h_2 have the same fitness, $p(h_1,t) = p(h_2,t) = \frac{1}{2}$ since $m(h_1,t) = m(h_2,t) = 1$ and $M = 2$. Therefore Equation (3.9) becomes:

$$E\left[\frac{m(h_1,t+1)}{2}\right] = (1-p_{xo})\frac{1}{2} + p_{xo}\frac{759}{4900}$$
$$= \frac{1}{2} - p_{xo}\frac{1691}{4900} \approx 0.5 - 0.345p_{xo}$$

If crossover is applied with 100% probability, the expected proportion of program h_1 at the next generation is approximately 15.5% (which represents a very large drop from its current proportion of 50%).

Equation (3.7) is a model for the propagation of programs under standard crossover assuming that only one offspring is produced as a result of each crossover operation. From this model, defining a schema as being a subexpression s, Altenberg obtained:

$$\bar{u}(s,t+1) = (1-p_{xo})\frac{f(s,t)}{\bar{f}(t)}\bar{u}(s,t) + \qquad (3.10)$$
$$p_{xo}\sum_{i,j,k\in\mathcal{P}} C(s\leftarrow i)\frac{f(j)f(k)}{\bar{f}^2(t)}\frac{m(j,t)}{M}\frac{m(k,t)}{M}\sum_{r\in\mathcal{S}} P(i\leftarrow j,r)C(r\leftarrow k)$$

where $\bar{u}(s,t) = \sum_{i\in\mathcal{P}} C(s\leftarrow i)\frac{m(i,t)}{M}$ is the average chance that crossover picks up expression s from a randomly chosen program at generation t, and

$$f(s,t) = \frac{\sum_{i\in\mathcal{P}} C(s\leftarrow i)f(i)\frac{m(i,t)}{M}}{\sum_{i\in\mathcal{P}} C(s\leftarrow i)\frac{m(i,t)}{M}}$$

is the schema fitness (remember that expression s is a schema).

This can be considered an exact microscopic subtree-schema theorem for GP with standard crossover and infinite populations. Altenberg did not analyse its components in more detail, but he indicated that a simpler schema theorem,

$$\bar{u}(s,t+1) \geq (1-p_{xo})\frac{f(s,t)}{\bar{f}(t)}\bar{u}(s,t)$$

could be obtained by neglecting the effects of crossover (represented by the triple summation in Equation (3.10)), which, however, models only the effects of selection.

Again, we generalise Equation (3.10) to finite populations and other selection methods to obtain:

$$E[\bar{u}(s,t+1)] = (1-p_{xo})\sum_{i\in\mathcal{P}} C(s \leftarrow i)p(i,t) + \\ p_{xo}\sum_{i\in\Sigma} C(s \leftarrow i) \sum_{j,k\in\mathcal{P}} p(j,t)p(k,t) \sum_{r\in\mathcal{S}} P(i \leftarrow j,r)C(r \leftarrow k) \tag{3.11}$$

where Σ is the set of all possible programs that can be created by crossing over the elements of $\mathcal{P}$.[5] In order to calculate the expected value of $\bar{u}(s,t+1)$ for a given expression s, one has to do calculations like the ones that led to Equation (3.9) but in this case for all possible members of Σ. These calculations, although relatively simple, may be exceedingly tedious and lengthy even for the simplest of cases.

Altenberg's notion of schema is that a schema is a subexpression. This is not exactly the same as the notion introduced by Koza, where a schema could be made up of multiple subexpressions.

3.6 O'Reilly's Genetic Programming Schemata

Koza's work on schemata was also formalised and refined by O'Reilly [O'Reilly, 1995, O'Reilly and Oppacher, 1995] who derived a schema theorem for genetic programming with fitness proportionate selection and crossover. The theorem was based on the idea of defining a schema as an unordered collection (a multiset) of subtrees and tree fragments. Tree fragments are trees with at least one leaf that is a "don't care" symbol ('`#`') which can be matched by any subtree (including subtrees with only one node). For example the schema H=`[(+ # x), (* x y), (* x y)]` represents all the programs including at least one occurrence of the tree fragment `(+ # x)` and at least *two* occurrences of `(* x y)`.[6] The tree fragment `(+ # x)` is present in all programs that include a `+`, the second argument of which is `x`. Figure 3.3 shows the instantiation of the schema H=`[(+ # x), x, x]` in the program `(+ (- 2 x) x)`. Like Koza's definition, O'Reilly's schema definition gives only the defining components of a schema, not their position. So again, a single schema can be matched (instantiated) by a program in different ways, and therefore multiple times.

O'Reilly's definition of schema allowed her to define the concepts of order and defining length for genetic programming schemata. In her definition the order of a schema is the number of non-# nodes in the expressions or fragments contained in the schema. The defining length is the number of links included in the expressions and tree fragments in the schema plus the links that connect them together. Unfortunately, the definition of defining length

[5] Alternatively, Σ can be seen as the set of all possible programs, i.e. the search space, explored by GP.

[6] We use here the standard notation for multisets, which is slightly different from the one used in O'Reilly's work.

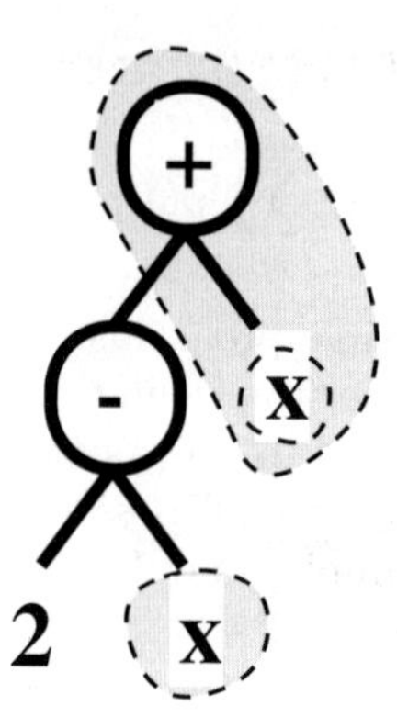

Fig. 3.3. O'Reilly's schema H=`[(+ # x), x, x ]`. The shaded regions show the locations where components of the schema match the program `(+ (- 2 x) x)`. The leaves `x` match the schema multiple times.

is complicated by the fact that the components of a schema can be embedded in different ways in different programs. Therefore, the defining length of a schema is not constant but depends on the way a schema is instantiated (matched) inside the programs sampling it. The probability of disruption $P_d(H, h, t)$ of a schema H contained in the program h at generation t due to crossover is the ratio between the defining length of H in h and the total number of crossover locations in h. This implies that $P_d(H, h, t)$ depends on the shape, size and composition of the tree h matching the schema. The schema theorem derived by O'Reilly,

$$E[i(H, t+1)] \geq i(H, t) \cdot \frac{f(H, t)}{\bar{f}(t)} \cdot \left(1 - p_{xo} \cdot \overbrace{\max_{h \in \text{Pop(t)}} P_d(H, h, t)}^{P_d(H,t)}\right)$$

overcame this problem by considering the maximum of such a probability, $P_d(H, t) = \max_{h \in \text{Pop(t)}} P_d(H, h, t)$. Pop(t) is the population at generation t. Naturally considering a maximum may lead to severely underestimating the number of occurrences of the given schema in the next generation. NB $i(H, t)$ is the number of *instances* of the schema H at generation t and $f(H, t)$ is the mean fitness of the instances of H. This is computed as the weighted sum of the fitnesses of the programs matching H, using as weights the ratios between the number of instances of H that each program contains and the total number of instances of H in the population. The theorem describes the way in which the components of the representation of a schema propagate from one generation to the next, rather than the way that the number of programs sampling a given schema changes during time. O'Reilly discussed the usefulness of her result and argued that the intrinsic variability of $P_d(H, t)$ from generation to generation is one of the major reasons why no hypotheses can be made on the real propagation and use of building blocks (by which she meant small relatively fit schemata containing few defining symbols (low-order)) in genetic programming. O'Reilly's schema theorem did not include the effects of mutation.

3.7 Whigham's Genetic Programming Schemata

In the framework of his genetic programming system based on context-free grammars (CFG-GP), Whigham produced a definition of schema for context-free grammars and the related schema theorem [Whigham, 1995, Whigham, 1996a, Whigham, 1996b]. In CFG-GP programs are the result of applying a set of rewrite rules taken from a pre-defined grammar to a starting symbol S. In order to create a program, CFG-GP starts with a derivation tree. The tree's internal nodes are rewrite rules from the user supplied grammar. The tree's terminals (leaf nodes) are the functions and terminals used in the program. The derivation tree can be readily converted into the corresponding program, which can then be run. In CFG-GP the individuals in the population are derivation trees (rather than the programs themselves) and the search proceeds using special crossover and mutation operators which always produce valid derivation trees.

Whigham defines a schema as a partial derivation tree rooted in some non-terminal node, i.e. as a collection of rewrite rules organised into a single derivation tree. Given that the terminals of a schema can be both terminal and non-terminal symbols of a grammar and that the root of a schema can be a symbol different from the starting symbol S, a schema represents all the programs that can be obtained by completing the schema (i.e. by adding other rules to its leaves until only terminal symbols are present) and all the programs represented by schemata that contain it as a component. When the root node of a schema is not S, the schema can occur multiple times in the derivation tree of the same program. This is the result of the absence of positional information in the schema definition. Figure 3.4 shows one of the two possible instantiations of the schema $H = (\mathtt{A} \stackrel{+}{\Rightarrow} \mathtt{FAA})$ in the derivation tree of the program `(+ (- 2 x) x)`.

Whigham's definition of schema leads to simple equations for the probability of disruption of schemata under crossover, $P_{d_c}(H, h, t)$, and mutation, $P_{d_m}(H, h, t)$ (see [Whigham, 1995, Whigham, 1996a, Whigham, 1996b] for more details). Unfortunately, as with O'Reilly's, these probabilities vary with the size of the tree h matching the schema. In order to produce a schema theorem for CFG-GP, Whigham used the average disruption probabilities of the instances of a schema under crossover and mutation, $\bar{P}_{d_c}(H, t)$ and $\bar{P}_{d_m}(H, t)$, and the average fitness $f(H, t)$ of such instances. The theorem is

$$E[i(H, t+1)] \geq i(H, t)\frac{f(H,t)}{\bar{f}(t)} \cdot \{[1 - p'_m \bar{P}_{d_m}(H, t)]\, [1 - p_{xo}\bar{P}_{d_c}(H, t)]\}$$

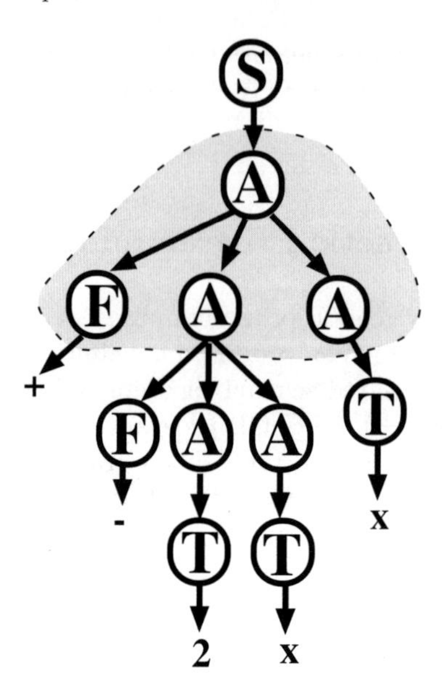

Fig. 3.4. Whigham's schema $H = (\mathtt{A} \stackrel{+}{\Rightarrow} \mathtt{FAA})$. The shaded region shows one of the cases where the schema matches the derivation tree. (In this example, the derivation tree creates the program `+ - 2 x x`, which is equivalent to $(2-x)+x$). Note that as with Koza (Figure 3.2) and O'Reilly (Figure 3.3), Whigham's schemata can match at any point of the tree, not only the root.

where p_{xo} and p'_m are the probabilities of applying crossover and mutation. (Note $p'_m \neq p_m$ since p_m is defined as the probability per location in the chromosome, while p'_m, like p_{xo} is defined per child.) By changing the grammar used in CFG-GP this theorem can be shown to be applicable both to GAs with fixed length binary strings and to standard GP, of which CFG-GP is a generalisation (see the GP grammar given in [Whigham, 1996b, page 130]). Like in O'Reilly's case, this theorem describes the way in which the components of the representation of a schema propagate from one generation to the next, rather than the way in which the number of programs sampling a given schema changes over time. The GP schema theorem obtained by Whigham is different from the one obtained by O'Reilly as the concept of schema used by the two authors is different. Whigham's schemata represent derivation-tree fragments which always represent single subexpressions, while O'Reilly's schemata can represent multiple subexpressions.

3.8 Summary

In this chapter we have presented early attempts to build a theory for genetic programming, based on the concept of schema. In these approaches schemata were mostly interpreted as components or groups of components which could propagate within the population. The emphasis of this work was to model how the number of instances of such components would vary over time. In the chapter we have also shown how Price's Theorem is a particularly general and useful instrument for this kind of analysis.

In the next chapter we will start considering a different type of schema theory, in which schemata are considered to be subsets of the search space and the emphasis is on modelling how the number of individuals in such subsets varies over time.

4. Pessimistic GP Schema Theories

In Chapter 3 we described early schema theories for genetic programming. These are based on the idea of schemata as components within the program tree. In this chapter we will consider two GP schema theories of a different kind. These schemata have the advantage that not only can they be considered as parts of programs but, also that they are true subsets of the search space. In Section 4.1 we summarise the schema theory proposed by Justinian Rosca [Rosca, 1997a] that is applicable to GP with standard crossover. However, the main focus of the chapter is on our schema theory. This is based on a concept of schema which is much closer to the original concept of schema in genetic algorithms and it is applicable to GP (and to different kinds of GAs) when one-point crossover is used. This is a generalisation of the corresponding GA operator in which *the same* crossover point is selected in both parents.

We present our notion of schema, we describe one-point crossover and show how a simple and natural schema theorem for GP can be derived. This GP schema theorem is the natural counterpart (and generalisation of) a version of Holland's schema theorem proposed in [Whitley, 1993, Whitley, 1994]. This result, like Holland's theorem, is obtained by focusing on the disruption and survival of schemata. This leaves out schema creation effects which are much harder to model. Therefore these versions of GP schema theorems, like most of those in Chapter 3, only give a lower bound for the expected number of individuals belonging to a given schema at the next generation.

Recent results for binary genetic algorithms have been able to model schema creation in an exact manner. This has led to an exact schema theorem for binary GAs. This result has been recently extended to genetic programming with one-point crossover and other GP operators thanks to some of the ideas presented in this chapter. These exact schema theories will be described in Chapter 5.

4.1 Rosca's Rooted Tree Schemata

[Rosca, 1997a] proposed rooted tree-schemata, where, syntactically, every schema is a contiguous tree fragment, which includes the root node of the

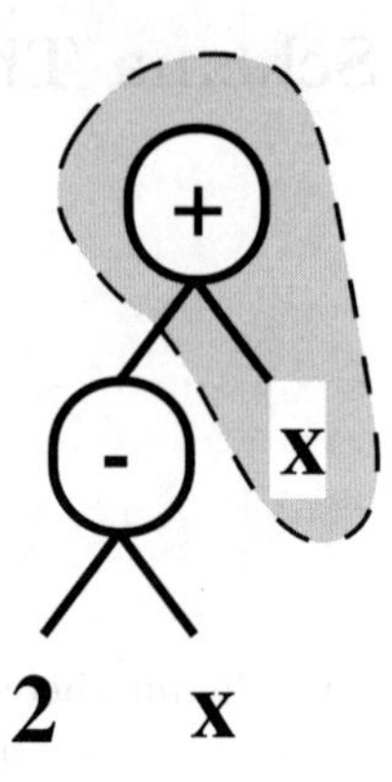

Fig. 4.1. Match of Rosca's schema H=(+ # x) onto program (+ (- 2 x) x).

tree. Semantically a rooted tree schema is a set of programs. For example, the rooted tree-schema H=(+ # x) represents all the programs whose root node is a +, and whose second argument is a x. Figure 4.1 shows this schema's match in (i.e. its instantiation in) the program (+ (- 2 x) x).

The rootedness of this schema representation, which was developed at the same time as and independently from ours, is very important as it reintroduces in the schema definition the positional information (Section 3.3) lacking in the previous definitions of schema for GP. As a consequence, a schema can be matched (instantiated) at most once within a program and studying the propagation of the components of the schema in the population is equivalent to analysing the way the number of programs sampling the schema change over time. Rosca derived the following schema theorem:[1]

$$E[m(H,t+1)] \geq m(H,t)\frac{f(H,t)}{\bar{f}(t)} \times \left[1 - (p'_m + p_{xo}) \underbrace{\sum_{h \in H \cap \mathrm{Pop(t)}} \overbrace{\frac{\mathcal{O}(H)}{N(h)}}^{P_d(H,h,t)} \frac{f(h)}{\sum_{h \in H \cap \mathrm{Pop(t)}} f(h)}}_{P_d(H,t)} \right] \tag{4.1}$$

where $N(h)$ is the size of a program h matching the schema H, $f(h)$ is its fitness, the order of a schema $\mathcal{O}(H)$ is the number of defining symbols it contains, Pop(t) is a multiset representing the population at generation t and

[1] We have rewritten the theorem in a form that is slightly different from the original one in order to highlight some of its features.

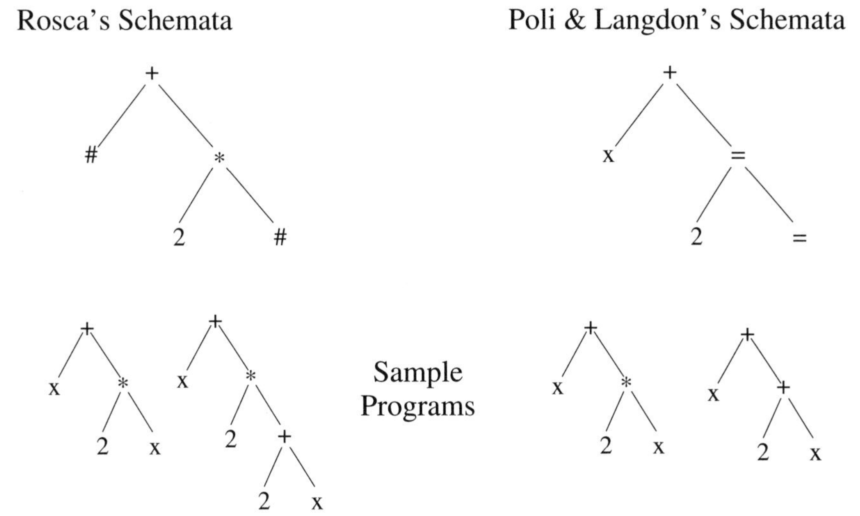

Rosca's Schemata

Poli & Langdon's Schemata

Sample Programs

Fig. 4.2. Examples of rooted schema (top) and some examples of programs sampling them (bottom).

$H \cap \text{Pop}(t)$ represents the multiset including all programs in Pop(t) which are also members of H.[2]

Rosca did not give a definition of the defining length for a schema. Rosca's schema theorem involves the evaluation of the weighted sum of the fragilities $\frac{\mathcal{O}(H)}{N(h)}$ of the instances of a schema within the population, using as weights the ratios between the fitness of the instances of H and the sum of the fitness of such instances. Another example of Rosca's schema and some of its instances are shown on the left-hand side of Figure 4.2.

In the definitions of GP schema mentioned in this section and in Sections 3.4–3.7, schemata divide the space of programs into subspaces containing programs of different sizes and shapes. In the next section we give a definition of schema which partitions the program space into subspaces of programs of fixed size and shape.

4.2 Fixed-Size-and-Shape Schemata in GP

If we take the viewpoint that a schema is a subspace of the space of possible solutions (Figure 1.5, page 8), then we can see schemata as mathematical tools to describe which areas of the search space are sampled by a population. For

[2] Multisets are extensions of the concept set in which elements can be present multiple times. So, for example, if the population contains five copies of a particular program, say $h=$`(+ 1 x)`, and this program is also a member of H, then the sum in Equation (4.1) will include 5 times the term relative to h.

schemata to be useful in explaining how GP searches, their definition must make the effects of selection, crossover and mutation comprehensible and relatively easy to calculate. The problem with some of the earlier definitions of schema for GP is that they make the effects on schemata of the genetic operators used in GP too difficult to evaluate. In the rest of this chapter we will provide a definition of schema for GP which has allowed us to overcome this problem [Poli and Langdon, 1997c, Poli and Langdon, 1998b].

Our definition of GP schema was inspired by the definition of schema for linear genetic algorithms (Section 3.2). In standard fixed length linear GAs, it is possible to obtain all the schemata contained in a string by simply replacing, in all possible ways, its symbols with "don't care" symbols which represent a *single character*. This process is shown in Figure 4.3(a) for a 6-bit string. Exactly the same approach can be used to obtain the GP schemata sampled by a program. It is sufficient to replace the nodes in a program with "don't care" nodes representing exactly *a single function or terminal* as in Figure 4.3(b). In the figure the equal sign "=" is used as a "don't care" symbol to emphasise the difference between our definition of schema and other GP definitions where the "don't care" symbol # is used to represent entire subtrees.

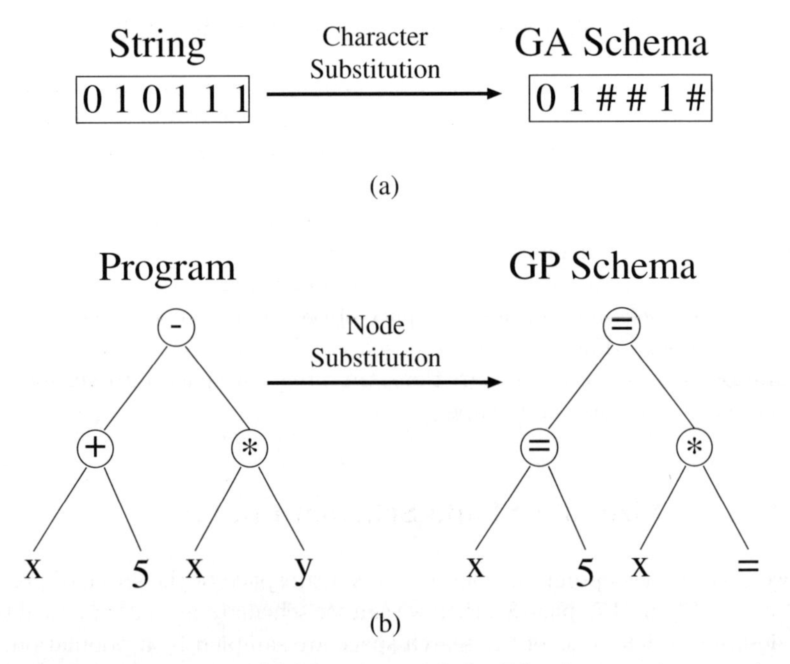

Fig. 4.3. Relation between (a) bit strings and the GA schemata they sample and (b) programs and the GP schemata they sample.

A natural extension to GP of the GA schema definition is:

Definition 4.2.1 (GP schema). *A* GP schema *is a rooted tree composed of nodes from the set* $\mathcal{F} \cup \mathcal{T} \cup \{=\}$*, where* $\mathcal{F}$ *and* $\mathcal{T}$ *are the function set and the terminal set used in a GP run, and the operator* = *is a polymorphic function with as many arities as the number of different arities of the elements of* $\mathcal{F} \cup \mathcal{T}$*, the terminals in* $\mathcal{T}$ *being 0-arity functions.*[3]

In the following we will sometimes refer to our schemata as *fixed-size-and-shape schemata.* An example of GP schema is shown on the right-hand side of Figure 4.2. There are cases in which a "don't care" symbol in one of our GP schemata can only stand for a particular function or terminal. This happens if there is only one function in $\mathcal{F} \cup \mathcal{T}$ with a particular arity (for an example see page 163). The same thing would happen in a multi-cardinality GA if a particular location (locus) could only take one value (allele). The theory presented in the rest of this chapter could readily be extended to cover this special case.

In line with the original definition of schema for genetic algorithms, a GP schema H represents multiple programs, all having the same shape as the tree representing H and the same labels for the non-= nodes. For example, if the function set $\mathcal{F}$={`+, -`} and the terminal set $\mathcal{T}$={`x, y`}, the schema H=`(+ (- = y) =)` would represent the four programs

```
(+ (- x y) x)
(+ (- x y) y)
(+ (- y y) x)
(+ (- y y) y)
```

We can now extend the concepts of order, length and defining length to our GP schemata.

Definition 4.2.2 (Order). *The number of non-= symbols is called the* order $\mathcal{O}(H)$ *of a schema* H*.*

Definition 4.2.3 (Length). *The total number of nodes in the schema is called the* length $N(H)$ *of a schema* H*.*[4]

Definition 4.2.4 (Defining Length). *The number of links in the minimum tree fragment including all the non-= symbols within a schema* H *is called the* defining length $\mathcal{L}(H)$ *of the schema.*

[3] We could have used a different "don't care" symbol for each arity instead of a single polymorphic "don't care" symbol. However, given that the arity of each = sign in a schema can be inferred by counting its arguments, for the sake of simplicity we decided to use a single symbol. The complications of the properties of the "don't care" symbol given in Definition 4.2.1 are not due to the variable size nature of the representation used in GP: they would be necessary also for linear genetic algorithms if different locations (loci) could take a different number of values (alleles), i.e. in a multi-cardinality GA.

[4] The length of a schema H is also equal to the size (number of nodes) of the programs matching H.

The defining length $\mathcal{L}(H)$ can be computed with a simple recursive procedure, similar to the one necessary to compute the size of the schema, $N(H)$. Figure 4.4 shows some of the 32 schemata matching the program `(+ (- 2 x) x)` along with their order $\mathcal{O}$ and defining length $\mathcal{L}$.

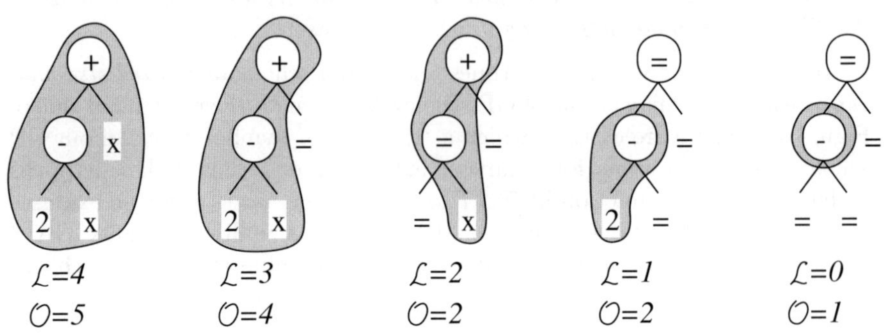

Fig. 4.4. Defining length $\mathcal{L}$ and order $\mathcal{O}$ of some some schemata sampled by the program `(+ (- 2 x) x)`. The minimum tree fragments used to compute $\mathcal{L}$ are enclosed in the shaded regions.

Like traditional binary genetic algorithms our definitions of order, length and defining length of a schema are totally independent of the shape and size of the programs in the actual population.

Schemata of low order and large length can represent a huge number of programs. For example, if there are two functions and two terminals, a schema of order $\mathcal{O} = 5$ and length $N = 30$ represents $2^{N-\mathcal{O}} = 2^{25} = 33,554,432$ different programs. While other definitions of GP schema are able to represent huge numbers of programs, the difference is that the number of programs represented by our schemata is *finite* and can easily be computed (if the function set and the terminal set are finite). In contrast the number is infinite with other schema definitions (unless the depth or size of the programs is limited, in which case the number is finite but cannot be expressed in a simple closed form).

Likewise, a program of length N includes 2^N different schemata. Therefore a modest-size population of programs can actually sample a huge number of schemata. Computing lower and upper bounds of such a number is trivial with Definition 4.2.1. There are between $2^{N_{\min}}$ and $M \cdot 2^{N_{\max}}$ different schemata in a population of size M whose smallest and largest trees contain $N_{\min}$ and $N_{\max}$ nodes, respectively. Calculating bounds like these is quite difficult with other definitions of schema.[5]

[5] This kind of calculations has traditionally been used to provide support for the implicit parallelism hypothesis. Roughly this hypothesis says that, because each individual in a GA belongs to a large number of schemata, the GA is implicitly processing huge numbers of schemata in parallel while (explicitly) operating on strings. This idea has been strongly criticised. Here, we provide these calculations only to emphasise the features of our definition of schema.

Our definition of schema is in some sense lower-level than those adopted by Koza, Altenberg, O'Reilly, Whigham and Rosca as a smaller number of trees can be represented by schemata with the same number of "don't care" symbols and it is possible to represent other kinds of schema by using a collection of ours. For example, assuming that the maximum allowed depth for trees is two and that all functions have arity 2, O'Reilly's schema H=[(+ # #)] can be represented by our schemata (+ = =), (= = (+ = =)), (= (+ = =) =), (= (+ = =) (= = =)), (= (= = =) (+ = =)), (+ = (= = =)), (+ (= = =) =) and (+ (= = =) (= = =)). Again, the converse is not true, since for example the similarity template (= x y) cannot be represented using O'Reilly's schemata. Similarly, it is easy to show that Rosca's schemata can be represented by sets of our schemata. The converse is not true. For example, schemata whose defining nodes do not form a compact subtree (where the minimum tree fragment linking all the defining node also has to contain "don't care" = nodes) cannot be represented using Rosca's schemata.

Our definition of schema, like Rosca's, reintroduces the positional information lacking in previous definitions of schema for GP. As a consequence, the number of instances of a schema in the population coincides with the number of programs sampling the schema.

In order to make it easier to understand the effects on schemata of one-point crossover (see next section) it is useful to introduce two more definitions which allow us to distinguish two broad classes of schemata.

Definition 4.2.5 (Hyperspaces and Hyperplanes). *A schema G is a* hyperspace *if it does not contain any defining nodes (i.e. order $\mathcal{O}(G) = 0$). A schema H is a* hyperplane *if it contains at least one defining node. The schema $G(H)$ obtained by replacing all the defining nodes in a hyperplane H with "don't care" symbols is called the* hyperspace *associated with H.*

Figure 4.5 represents the relation between programs, hyperplanes (ordinary schemata) and hyperspaces of programs. Each hyperspace represents all the programs with a given shape. Hyperspaces subdivide the space of possible programs into non-overlapping subsets whose union is the original space. That is, they form a tessellation of it. Unless a depth or size limit is imposed on the programs being considered, there are infinitely many hyperspaces. However, the number of programs belonging to each hyperspace is finite (assuming that both the function set $\mathcal{F}$ and the terminal set $\mathcal{T}$ are finite). Hyperplanes are smaller subsets of programs. Their name derives from the geometric interpretation illustrated in Figure 4.6 in which schemata of order 0 (hyperspaces) are represented as multidimensional parallelepipeds, the edges and faces of which represent schemata with order greater than 0 (i.e. hyperplanes). This interpretation is a direct extension of the usual interpretation of GA schemata as faces and edges of the unit hypercube, the vertices of which are all the possible bit strings of a prefixed length.

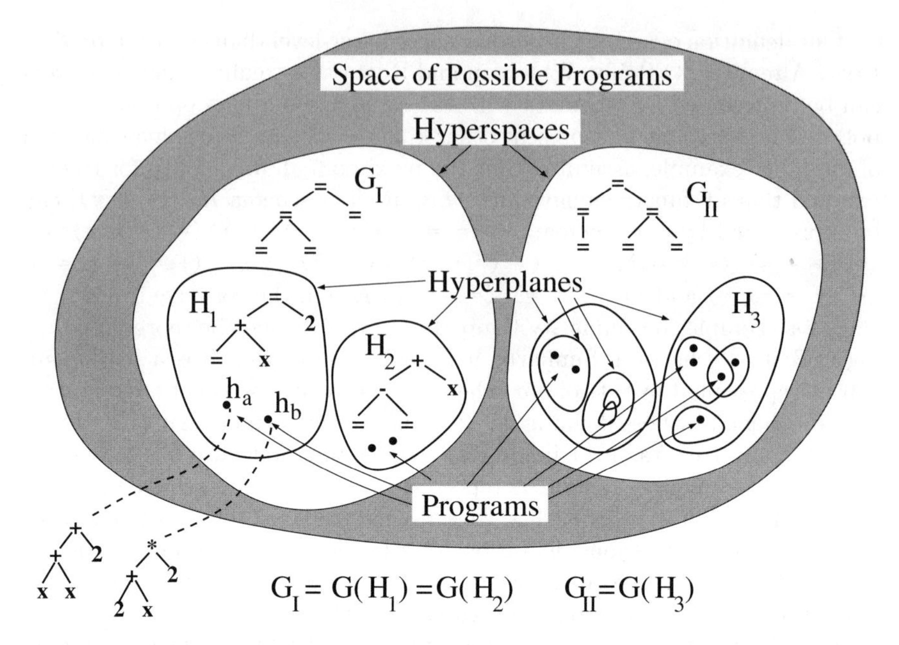

Fig. 4.5. Interpretation of hyperspaces, hyperplanes and programs as sets. Hyperspaces G are the set of all programs of a given shape. They do not overlap with one another. Hyperplanes H are subsets of a hyperspace. They can overlap. Individual programs are elements of the sets represented by hyperplanes and hyperspaces.

4.3 Point Mutation and One-Point Crossover in GP

The concept of schema just introduced is very simple and natural. However, like most previous schema definitions it does not make the effects of standard GP crossover easy to calculate. Given the similarity of our definition with the one of GA schema, we started wondering whether it would have been possible to get again inspiration from Holland's work and define more natural crossover and mutation operators.

The obvious analogue of bit-flip mutation is the substitution of a function in the tree with another function with the same arity, or the substitution of a terminal with another terminal: a technique that has been sometimes used in the GP literature [McKay *et al.*, 1995]. We will call this operation *point mutation*. The perhaps less obvious equivalent of one-point crossover for bit strings is a crossover operator that we also call *one-point crossover*.

The way one-point crossover works was suggested by considering the three steps involved in one-point crossover in binary GAs. These steps are shown in Figure 4.7(a). They are: (1) alignment of the parent strings, (2) selection of a common crossover point, and (3) swapping of the right ends of the strings to obtain two offspring (one of which is often discarded). If all the trees in a GP population had exactly the same shape and size, then a crossover

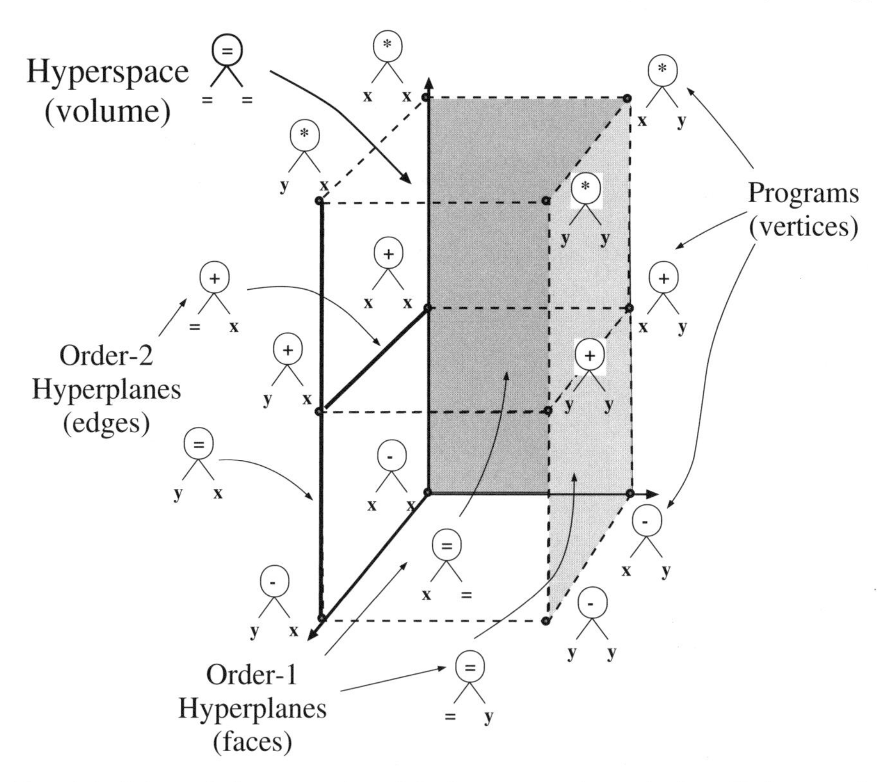

Fig. 4.6. Geometric interpretation of the hyperspace `(= = =)` when $\mathcal{F}$={`-`, `+`, `*`} and $\mathcal{T}$={`x`, `y`}. Individual points in the space represent programs, e.g. `(* x y)`. Edges are order 2 hyperplanes (schemata of order 2). For example, the vertical edge `(= x y)` links all three programs (`(* x y)`, `(+ x y)` and `(- x y)`) which match schema `(= x y)`. In this example order 1 schemata have only one defining component (e.g. `(= = y)`) and they define planes. The hyperspace `(= = =)` has no defining components and defines a volume (space).

operator involving exactly the same three steps could easily be implemented. The nodes in the trees would play the role of the digits in the bit strings, and the links would play the role of the gaps between digits. Figure 4.7(b) depicts this situation. In this figure the alignment process corresponds to translating the graphical representation of one of the parents until its root node coincides with the root node of the other.

What would happen if one tried the same procedure with trees having different shapes? To a certain extent it would still be possible to align the trees. After the alignment it would be easy to identify the links that overlap, to select a common crossover point and swap the corresponding subtrees. Our one-point crossover operator for GP is based on these simple steps. They are sketched in Figure 4.7(c).

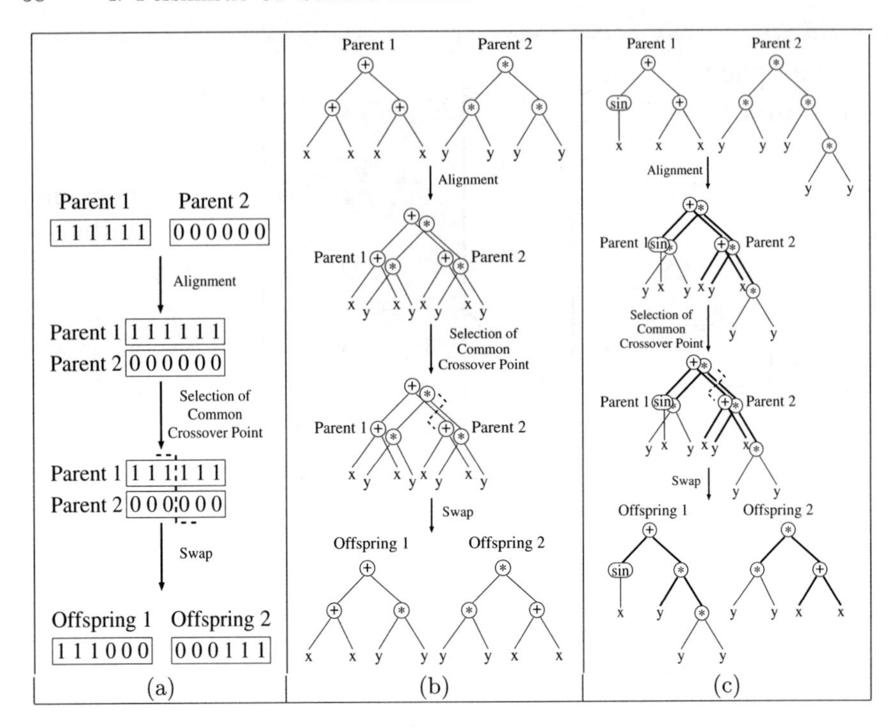

Fig. 4.7. (a) One-point crossover in binary GAs. (b) One-point crossover for GP when both parents have the same shape. (c) General form of one-point crossover for GP. In (c), the links that can be selected as common crossover points are drawn with thick lines.

From the implementation point of view, the three phases involved in one-point crossover are as follows:

1. *Alignment.* Copies of the two parent trees are recursively (jointly) traversed starting from the root nodes to identify the parts with the same shape, i.e. with the same arity in the nodes visited. Recursion is stopped as soon as an arity mismatch between corresponding nodes in the two trees is present. All the links encountered are stored.
2. *Crossover point selection.* A random crossover point is selected with a uniform probability among the stored links.
3. *Swap.* The two subtrees below the common crossover point are swapped in exactly the same way as in standard crossover.

One-point crossover has several interesting features that make it different from standard crossover.

First, in the absence of mutation one-point crossover allows the population to converge (like crossover in binary GAs), possibly with help from genetic drift. That is, the members of the population become more similar, and eventually every program becomes identical to every other one. This can be understood by considering the following. Until a large-enough proportion of the population has exactly the same structure in the upper parts of the tree, the probability of selecting a crossover point in the lower parts is relatively small. This effectively means that until a common upper structure emerges, one-point crossover is actually searching a much smaller space of approximately fixed-size (upper) structures. Obviously the lower parts of the subtrees moved around by crossover influence the fitness of the fixed-size upper parts, but they are not modified by crossover at this stage. Therefore, GP behaves like a GA searching for a partial solution (i.e. a good upper part) in a relatively small search space. This means that the algorithm quickly converges towards a common upper part, which cannot later be modified unless mutation is present. At that point the search concentrates on slightly lower levels in the tree with a similar behaviour, until level-after-level the entire population has completely converged. An important consequence of the convergence property of GP with one-point crossover is that, like in GAs, mutation becomes a very important operator to prevent premature convergence and to maintain diversity.

Secondly, one-point crossover does not increase the depth of the offspring beyond that of their parents, and therefore beyond the maximum depth of the initial random population. Similarly, one-point crossover will not produce offspring whose depth is smaller than that of the shallowest branch in their parents and therefore than the smallest of the individuals in the initial population. This means that the search performed by GP with one-point crossover and point mutation is limited to a subspace of programs defined by the initial population. Therefore, the initialisation method and parameters chosen for the creation of the initial population can modify significantly the behaviour of the algorithm. For example, if all functions in $\mathcal{F}$ have the same arity and one uses the "full" initialisation method [Koza, 1992] which produces balanced trees with a fixed depth, then the search will be limited to programs with a fixed size and shape. If on the contrary the "ramped half-and-half" initialisation method is used [Koza, 1992], which produces trees of variable shape and size with depths ranging from 0 to the prefixed maximum initial tree depth D, then the entire space of programs with maximum depth D will be searched (at least if the population is big enough). (In [Langdon, 2000] we define an alternative random initialisation, that produces programs of random shapes).

Thirdly, the offspring produced by one-point crossover inherit the common structure (emphasised with thick lines in Figure 4.7(c)) of the upper parts of their parents. So, crossing over a program with itself produces the original program. This property is an instance of a more general property: *GP schemata are closed under crossover* (i.e. when one-point crossover is applied to two programs sampling the same schema, the offspring produced will also sample the same schema). This property, which is also known as *respect* [Radcliffe, 1994], is considered to be very important in GAs. A respectful recombination operator makes sure that the offspring inherit common characteristics present in the parents. This has the effect of reducing the size of the search space explored by a GA over time. In the presence of selection, this gives a GA the ability to focus the search towards regions of the search space which have shown a consistently higher fitness.

Fourthly, and perhaps most importantly, one-point crossover makes the calculations necessary to model the disruption of GP schemata feasible. This means that it is possible to study in detail its effects on different kinds of schemata and to obtain a schema theorem as described in the next section.

One-point crossover has some similarity to strong context preserving crossover operator [D'haeseleer, 1994]. However, one-point crossover constraints crossover points more tightly than strong context preserving crossover.

Other operators, like two-point crossover and uniform crossover [Poli and Langdon, 1998a, Page *et al.*, 1999, Poli and Page, 2000], can easily be defined by extending the basic mechanisms of one-point crossover.

4.4 Disruption-Survival GP Schema Theorem

Our schema theorem (page 65) describes how the number of programs contained in a schema is expected to vary from one generation to the next as a result of fitness proportionate selection, one-point crossover and point mutation.

4.4.1 Effect of Fitness Proportionate Selection

The effect of fitness proportionate selection on our schemata can be obtained by performing exactly the same calculations as for the GA schema theorem (see Inequality (3.6) page 33 [Holland, 1975, Goldberg, 1989b, Whitley, 1993, Whitley, 1994, Whigham, 1996b, Appendix C]). If we assume that programs are selected using fitness proportionate selection and we denote with $\{h \in H\}$ the event "a program h sampling the schema H is selected" then:

$$\Pr\{h \in H\} = \frac{m(H,t)f(H,t)}{M\bar{f}(t)} \tag{4.2}$$

where all the symbols have the same meaning as in Inequality (3.6) (page 33).

As usual, if the population could include a large number of individuals and the ratio between schema fitness and population fitness was constant, this result would support the claim that schemata with above-average (below-average) fitness will tend to get an exponentially increasing (decreasing) number of programs sampling them [Goldberg, 1989b]. However, this can only be considered as an approximation since, in real-life GAs and GP, populations are always finite and the fitness of schemata changes since the population changes from generation to generation.

4.4.2 Effect of One-Point Crossover

One-point crossover can affect the number of instances of a schema in the next generation (its propagation) by disrupting some of them or creating some new schemata. Let us term $D_c(H)$ as the event: "H is disrupted when a program h matching H is crossed over with a program $\hat{h}$". (By "disrupted" we mean that the offspring produced does not match H).

There are two ways in which H can be disrupted. First, H can be disrupted when program h is crossed over with a program $\hat{h}$ with a different shape. (We discuss below the second way, when $\hat{h}$ has the same shape as h). If $G(H)$ is the hyperspace associated with H (i.e. $G(H)$ is the shape of all programs matching schema H), this can be expressed as the joint event $D_{c_1}(H) = \{D_c(H), \hat{h} \notin G(H)\}$. An example of this is shown in Figure 4.8. As another example of schema disruption of the first kind, consider the schema H=`(+ = =)` which program h=`(+ x y)` matches. If we cross h over with $\hat{h}$=`(+ z (+ 2 3))`, swapping the subtree `(+ 2 3)` with `y` produces the program `(+ x (+ 2 3))`. This has a different shape, i.e. it does not sample $G(H)$=`(= = =)`, and therefore cannot match H.

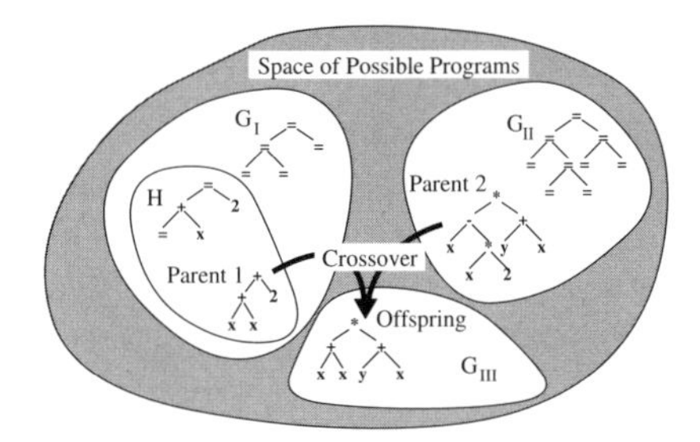

Fig. 4.8. Schema disruption of the first type, $D_{c_1}(H)$. Crossover between programs (Parent 1 and Parent 2) of different shapes (G_{I} and G_{II}) produces a program (Offspring) whose shape (G_{III}) is different and so cannot match any of the schemata that its parents do.

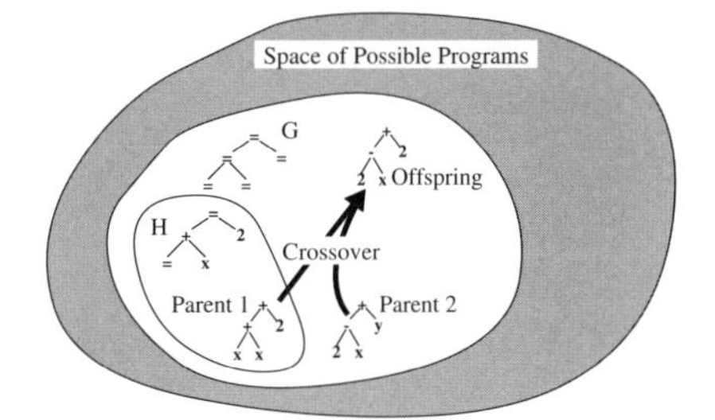

Fig. 4.9. Schema disruption of the second type, $D_{c_2}(H)$. Crossover of Parent 1 and Parent 2 of the same shape (G) produces an Offspring program of the same shape which does not match schema H. (Parent 1 matches H but Parent 2 does not).

The probability of the event D_{c_1} is given by:

$$\Pr\{D_{c_1}(H)\} = \Pr\{D_c(H)|\hat{h} \notin G(H)\} \Pr\{\hat{h} \notin G(H)\} \tag{4.3}$$

Consider the second term first. The probability of selecting a program $\hat{h}$ that has a different shape from h ($\Pr\{\hat{h} \notin G(H)\}$) is simply one minus the probability that it has the same shape: $1-\Pr\{\hat{h} \in G(H)\}$. Since parents are selected independently the probabilities associated with the second parent $\hat{h}$ are the same as those for the first h. As we are using fitness proportionate reproduction, $\Pr\{\hat{h} \in G(H)\} = \Pr\{h \in G(H)\}$ is given by applying Equation (4.2) to the schema $G(H)$. That is,

$$\Pr\{\hat{h} \notin G(H)\} = 1 - \Pr\{\hat{h} \in G(H)\} = 1 - \frac{m(G(H),t)f(G(H),t)}{M\bar{f}(t)} \quad (4.4)$$

The term $\Pr\{D_c(H)|\hat{h} \notin G(H)\}$, which we will denote by $p_{\text{diff}}(t)$ for brevity, is harder to quantify as not all crossovers between two parents h and $\hat{h}$ with different structure produce offspring which do not sample H. In fact if the common part between h and $\hat{h}$ where the crossover point can be placed includes terminals then crossover can swap subtrees with exactly the same shape and size. This situation is shown in Figure 4.10, where crossover points placed in the shaded areas produce subtrees with identical shapes. Swapping such subtrees will produce a program with the same shape as h, i.e. which matches $G(H)$. Whether this will sample also H depends on the actual node composition of the subtrees swapped and on the characteristics of the corresponding subtree in H. For this reason we will leave the term $p_{\text{diff}}(t)$ in symbolic form to avoid using the tempting, but over simplistic assumption that, when $\hat{h} \notin G(H)$, all crossover operations produce programs which do not match H, i.e. that $p_{\text{diff}}(t) = 1$. $p_{\text{diff}}(t)$ can be interpreted as the *fragility of the shape* of a schema with respect to the shapes of the other schemata in the population.

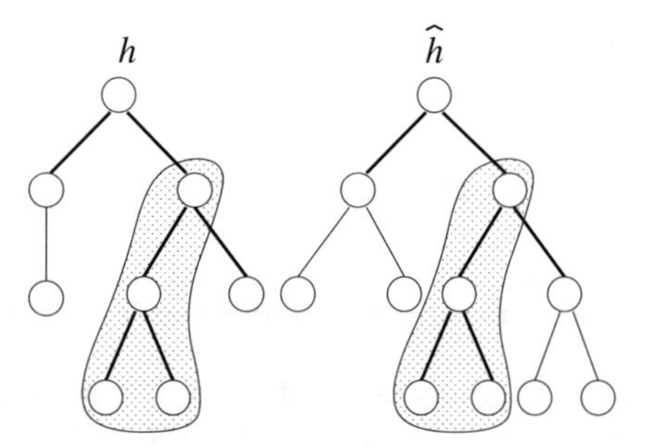

Fig. 4.10. Crossover between two individuals, h and $\hat{h}$, with different shapes. The links where the common crossover point can be placed are drawn with thick lines. The shaded areas enclose the links that, if selected as crossover points, need not lead to the disruption of the schemata in h.

The second way in which H can be disrupted is when h is crossed over with a program $\hat{h}$ with the same structure as H (i.e. $\hat{h} \in G(H)$) but which does not sample H (i.e. $\hat{h} \notin H$). Note that since our schemata are closed under one-point crossover, if both parents match schema H then (if there is no mutation) their offspring will always match schema H. The second type of disruption can be expressed as the joint event $D_{c_2}(H) = \{D_c(H), \hat{h} \notin H, \hat{h} \in G(H)\}$. This is shown in Figure 4.9.

The probability of $D_{c_2}(H)$ is given by:

$$\Pr\{D_{c_2}(H)\} = \Pr\{D_c(H)|\hat{h} \notin H, \hat{h} \in G(H)\} \Pr\{\hat{h} \notin H, \hat{h} \in G(H)\} \quad (4.5)$$

Consider the second term first. $\Pr\{\hat{h} \notin H, \hat{h} \in G(H)\}$ is the probability that the second parent $\hat{h}$ does not match the schema H but has the same shape $G(H)$ as the first parent. Given that the events $\{\hat{h} \in H\}$ are a subset of the events $\{\hat{h} \in G(H)\}$, as $H \subset G(H)$ (see Figure 4.11), we can write

$$\Pr\{\hat{h} \notin H, \hat{h} \in G(H)\} = \Pr\{\hat{h} \in G(H)\} - \Pr\{\hat{h} \in H\} \quad (4.6)$$

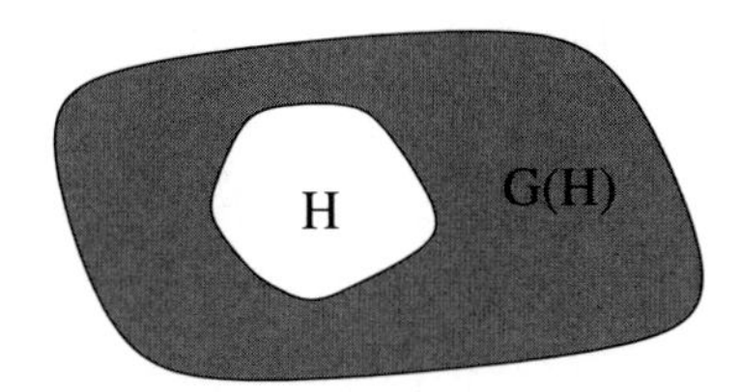

Fig. 4.11. H is a subset of $G(H)$ therefore the chance of selecting a parent which matches $G(H)$ but does not match H is the probability of matching $G(H)$ minus that of matching H.

Since we are using fitness proportionate selection, we can apply Equation (4.2) as before:

$$\Pr\{\hat{h} \in G(H)\} = \frac{m(G(H),t)f(G(H),t)}{M\bar{f}(t)} \quad (4.7)$$

and

$$\Pr\{\hat{h} \in H\} = \frac{m(H,t)f(H,t)}{M\bar{f}(t)} \quad (4.8)$$

Consider now the first part of Equation (4.5). A necessary condition for disruption of schema H by one-point crossover given that the second parent $\hat{h}$ has the same shape as the first (but $\hat{h}$ does not match H), i.e. the event $\{D_c(H)|\hat{h} \notin H, \hat{h} \in G(H)\}$, is that the crossover point is between the defining nodes (the non-= nodes) of H. We call this event $B(H)$. For example, Table 4.1 shows which schemata of h=`(+ x y)` could be disrupted depending on which of the two possible crossover points in h is chosen. The probability of $B(H)$ is the defining length of H divided by the total number of possible crossover points:

$$\Pr\{B(H)\} = \frac{\mathcal{L}(H)}{N(H)-1} \quad (4.9)$$

This probability can be interpreted as the intrinsic *fragility of the node composition* of a schema.

Since $B(H)$ is necessary for this type of disruption (but some crossovers between the defining nodes of H will not cause it to be disrupted), we have the inequality

$$\Pr\{D_c(H)|\hat{h} \notin H, \hat{h} \in G(H)\} \leq \Pr\{B(H)\} \tag{4.10}$$

Given Equations (4.5)–(4.9) and the inequality given by (4.10), we obtain:

$$\Pr\{D_{c_2}(H)\} \leq \frac{\mathcal{L}(H)}{N(H)-1} \frac{m(G(H),t)f(G(H),t) - m(H,t)f(H,t)}{M\bar{f}(t)} \tag{4.11}$$

Table 4.1. Possible effects of crossover on the schemata sampled by the program (+ x y) as a function of the crossover point selected, when the second parent has the same shape (= = =).

		Crossover Point	
Schema	$\Pr\{B(H)\}$	Between + and x	Between + and y
(= = =)	0	Unaffected	Unaffected
(= = y)	0	Unaffected	Unaffected
(= x =)	0	Unaffected	Unaffected
(+ = =)	0	Unaffected	Unaffected
(= x y)	1	Disrupted?	Disrupted?
(+ = y)	$\frac{1}{2}$	Unaffected	Disrupted?
(+ x =)	$\frac{1}{2}$	Disrupted?	Unaffected
(+ x y)	1	Disrupted?	Disrupted?

Since the events $D_{c_1}(H)$ and $D_{c_2}(H)$ are mutually exclusive, the probability of disruption of a schema H due to crossover is simply the sum of $\Pr\{D_{c_1}(H)\}$ and $\Pr\{D_{c_2}(H)\}$. Given the previous results (Equations (4.3), (4.4), the definition of $p_{\text{diff}}(t)$ and Inequality (4.11)), the following formula provides an upper bound for the probability of disruption due to one-point crossover:

$$\Pr\{D_c(H)\} \leq p_{\text{diff}}(t) \left(1 - \frac{m(G(H),t)f(G(H),t)}{M\bar{f}(t)}\right) + \frac{\mathcal{L}(H)}{N(H)-1} \frac{m(G(H),t)f(G(H),t) - m(H,t)f(H,t)}{M\bar{f}(t)} \tag{4.12}$$

When crossover is applied with a probability $p_{xo} < 1$, the actual probability of disruption of a schema H due to crossover is $p_{xo} \cdot \Pr\{D_c(H)\}$.

4.4.3 Effect of Point Mutation

Point mutation consists of changing the label of a node into a different one. We assume that it is applied to the nodes of the programs created using selection and crossover with a probability p_m per node. A schema H will survive mutation only if *all* its $\mathcal{O}(H)$ defining nodes are unaltered. (Thus a higher order schema, with more defining nodes, will be affected by point mutation more than a lower order one). The probability that each node is not altered is $1 - p_m$. So, a schema will be disrupted by mutation with a probability:

$$\Pr\{D_m(H)\} = 1 - (1 - p_m)^{\mathcal{O}(H)} \tag{4.13}$$

4.4.4 GP Fixed-size-and-shape Schema Theorem

By considering all the various effects discussed above, we can write:

$$E[m(H, t+1)] \geq M \Pr\{h \in H\}(1 - \Pr\{D_m(H)\})(1 - p_{xo} \Pr\{D_c(H)\}) \tag{4.14}$$

where the inequality is due to the fact that we ignore creation events due to crossover and mutation.

By substituting (4.2), (4.12) and (4.13) into expression (4.14) we obtain a lower bound for the expected number of individuals sampling a schema H at generation $t + 1$:

Theorem 4.4.1 (Survival-Disruption GP Schema Theorem). *In a generational GP with fitness proportionate selection, one-point crossover and point mutation:*

$$E[m(H, t+1)] \geq m(H, t)\frac{f(H, t)}{\bar{f}(t)} \times (1 - p_m)^{\mathcal{O}(H)} \times \left\{ 1 - p_{xo} \left[\begin{array}{c} p_{\text{diff}}(t) \left(1 - \dfrac{m(G(H), t) f(G(H), t)}{M\bar{f}(t)}\right) \\ + \\ \dfrac{\mathcal{L}(H)}{(N(H) - 1)} \dfrac{m(G(H), t) f(G(H), t) - m(H, t) f(H, t)}{M\bar{f}(t)} \end{array} \right] \right\} \tag{4.15}$$

The theorem can be easily generalised to other types of selection with replacement, obtaining:

$$E[m(H, t+1)] \geq M p(H, t) \times (1 - p_m)^{\mathcal{O}(H)} \times \left\{1 - p_{xo} \left[p_{\text{diff}}(t)\,(1 - p(G(H), t)) + \frac{\mathcal{L}(H)}{(N(H) - 1)}(p(G(H), t) - p(H, t))\right]\right\} \tag{4.16}$$

4.4.5 Discussion

The GP schema theorem given in Inequality (4.15) is considerably more complicated than the corresponding version for GAs in Inequality (3.6) (page 33). This is because in GP the trees undergoing optimisation have variable size and shape. In Inequality (4.15) the different sizes and shapes are accounted for by the presence of the terms $m(G(H),t)$ and $f(G(H),t)$, which summarise the characteristics of all the programs in the population having the same shape and size as the schema H whose propagation is being studied. Therefore, in GP the propagation of a schema H seems to depend not only on the features of the programs sampling it but also on the features of the programs having the same structure as H. However, the term $p_{\text{diff}}(t)$, which represents the fragility of the shape of the schema H in the population, also plays an important role.

In the following we will discuss how these terms change over time in a run and show that GP with one-point crossover tends asymptotically to behave like a GA and that, for large enough values of t, Inequality (4.15) transforms into Inequality (3.6) (page 33).[6] In particular we will consider the situation at the early and late stages of a run.

4.4.6 Early Stages of a GP Run

If we consider the typical case in which functions with different arities are present in the function set and the initialisation method is "ramped half-and-half" or "grow" [Koza, 1992]. *At the beginning of a run* it is unlikely that two trees h and $\hat{h}$ with *different* shapes undergoing crossover will swap a subtree with exactly the same shape. This is because given the diversity in the initial population, only a small number of individuals will have common parts including terminals (refer to Figure 4.10). Therefore, we can safely assume that $p_{\text{diff}}(t) \approx 1$.

The term $m(G(H),t)f(G(H),t)$ is the total fitness allocated to the programs with structure $G(H)$ (by definition it is equal to $\sum_{h\in \text{Pop(t)}\cap G(H)} f(h)$, where $f(h)$ is the fitness of individual h). Likewise, $M\bar{f}(t)$ is the total fitness in the population (i.e. $\sum_{h\in \text{Pop(t)}} f(h)$). Therefore the diversity of the population at the beginning of a run implies also that $m(G(H),t)f(G(H),t) \ll M\bar{f}(t)$ and that *the probability of schema disruption is very large (i.e. close to 1) at the beginning of a run*, even without considering disruption between programs of the same shape ($D_{c_2}(H)$ events). On average, a similarly high level of construction must correspond to this high level of disruption (since the

[6] Some of the arguments presented in the rest of this section are the result of observing the actual behaviour of runs of GP with one-point crossover (some of which are described in [Poli and Langdon, 1997b]) and also of the experiments reported in [Poli and Langdon, 1997a] in which we have counted the schemata present in small populations and studied their creation, disruption and propagation.

disrupted instances of a schema must still belong to some other schema). However, our schema theorem does not account for creation events. Therefore, we have to expect that the lower bound provided considering survival and disruption only will be in general quite pessimistic in the early stages of a GP run.[7]

4.4.7 Late Stages of a GP Run

If the mutation rate is small and we use only one-point crossover, after a while the population will start converging, like in a GA, and the diversity of shapes and sizes will decrease. This in turn will make the common part between pairs of programs undergoing crossover grow and include more and more terminals. As a result, the probability of swapping subtrees with the same structure, even when crossing over programs with different shapes, will increase and the fragility of the shape of the schema $p_{\text{diff}}(t)$ will decrease. Losing diversity also means that a relatively small number of different program structures $G(H)$ will survive in the population and that the factor multiplying $p_{\text{diff}}(t)$ in Inequality (4.15) will become smaller.

Therefore, in a second phase of a run the probability of disruption when crossover is performed between trees with the same shape and size, $\Pr\{D_{c_2}(H)\}$, will start becoming a more important component in overall disruption probability.

After many generations the probability of swapping subtrees with the same structure tends to 1, and $p_{\text{diff}}(t)$ tends to become unimportant. In this case (assuming H is in the population) $m(G(H),t) = M$ and $f(G(H),t) = \bar{f}(t)$ so Inequality (4.15) becomes exactly the same as Inequality (3.6) (page 33). That is, the survival-disruption GP schema theorem becomes the same as the GA schema theorem. This is consistent with the intuition that if all programs have the same structure then their nodes can be described with fixed-length strings, and GP with one-point crossover is really nothing more than a GA. (In less typical conditions, where the "full" initialisation method is used and all functions have the same arity, every individual in the population has the same shape and Inequality (4.15) is the same as Inequality (3.6) from the beginning of a run). This asymptotic behaviour of GP with one-point crossover can be summarised with the slogan:

$$\lim_{t\to\infty} \text{GP}_{\text{1pt}}(t) = \text{GA}.$$

[7] This view on the effects of disruption differs considerably from the corresponding views in [Poli and Langdon, 1997c, Poli and Langdon, 1998b], which now seem dated.

4.4.8 Interpretation

We can interpret the presence of the two parts in the term multiplying p_{xo} in Inequality (4.15) by saying that in GP with one-point crossover there are two competitions going on. The first one happens at the beginning of a run when different *hyperspaces* $G(H)$, representing all programs with a given shape and size, compete. The second competition starts when the first one starts settling. In this second phase the hyperplanes within the few remaining hyperspaces start competing. In this phase GP tends to behave more and more like a standard GA.

As discussed in [O'Reilly and Oppacher, 1995], in general no schema theorem alone can support conclusions on the correctness of the building block hypothesis. However, our definition of schema and the characterisation of $\Pr\{D_c(H)\}$ presented above seem to suggest that the building block hypothesis is much less troublesome in GP with one-point crossover than in standard GP. It is possible, however, that two kinds of building blocks are needed to explain how GP builds solutions: building blocks representing hyperspaces and building blocks representing hyperplanes within hyperspaces. We will come back to the building-block issue in Chapter 5 after introducing two exact schema theories for GP.

4.5 Summary

While we started in Section 4.1 by looking at Rosca's GP schemata and schema theorem, mostly this chapter has described our fixed-size-and-shape GP schemata. Using these we presented the first respectful crossover operator for tree GP: one-point crossover. By generalising a version of Holland's GA schema theorem, we have presented a GP schema theorem for one-point crossover and point mutation. Like Holland's GA schema theorem, (4.15) gives a pessimistic lower bound on the expected number of copies of a schema in the next generation. In the next chapter we will consider schema theorems which give an exact expected number rather than a bound.

5. Exact GP Schema Theorems

In Chapter 4 we introduced pessimistic schema theories for genetic programming in which schemata are treated as subsets of the search space. These two theories (Rosca's and ours) do not take schema creation into account and consider only the worst-case scenario for schema disruption. As a result they only provide lower bounds for the expected number of members of the population matching a given schema at the next generation.

We start this chapter by discussing the weaknesses presented by worst-case scenario GP schema theorems. (These are similar to those of many GA schema theorems presented in the past). Then in Section 5.2 we consider schema creation. Section 5.3 reviews recent results for binary GAs that have been able to model exactly schema creation thus leading to an exact macroscopic schema theorem. Section 5.4 shows how this result can be extended to genetic programming with one-point and standard crossover. This is followed by a worked example, Section 5.5. We conclude in Section 5.6 with an exact schema theorem for GP using subtree swapping crossover.

5.1 Criticisms of Schema Theorems

The usefulness of Holland's schema theorem has been widely criticised [Chung and Perez, 1994, Altenberg, 1995, Fogel and Ghozeil, 1997, Fogel and Ghozeil, 1998, Vose, 1999], and many people in the evolutionary computation field nowadays seem to believe that the theorem is nothing more than a trivial tautology of no use whatsoever (see for example [Vose, 1999, preface]). So recently the attention of GA theorists has moved away from schemata to land onto Markov chain models [Nix and Vose, 1992, Davis and Principe, 1993, Rudolph, 1994, Rudolph, 1997a, Rudolph, 1997b, Rudolph, 1997c]. However, many of the problems attributed to the schema theorem are probably not due to the theorem itself, but rather to its over-interpretations [Radcliffe, 1997].

The main criticism of schema theorems is that they cannot be used easily for predicting the behaviour of a GA over multiple generations. Two reasons for this are that schema theorems firstly give only a *lower bound* and, secondly, this bound is for the *expected value* of the number of instances of a schema H at the next generation $E[m(H, t + 1)]$.[1] Unless one assumes that the population is infinite, the expectation operator means that it is not easy to use schema theorems recursively to predict the behaviour of a genetic algorithm over multiple generations (see [Poli, 2000c] for a taste of the complexities involved in performing this kind of calculations for finite populations). In addition, since the schema theorem provides only a lower bound, some people argue that the predictions of the schema theorem are not very useful even for a single generation ahead.

Clearly there is some truth in these criticisms. However, this does not mean that schema theorems are useless. As shown by our own recent work and that of others [Stephens and Waelbroeck, 1997, Stephens and Waelbroeck, 1999, Stephens and Vargas, 2000, Poli, 1999b, Poli, 2000b, Poli, 2000c, Poli, 2001a, Poli and McPhee, 2001c, Poli and McPhee, 2001a] schema theorems have not been fully exploited nor fully developed, and when this is done they become very useful.

In the rest of the chapter we will present new theoretical results about GP and GA schemata which largely overcome some of the weaknesses discussed above. First, unlike previous results which concentrated on schema survival and disruption, our results extend recent work on GA theory by Stephens and Waelbroeck to GP, and so make the effects of schema creation explicit. This allows us to give an exact microscopic formulation (rather than a lower bound) for the expected number of individuals in the population matching schema H at the next generation, $E[m(H, t+1)]$. We are then able to provide an improved version for the GP schema theorem in Equation 4.15 (page 65) where some schema creation events are accounted for, thus obtaining a tighter bound for $E[m(H, t + 1)]$. By properly manipulating the terms in the exact microscopic schema theorem mentioned above, we are then able to obtain an exact macroscopic equation to calculate $E[m(H, t + 1)]$. The equation indicates that $E[m(H, t+1)]$ is a function of the selection probabilities of the schema itself and those of an associated set of lower-order schemata that one-point crossover uses to build instances of the schema. This result supports the existence of building blocks in GP which, however, are not necessarily all short, low-order or highly fit. Finally, by further extending our notation, we are able to present an exact macroscopic schema theorem for GP with subtree crossover.

We start this chapter (Section 5.2) by showing the kind of things one can do when schema creation is incorporated in the analysis, like in the case of Stephens and Waelbroeck's GA results and the new GP results described in this chapter. We will show how exact schema theorems can be used to

[1] Holland's original theorem was formulated for a GA in which the second parent is not selected on the basis of fitness, like the first parent, but it is randomly chosen from the population. Recently it has been shown that the theorem offers a slightly incorrect bound, due to an approximation error [Menke, 1997]. However, the version of Holland's schema theorem presented in this book does not seem to be affected by the error.

evaluate schema variance, signal-to-noise ratio and, in general, the probability distribution of the number of programs matching schema H, i.e. $m(H, t+1)$. It is then possible to predict, with a known certainty, whether $m(H, t+1)$ is going to be above a given threshold. That is the expectation operator can be removed from the schema theorem, giving results about $m(H, t+1)$ rather than $E[m(H, t+1)]$.

Then, in Section 5.4, we introduce the hyperschema theory for GP, indicating its uses and interpretations. Since this work extends both our and other people's work, in particular the recent GA results by Stephens and Waelbroeck, we will extensively review them in Section 5.3.

5.2 The Role of Schema Creation

In a GA various processes contribute to the overall production and loss of individuals matching a schema H. Crossover will possibly cause some of the individuals matching a schema H selected to have children to produce offspring which do not match H, and will possibly create some children which match H when their parents do not. Since typically some individuals are simply cloned, when selection plus cloning is applied to individuals matching H, it is guaranteed to "produce" individuals matching H. If mutation is applied to the individuals thus produced, things are even more complicated. In general we can summarise all these effects using a single parameter: the total schema transmission probability for the schema H, $\alpha(H, t)$. This represents the probability that, at generation t, the individuals produced by selection, crossover, cloning and mutation will match H.

As shown in the previous chapters, identifying all possible ways in which schemata can survive, be destroyed or be created, and calculating the probabilities of each type of event, is quite difficult. So, calculating α may generally be quite difficult, and it would not be possible to do so using the analyses presented in the previous chapters. However, let us for a moment assume that we know the exact expression for $\alpha(H, t)$ for a given schema and a given set of operators, and let us investigate the kind of things we could do with such an expression.

If each selection/crossover is independent (as is usually the case in GAs and GP, e.g. in roulette wheel selection), then the chance of each child matching H will be the same, so the total $m(H, t+1)$ will be binomially distributed. This is really not surprising. In fact it is a simple extension of the ideas, formulated mathematically in [Wright, 1931], which form the basis of the Wright-Fisher model of reproduction for a gene in a finite population with non-overlapping generations. The binomial distribution gives

$$\Pr\{m(H, t+1) = k\} = \binom{M}{k} \alpha(H, t)^k (1 - \alpha(H, t))^{M-k} \tag{5.1}$$

Since the probability distribution of $m(H,t+1)$ is a Binomial, its mean (which is $E[m(H,t+1)]$) and variance can be readily calculated:

$$E[m(H,t+1)] = M\alpha(H,t), \qquad (5.2)$$
$$\mathrm{Var}[m(H,t+1)] = M\alpha(H,t)(1-\alpha(H,t))$$

So, the total schema transmission probability α at time t corresponds to the expected proportion of population sampling H at generation $t+1$.

We define the signal-to-noise ratio of the schema as the mean divided by the standard deviation:

$$\left(\frac{S}{N}\right) \stackrel{\mathrm{def}}{=} \frac{E[m(H,t+1)]}{\sqrt{\mathrm{Var}[m(H,t+1)]}} = \sqrt{M}\sqrt{\frac{\alpha(H,t)}{1-\alpha(H,t)}}$$

When the signal-to-noise ratio is large, the propagation of a schema will occur nearly exactly as predicted. When the signal-to-noise ratio is small, the actual number of instances of a schema in the next generation is essentially random with respect to its predicted value.

From Equation (5.1) it is also possible to compute the probability of schema extinction in one generation:

$$\Pr\{m(H,t+1)=0\} = (1-\alpha(H,t))^M$$

Also, given $\alpha(H,t)$ we can calculate exactly the probability, $\Pr\{m(H,t+1) \geq x\}$, that the schema H will have at least x instances at generation $t+1$, for any given x. Unfortunately, the result of this calculation is difficult to use [Poli, 2000c].

One way to remove this problem is to *not* fully exploit our knowledge about the probability distribution of $m(H,t+1)$ when computing $\Pr\{m(H,t+1) \geq x\}$. Instead we could use Chebyshev's inequality: $\Pr\{|X-\mu| < k\sigma\} \geq 1-\frac{1}{k^2}$ where X is a random variable (with *any* probability distribution), μ is the mean of X, σ is its standard deviation and k is an arbitrary positive number [Spiegel, 1975].

Since $m(H,t+1)$ is binomially distributed, $\mu = M\alpha$ and $\sigma = \sqrt{M\alpha(1-\alpha)}$, where α is a shorthand notation for $\alpha(H,t)$ and M is the population size. By substituting these equations into Chebyshev's inequality we obtain:

Theorem 5.2.1 (Two-sided Probabilistic Schema Theorem). *For any given constant $k > 0$,*

$$\Pr\{|m(H,t+1) - M\alpha| \leq k\sqrt{M\alpha(1-\alpha)}\} \geq 1-\frac{1}{k^2}$$

Also, since $\Pr\{m(H,t+1) > M\alpha - k\sqrt{M\alpha(1-\alpha)}\} \geq \Pr\{|m(H,t+1) - M\alpha| \leq k\sqrt{M\alpha(1-\alpha)}\}$, we obtain

Theorem 5.2.2 (Probabilistic Schema Theorem). *For any given constant $k > 0$,*

$$\Pr\{m(H,t+1) > M\alpha - k\sqrt{M\alpha(1-\alpha)}\} \geq 1-\frac{1}{k^2}$$

Because the results reported above only require the knowledge of the transmission probability α, the results in this section are valid whatever the representation of the individuals, the operators used and the definition of schema. Therefore, they are valid for genetic algorithms as well as for genetic programming.

In [Poli, 2000c] we used the previous theorem in conjunction with the Stephens and Waelbroeck's theory presented in the next section to study the convergence of genetic algorithms. Knowing an expression for α also allows one to study the propagation of schemata in the presence of stochastic effects [Poli, 2000d].

5.3 Stephens and Waelbroeck's GA Schema Theory

The results in the previous section assume that we know the total transmission probability, α. That is, they require that schema survival, disruption and creation events be modelled mathematically. This is not an easy task, especially if one wants to use *only* the properties of the schema H (such as the number of instances of H and the fitness of H). Indeed, until now none of the schema theorems have succeeded in doing this. This is the reason why all of schema theorems provide only lower bounds. However it is now possible, thanks to the recent work in [Stephens and Waelbroeck, 1997, Stephens and Waelbroeck, 1999], to express $\alpha(H,t)$ exactly in the case of genetic algorithms operating on fixed-length bit strings with one-point crossover. Stephens and Waelbroeck describe $\alpha(H,t)$ in terms of the properties of lower-order schemata that are related to H.

For a bit string genetic algorithm with one-point crossover applied with a probability p_{xo}, $\alpha(H,t)$ is given by:[2]

$$\alpha(H,t) = (1 - p_{xo})p(H,t) + \frac{p_{xo}}{N-1} \sum_{i=1}^{N-1} p(L(H,i),t)p(R(H,i),t), \quad (5.3)$$

where:

$L(H,i)$ is the left part of schema H, i.e. the schema obtained by replacing with "don't care" symbols all the elements of H from position $i+1$ to position N.

$R(H,i)$ is the right part of schema H, i.e. the schema obtained by replacing with "don't care" symbols all the elements of H from position 1 to position i.

$p(H,t)$ is again the probability of selecting an individual matching schema H to be a parent.

Note the summation is over all the valid crossover points.

[2] Equation (5.3) is equivalent to the results in [Stephens and Waelbroeck, 1997, Stephens and Waelbroeck, 1999], assuming $p_m = 0$.

Some examples may clarify the meaning of $L(H, i)$ and $R(H, i)$:

H	$=$ `1#111`
$L(H,1)$	$=$ `1####`
$R(H,1)$	$=$ `##111`
$L(H,2)$	$=$ `1####`
$R(H,2)$	$=$ `##111`
$L(H,3)$	$=$ `1#1##`
$R(H,3)$	$=$ `###11`
$L(H,4)$	$=$ `1#11#`
$R(H,4)$	$=$ `####1`

5.4 GP Hyperschema Theory

A question that immediately comes to mind is whether it would be possible to extend the GP schema theories described in the previous chapter to obtain an exact schema theorem for GP. As will be described below, this is possible thanks to the introduction of a generalisation of the definition of GP schema: the hyperschema.

5.4.1 Theory for Programs of Fixed Size and Shape

When the population of programs all have exactly the same size and shape, it is possible to express the total transmission probability of a fixed-size-and-shape schema H, $\alpha(H, t)$, in the presence of one-point crossover, in the same way as in Equation 5.3; i.e.

$$\alpha(H,t) = (1 - p_{xo})p(H,t) + \frac{p_{xo}}{N(II)} \sum_{i=0}^{N(H)-1} p(l(H,i),t)p(u(H,i),t), \quad (5.4)$$

where:

$N(H)$ is the number nodes in the schema H (which is assumed to have the same size and shape as the programs in the population).

$l(H, i)$ is the schema obtained by replacing all the nodes above crossover point i with `=` nodes.

$u(H, i)$ is the schema obtained by replacing all the nodes below crossover point i with `=` nodes.

Note the summation is over all the $N(H)$ valid crossover points.

The symbol l stands for "lower part of", while u stands for "upper part of". For example, Figure 5.1 (second to fifth columns) shows how $l(H, 1)$, $u(H, 1)$, $l(H, 3)$ and $u(H, 3)$ are obtained from $H =$`(* = (+ x =))`. The numbering of the crossover points is given in the first column of Figure 5.1.

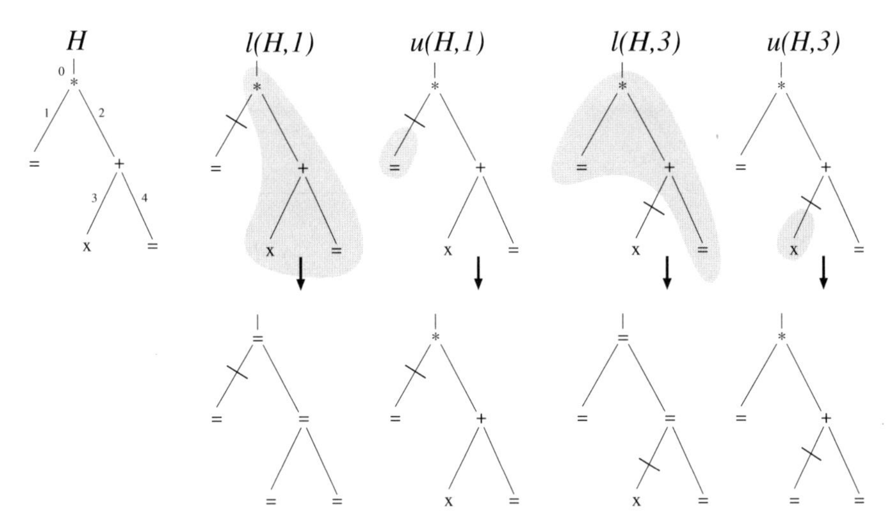

Fig. 5.1. Example of schema and some of its potential fixed-size-and-shape building blocks. The top row of trees are all the same schema H. The cross bars (except leftmost tree) indicates a crossover point. The shaded area (top row) shows the part of H to be replaced by "don't care" symbols. The lower row of trees are the resulting l and u schemata.

We obtain Equation (5.4) as follows. First we assume that while producing each individual for a new generation, one first decides whether to apply selection followed by cloning (probability $1 - p_{xo}$) or selection followed by crossover (probability p_{xo}). If selection followed by cloning is applied, the new individual created samples H with a probability $p(H,t)$, hence the first term in Equation (5.4). If selection followed by crossover is chosen, we first select the crossover point (to be interpreted as a link between nodes) randomly out of the $N(H)$ crossover points available. Then we select two parents and perform crossover. This will result in an individual that samples H only if the first parent has the correct lower part (with respect to the crossover point) *and* the second parent has the correct upper part. Assuming that crossover point i has been selected, the parents create a child that matches H only if they belong to $l(H,i)$ and $u(H,i)$ respectively. This gives the terms in the summation. By combining the probabilities of all these events we obtain Equation (5.4).

Equation (5.4) is applicable to any kind of tree-like structures of fixed size and shape, so although Equation (5.4) seems to have exactly the same form as Stephens and Waelbroeck's result for binary GAs in Equation (5.3), the GP result (5.4) is a generalisation of the GA result (5.3). If one considers program trees made from only unary functions, the two results coincide[3].

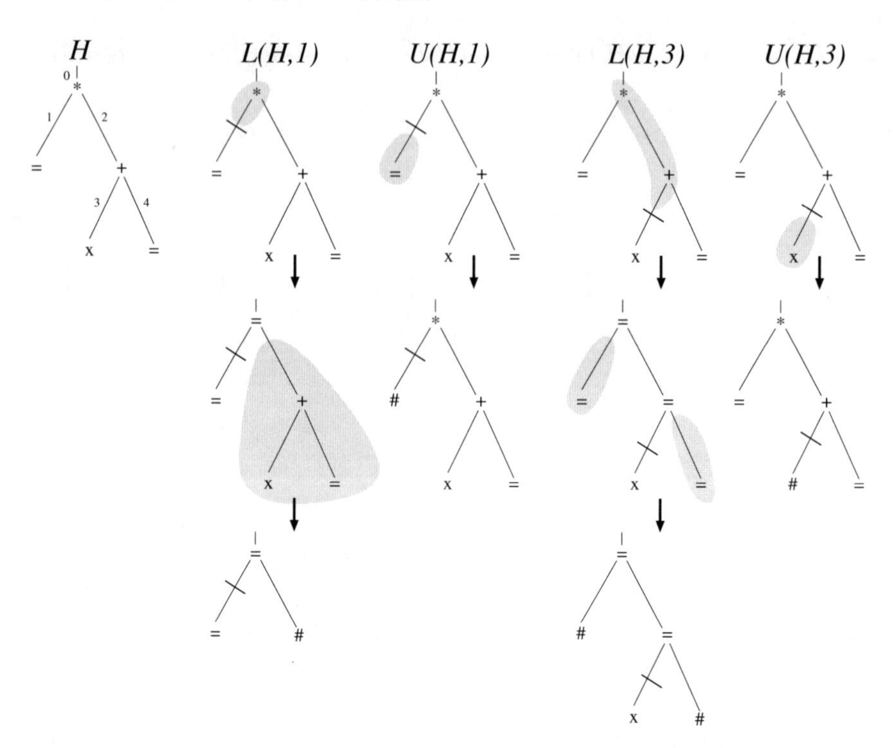

Fig. 5.2. Example of a schema and some of its potential hyperschema building blocks. The top row of trees are all the same schema H. The cross bars (except leftmost tree) indicates a crossover point. (H and the crossovers are the same as Figure 5.1). The shaded area (top row) shows the first part of H to be replaced by "don't care" symbols. In the U hyperschemata, the shaded region is replaced by a # (match any subtree), shown in the second row. While in L hyperschemata, the shaded region is replaced by = (match exactly one symbol). The third row are the final L hyperschemata produced by the second stage, where side subtrees (shown shaded in the middle row) are replaced by # "don't care" symbols.

[3] This is because GP one-point crossover and GA one-point crossover behave in exactly the same way in these conditions. Since Equation (5.4) is applicable only to structures of a fixed size (and shape), one could think of the action of GP one-point crossover on non-linear trees as the action of a non-standard GA crossover acting on the flattened representation of such trees. For example, GP one-point crossover operating on structures matching the schema $(== \cdots =)$ of size N (a tree including only a root node whose arguments are $N-1$ terminals) would behave like a GA crossover operator acting on strings of length N which, with equal probability, does one of the following actions: swaps all the bits in one parent with all the bits in the other, swaps bit 2 only, swaps bit 3 only, ... , swaps bit N only.

5.4.2 Hyperschemata

To generalise Equation (5.4) to populations including programs of different sizes and shapes, we need to introduce a new, more general definition of schema. The definition is:

Definition 5.4.1 (GP hyperschema). *A* GP hyperschema *is a rooted tree composed of internal nodes from the set $\mathcal{F} \cup \{=\}$ and leaves from $\mathcal{T} \cup \{=, \#\}$. That is the hyperschema function set is the function set used in a GP run plus $=$. The hyperschema terminal set is the GP terminal set plus $=$ and $\#$. The operator* `=` *is a "don't care" symbol which stands for exactly one node, while the new operator* `#` *stands for any valid subtree.*

Perhaps an example of hyperschema will make this clearer. Hyperschema `(* # (+ x =))` represents all the programs with the following characteristics: (1) the root node is a product (`*`), (2) the first argument of the root node is any valid subtree (`#`), (3) the second argument of the root node is `+`, (4) the first argument of the `+` is the variable `x` and (5) the second argument of the `+` is any valid node (`=`) in the terminal set.

Definition 5.4.1 has some of the features of the fixed-size-and-shape schema definition we used for the GP schema theory in [Poli and Langdon, 1997c, Poli and Langdon, 1998b] and summarised in Section 4.4.4: hyperschemata are rooted trees, and they include "don't care" symbols which stand for one node only. However Definition 5.4.1 also includes one of the features of the schema definitions proposed by other authors [O'Reilly and Oppacher, 1995, Rosca, 1997a]: hyperschemata also include "don't care" symbols which stand for entire subtrees. Indeed, the notion of hyperschema is a generalisation of both Rosca's schemata (which are hyperschemata without `=` symbols) and fixed-size-and-shape schemata (which are hyperschemata without `#` symbols). So a hyperschema can represent a group of schemata in essentially the same way as such schemata represent groups of program trees (hence the name "hyperschema").

5.4.3 Microscopic Exact GP Schema Theorem

Thanks to hyperschemata, it is possible to obtain a general result which is valid for populations of programs of any size and shape. However, this is a *microscopic* schema theorem in the sense that it is necessary to consider in detail each member of the population rather than average properties:

Theorem 5.4.1 (Microscopic Exact GP Schema Theorem). *The total transmission probability for a fixed-size-and-shape GP schema H under one-point crossover and no mutation is*

$$\alpha(H,t) = (1 - p_{xo})p(H,t) + \\ p_{xo} \sum_{h_1} \sum_{h_2} \frac{p(h_1,t)p(h_2,t)}{\mathbf{NC}(h_1,h_2)} \sum_{i \in C(h_1,h_2)} \delta(h_1 \in L(H,i))\delta(h_2 \in U(H,i)) \tag{5.5}$$

where: the first two summations are over all the individuals in the population.

NC(h_1, h_2) *is the number of nodes in the tree fragment representing the common region between program* h_1 *and program* h_2.
$C(h_1, h_2)$ *is the set of indices of the crossover points in the common region.*
$\delta(x)$ *is a function than returns 1 if* x *is true and 0 otherwise.*
$L(H, i)$ *is the hyperschema obtained by replacing all the nodes on the path between crossover point* i *and the root node with* `=` *nodes, and all the subtrees connected to those nodes with* `#` *nodes.*
$U(H, i)$ *is the hyperschema obtained by replacing the subtree below crossover point* i *with a* `#` *node.*

The hyperschemata $L(H, i)$ and $U(H, i)$ are generalisations of the schemata $l(H, i)$ and $u(H, i)$ used in Equation (5.4) (compare Figures 5.1 and 5.2). If one crosses over at point i *any* individual in $L(H, i)$ and with *any* individual in $U(H, i)$, the resulting offspring is always an instance of H.

Before we proceed with the proof of the theorem, let us try to understand how $L(H, i)$ and $U(H, i)$ are built with an example. We take the same example as in Section 5.4.1, i.e. $H =$`(* = (+ x =))`; cf. Figure 5.2. $L(H, 1)$ (second column of Fig 5.2) is obtained by first replacing the root node with a `=` symbol and then replacing the subtree connected to the right-hand side of the root node with a `#` symbol obtaining `(= = #)`. The hyperschema $U(H, 1)$ (third column) is constructed by replacing the subtree below the crossover point with a `#` symbol obtaining `(* # (+ x =))`. The fourth and fifth columns of Figure 5.2 show how $L(H, 3) =$ `(= # (= x #))` and $U(H, 3) =$ `(* = (+ # =))` are obtained.

Once the concepts of $L(H, i)$ and $U(H, i)$ are available, Theorem 5.4.1 can easily be proven by: (1) considering all the possible ways in which parents can be selected for crossover and in which the crossover points can be selected in such parents, (2) computing the probabilities that each of those events happens *and* the first parent has the correct lower part (with respect to the crossover point) to create H while the second parent has the correct upper part, and (3) adding up such probabilities (this gives the three summations in Equation 5.5). Obviously, in order for this equation to be valid it is necessary to assume that if a crossover point i is in the common region between two programs but outside the schema H under consideration, then $L(H, i)$ and $U(H, i)$ are empty sets (i.e. they cannot be matched by any individual).

A more formal way to prove the result is the following.

Proof. Let $p(h_1, h_2, i, t)$ be the probability that, at generation t, the selection/crossover process will choose parents h_1 and h_2 and crossover point i. Then, let us consider the function:

$$g(h_1, h_2, i, H) = \delta(h_1 \in L(H, i))\delta(h_2 \in U(H, i))$$

Given two parent programs, h_1 and h_2, and a schema of interest H, g returns the value 1 if crossing over h_1 and h_2 at position i yields an offspring in H; it returns 0 otherwise. (g is a measurement function in the sense defined in

[Altenberg, 1995]). If h_1, h_2 and i are stochastic variables with joint probability distribution $p(h_1, h_2, i, t)$, the expected value of $g(h_1, h_2, i, H)$ is the proportion of times the offspring of h_1 and h_2 are in H. The expected value of g is:

$$\sum_{h_1} \sum_{h_2} \sum_{i \in C(h_1,h_2)} g(h_1, h_2, i, H) p(h_1, h_2, i, t) \tag{5.6}$$

By using the definition of conditional probability and assuming that the probability of selecting a particular crossover point depends only on the parents, not the actual generation at which such parents were selected, we obtain:

$$p(h_1, h_2, i, t) = p(i|h_1, h_2) \times p(h_1, h_2, t)$$

where $p(i|h_1, h_2)$ is the conditional probability that crossover point i will be selected when the parents are h_1 and h_2. Because each parent is selected independently, this becomes

$$p(h_1, h_2, i, t) = p(i|h_1, h_2) \times p(h_1, t) \times p(h_2, t)$$

where $p(h_1, t)$ and $p(h_2, t)$ are the selection probabilities for the parents. In one-point crossover

$$p(i|h_1, h_2) = 1/\mathbf{NC}(h_1, h_2)$$

So,

$$p(h_1, h_2, i, t) = \frac{p(h_1, t) p(h_2, t)}{\mathbf{NC}(h_1, h_2)}$$

which substituted into Equation (5.6) leads to the term due to crossover in the right hand side of Equation (5.5). By multiplying this by p_{xo} (the probability of doing selection followed by crossover) and adding the selection-followed-by-cloning term $(1-p_{xo})p(H, t)$ one obtains the right hand side of Equation (5.5). □

Equation (5.5) allows one to compute the exact total transmission probability $\alpha(h, t)$ of a GP schema. In the same way as [Poli *et al.*, 1998, Poli, 2000a] a series of new results for GP schemata can be obtained which provide important statistical information on schema behaviour (e.g. schema probability distribution, schema variance, schema signal-to-noise ratio and extinction probability). However, in this chapter we will not study these results, instead in the next two sections we will concentrate on understanding Equation (5.5) in greater depth and on transforming it into a macroscopic model. That is dealing with global properties in the population rather than the intimate details of each individual within it.

5.4.4 Macroscopic Schema Theorem with Schema Creation

If one restricts the first two summations in Equation (5.5) to include only the individuals having the same shape and size of the schema H, i.e. which belong to $G(H)$, one obtains a lower bound for $\alpha(H,t)$:

$$\alpha(H,t) \geq (1-p_{xo})p(H,t) + \tag{5.7}$$

$$\frac{p_{xo}}{N(H)} \sum_{i=0}^{N(H)-1} \sum_{h_1 \in G(H)} p(h_1,t)\delta(h_1 \in L(H,i)) \sum_{h_2 \in G(H)} p(h_2,t)\delta(h_2 \in U(H,i))$$

where we used the following three facts: (1) $\mathbf{NC}(h_1,h_2) = N(H)$ since $h_1, h_2 \in G(H)$, (2) all common regions have the same shape and size as H, and so (3) assuming that the crossover points in H are numbered 0 to $N(H)-1$, $C(h_1,h_2) = \{0,1,\ldots,N(H)-1\}$.

Equation (5.7) can easily be transformed into:

Theorem 5.4.2 (GP Schema Theorem with Schema Creation Correction). *A lower bound for the chance of a child matching a fixed-size-and-shape GP schema H (total transmission probability) with one-point crossover and no mutation is*

$$\alpha(H,t) \geq (1-p_{xo})\,p(H,t) + \tag{5.8}$$

$$\frac{p_{xo}}{N(H)} \sum_{i=0}^{N(H)-1} p(L(H,i)\cap G(H),t)\,p(U(H,i)\cap G(H),t)$$

If all the programs in the population have the same shape $G(H)$ then (5.8) becomes an equality rather than a lower bound.

Let us compare this result to the one described in Section 4.4.4.

It is easy to see that by dividing the right-hand side of Inequality (4.16) (page 65) by M one obtains a lower bound for $\alpha(H,t)$. If we consider the case in which $p_m = 0$ and assume the worst-case scenario for $p_{\text{diff}}(t)$, i.e. $p_{\text{diff}}(t) = 1$, we obtain

$$\alpha(H,t) \geq p(H,t)\left\{1 - p_{xo}\left[1 - p(G(H),t) + \frac{\mathcal{L}(H)}{N(H)}\Big(p(G(H),t) - p(H,t)\Big)\right]\right\}$$

We define $\Delta\alpha(H,t)$ by subtracting the right-hand side of this equation from the lower bound provided by Inequality (5.8). If we can show $\Delta\alpha(H,t)$ is positive, then (5.8) yields a tighter bound than the GP fixed-size-and-shape schema theorem, Inequality (4.16) (page 65):

$$\begin{aligned}\Delta\alpha(H,t) &= (1-p_{xo})\,p(H,t)\\ &+\frac{p_{xo}}{N(H)} \sum_{i=0}^{N(H)-1} p(L(H,i)\cap G(H),t)\,p(U(H,i)\cap G(H),t)\\ &-p(H,t) + p_{xo}p(H,t)\left[1 - p(G(H),t) + \frac{\mathcal{L}(H)}{N(H)}\Big(p(G(H),t) - p(H,t)\Big)\right]\end{aligned}$$

We then split the first summation into two separate summations: those where the crossover point i lies within the defining nodes of H, i.e. $i \in B(H)$, and the rest, i.e. $i \notin B(H)$. Then expand the brackets and cancel $(1 - p_{xo}\, p(H,t))$:

$$\begin{aligned}
\Delta\alpha(H,t) = \\
&\frac{p_{xo}}{N(H)} \sum_{i \in B(H)} p(L(H,i) \cap G(H),t)\, p(U(H,i) \cap G(H),t) \\
&+\frac{p_{xo}}{N(H)} \sum_{i \notin B(H)} p(L(H,i) \cap G(H),t)\, p(U(H,i) \cap G(H),t) \\
&-p_{xo} p(H,t) p(G(H),t) + p_{xo} p(H,t) \frac{\mathcal{L}(H)}{N(H)} p(G(H),t) \\
&-p_{xo} p(H,t)^2 \frac{\mathcal{L}(H)}{N(H)}
\end{aligned}$$

If the crossover point does not lie within the defining nodes of H, then either the crossover point lies between the defining nodes and the root, or it does not. In the first case, every program that matched H also matches $L(H,i)$, while in the second case, every program that matched H also matches $U(H,i)$. Since we are only considering individuals with the same size and shape as H, $L(H,i) \cap G(H) = H$ or $U(H,i) \cap G(H) = H$. So $\sum_{i \notin B(H)} \cdots = (N(H) - \mathcal{L}(H)) p(G(H),t) p(H,t)$, since every item in the summation is equal to $p(G(H),t)p(H,t)$ and the number of items to be summed is the total size of H less the size of $B(H)$ (i.e. H's defining length $\mathcal{L}(H)$). Replacing $\sum_{i \notin B(H)} \cdots$ with this result and cancelling terms in $p(G(H),t)p(H,t)$, yields:

$$\begin{aligned}
\Delta\alpha(H,t) &= \frac{p_{xo}}{N(H)} \sum_{i \in B(H)} p(L(H,i) \cap G(H),t)\, p(U(H,i) \cap G(H),t) \\
&\quad - p_{xo} p(H,t)^2 \frac{\mathcal{L}(H)}{N(H)} \\
&= \frac{p_{xo}}{N(H)} \left(\sum_{i \in B(H)} p(L(H,i),t) p(U(H,i),t) - \mathcal{L}(H) p(H,t)^2 \right)
\end{aligned}$$

where $B(H)$ is the set of crossover points in the tree fragment used in the definition of the defining length of H, $\mathcal{L}(H)$. Since such a tree fragment contains exactly $\mathcal{L}(H)$ crossover points, we can rewrite this equation as

$$\Delta\alpha(H,t) = \frac{p_{xo}}{N(H)} \sum_{i \in B(H)} \Big(p(L(H,i),t) p(U(H,i),t) - p(H,t)^2 \Big). \tag{5.9}$$

Now $L(H,i)$ and $U(H,i)$ are less specific than schema H so they contain all programs in the population matching H plus possibly some more. So, it follows that $p(L(H,i),t) \geq p(H,t)$ and $p(U(H,i),t) \geq p(H,t)$. Every term in the summation is non-negative, therefore $\Delta\alpha(H,t) \geq 0$. So, Theorem 5.4.2

provides a better estimate of the true transmission probability of a schema on the assumption that $p_{\text{diff}}(t) = 1$. This is why we called Theorem 5.4.2 the "GP schema theorem with schema creation correction".

5.4.5 Macroscopic Exact GP Schema Theorem

In order to transform Equation (5.5) into an exact macroscopic description of schema propagation, let us start by numbering all the possible program shapes, i.e. all the possible fixed-size-and-shape schemata of order 0. Let us denote such schemata as $G_1, G_2, \cdots$. These schemata represent disjoint sets of programs. Their union represents the whole search space. For these reasons we have

$$\sum_j \delta(h_1 \in G_j) = 1.$$

We append the left hand side of this expression and of an analogous expression for $\delta(h_2 \in U(H, i))$ to the terms in the triple summation in Equation (5.5) and reorder the terms to obtain:

$$\begin{aligned}
&\sum_{h_1}\sum_{h_2} \frac{p(h_1,t)p(h_2,t)}{\mathbf{NC}(h_1,h_2)} \\
&\qquad \sum_{i\in C(h_1,h_2)} \sum_j \delta(h_1 \in L(H,i))\delta(h_1 \in G_j) \sum_k \delta(h_2 \in U(H,i))\delta(h_2 \in G_k) \\
&= \sum_j\sum_k\sum_{h_1}\sum_{h_2} \frac{p(h_1,t)p(h_2,t)}{\mathbf{NC}(h_1,h_2)} \\
&\qquad \sum_{i\in C(h_1,h_2)} \delta(h_1 \in L(H,i))\delta(h_1 \in G_j)\delta(h_2 \in U(H,i))\delta(h_2 \in G_k) \\
&= \sum_j\sum_k\sum_{h_1\in G_j}\sum_{h_2\in G_k} \frac{p(h_1,t)p(h_2,t)}{\mathbf{NC}(h_1,h_2)} \\
&\qquad \sum_{i\in C(h_1,h_2)} \delta(h_1 \in L(H,i))\delta(h_2 \in U(H,i)) \\
&= \sum_j\sum_k\sum_{h_1\in G_j}\sum_{h_2\in G_k} \frac{p(h_1,t)p(h_2,t)}{\mathbf{NC}(G_j,G_k)} \\
&\qquad \sum_{i\in C(G_j,G_k)} \delta(h_1 \in L(H,i))\delta(h_2 \in U(H,i)) \\
&= \sum_j\sum_k \frac{1}{\mathbf{NC}(G_j,G_k)} \\
&\qquad \sum_{i\in C(G_j,G_k)} \sum_{h_1\in G_j} p(h_1,t)\delta(h_1 \in L(H,i)) \sum_{h_2\in G_k} p(h_2,t)\delta(h_2 \in U(H,i))
\end{aligned}$$

From this one obtains the following:

Theorem 5.4.3 (Macroscopic Exact GP Schema Theorem for One-point Crossover). *The total transmission probability for a fixed-size-and-shape GP schema H under one-point crossover and no mutation is*

$$\alpha(H,t) = (1 - p_{xo})p(H,t) + p_{xo} \sum_j \sum_k \frac{1}{\mathbf{NC}(G_j, G_k)} \sum_{i \in C(G_j,G_k)} p(L(H,i) \cap G_j, t) p(U(H,i) \cap G_k, t) \quad (5.10)$$

The sets $L(H,i) \cap G_j$ and $U(H,i) \cap G_k$ either are (or can be represented by) fixed-size-and-shape schemata, or are the empty set $\emptyset$. So, the theorem expresses the total transmission probability of H only using the selection probabilities of a set of lower- (or same-) order schemata.

This theorem is a refinement of Equation (5.8) which can be obtained from Equation (5.10) by considering only one term in the summations in j and k (the term for which $G_j = G_k = G(H)$). The theorem is a generalisation of Equation (5.4) which, as noted earlier, is a generalisation of Equation (5.3) (which is equivalent to Stephens and Waelbroeck's schema theorem in the absence of mutation). That in turn is a refinement of Whitley's version of Holland's schema theorem [Whitley, 1993, Whitley, 1994]. So, in the absence of mutation, Equation (5.10) generalises and refines not only earlier GP schema theorems but also old and modern GA schema theories for one-point crossover.

5.5 Examples

This section gives a few examples that show how to use the schema theory developed in this chapter in practice and illustrate some of its benefits.

5.5.1 Linear Trees

This example describes how to apply the exact schema theorems (both microscopic and macroscopic; cf. Equations (5.5) and (5.10)). We assume there is no mutation ($p_m = 0$) and 100% one-point crossover ($p_{xo} = 1$). Since the calculations involved may be quite lengthy, we will consider one of the simplest non-trivial examples.

In the first example there are populations of trees without any branches; i.e. they consist of a single trunk. Each of the functions takes exactly one argument (i.e. they are unary functions). As they contain no branches the programs are linear trees. Let us imagine that we have a function set $\mathcal{F} = \{A_f, B_f, C_f, D_f, E_f\}$ including only unary functions and the terminal set $\mathcal{T} = \{A_t, B_t, C_t, D_t, E_t\}$. For example, $(A_f(B_f B_t))$ is a program with two functions and one terminal. Since we know the arity of all functions, we can remove the parentheses from the expression, to obtain $A_f B_f B_t$. In addition, since the only terminal in each expression is the rightmost node, we can remove the subscripts without generating any ambiguity, to obtain ABB. This can be done for every member of the search space, which can be seen as the space of variable-length strings over the alphabet $\{A, B, C, D, E\}$. So in this example, GP with one-point crossover is really a non-binary variable-length GA.

Let us now consider the schema `AB=`. We want to measure its total transmission probability under fitness proportionate selection and one-point crossover (with $p_{xo} = 1$) in two slightly different populations (note the underlined individuals):

Population 1	Fitness	$p(h,t)$	Population 2	Fitness	$p(h,t)$
AB	2	2/14	AB	2	2/14
BCD	2	2/14	BCD	2	2/14
ABC	4	4/14	ABC	4	4/14
ABCD	6	6/14	*BCDE*	6	6/14

Since we use fitness proportionate selection, the chance of selecting any program h to be a parent, $p(h,t)$, is its fitness divided by the sum of all the fitnesses in the population (14 in both Population 1 and 2). Then we need to compute the "lower part" and "upper part" building blocks of `AB=` (cf. Section 5.4.3). These are:

i	L(AB=, i)	U(AB=, i)
0	AB=	#
1	=B=	A#
2	===	AB#
3	$\emptyset$	$\emptyset$

($\emptyset$ is the null or empty set which contains no members).

We can now start calculating α(AB=, t) for Population 1, using the exact microscopic schema theorem Equation (5.5). (Remember that $\mathbf{NC}(h_1, h_2)$ is the number of available crossover points when crossing programs h_1 and h_2):

$$\alpha(\texttt{AB=},t) = \sum_{h_1,h_2} \frac{p(h_1,t)p(h_2,t)}{\mathbf{NC}(h_1,h_2)} \sum_{i\in C(h_1,h_2)} \delta(h_1 \in L(\texttt{AB=},i))\delta(h_2 \in U(\texttt{AB=},i))$$

$$= \underbrace{\overbrace{\frac{2}{14}}^{p(h_1,t)} \times \overbrace{\frac{2}{14}}^{p(h_2,t)} \times \overbrace{\frac{1}{2}}^{1/\mathbf{NC}(h_1,h_2)} \times (0\times 1 + 0 \times 1)}_{h_1=AB,h_2=AB,i=0,1}$$

$$+ \underbrace{\frac{2}{14} \times \frac{2}{14} \times \frac{1}{2} \times (0\times 1 + 0 \times 0)}_{h_1=AB,h_2=BCD,i=0,1}$$

$$\cdots$$

$$\text{(13 terms are omitted)}$$

$$\cdots$$

$$+ \underbrace{\frac{6}{14} \times \frac{6}{14} \times \frac{1}{4} \times (0\times 1 + 0 \times 1 + 0 \times 1 + 0 \times 0)}_{h_1=ABCD,h_2=ABCD,i=0,1,2,3} \tag{5.11}$$

$$= \frac{43}{147} \approx 0.2925$$

This equation contains a term for each pair of parents and for each crossover point. So clearly this is a lengthy calculation, which can only produce a numerical result. It cannot really be used to understand how instances of AB= are created in different populations since it includes only microscopic quantities, i.e. detailed calculation about everyone in the population.

Let us now use the macroscopic theorem Equation (5.10) to do the same calculation. First we need to number all the possible program shapes. Let G_1 be =, G_2 be ==, G_3 be === and G_4 be ====. We do not need to consider other, bigger shapes because the population does not contain any larger programs (i.e. $G_l = \emptyset$ for $l > 4$). Then we need to identify the schemata created by the intersection of the "building blocks" of schema AB= and each shape (hyperschema) G_j, i.e. $L(\texttt{AB=},i) \cap G_j$ and $U(\texttt{AB=},i) \cap G_k$:

Lower building blocks for each shape $L(\texttt{AB=},i) \cap G_j$

i	L(AB=,i)	G_1	G_2	G_3	G_4
0	AB=	$\emptyset$	$\emptyset$	AB=	$\emptyset$
1	=B=	$\emptyset$	$\emptyset$	=B=	$\emptyset$
2	===	$\emptyset$	$\emptyset$	===	$\emptyset$
3	$\emptyset$	$\emptyset$	$\emptyset$	$\emptyset$	$\emptyset$

Upper building block for each shape $U(\texttt{AB=}, i) \cap G_k$

i	$U(\texttt{AB=}, i)$	G_1	G_2	G_3	G_4
0	#	=	==	===	====
1	A#	∅	A=	A==	A===
2	AB#	∅	∅	AB=	AB==
3	∅	∅	∅	∅	∅

Finally we need to evaluate the shape of the common regions to determine the possible crossover locations, $C(G_j, G_k)$, and the number of them, $\mathbf{NC}(G_j, G_k)$, for all valid values of j and k. In general common regions can naturally be represented using the program shapes G_1, G_2, etc.

Shape of common region $C(G_j, G_k)$

	k			
j	1	2	3	4
1	G_1	G_1	G_1	G_1
2	G_1	G_2	G_2	G_2
3	G_1	G_2	G_3	G_3
4	G_1	G_2	G_3	G_4

Taking just the first term in Equation (5.10) (i.e. $j = 1, k = 1$),

$$\frac{1}{\mathbf{NC}(G_j, G_k)} \sum_i p(L(\texttt{AB=}, i) \cap G_j, t) p(U(\texttt{AB=}, i) \cap G_k, t)$$

$$= \underbrace{\overbrace{\frac{1}{1}}^{1/\mathbf{NC}(G_j,G_k)} \times \overbrace{[\ 0 \times p(\texttt{=})\ ]}^{\sum_i p(L(\texttt{AB=},i)\cap G_j,t)p(U(\texttt{AB=},i)\cap G_k,t)}}_{j=1,k=1,i=0}$$

we see it is zero, since $p(L(\texttt{AB=}, i) \cap G_j, t) = p(\emptyset) = 0$. In fact (as the following table shows) many j, k terms are equal to zero

Non-null building blocks contributing to Equation (5.10)

	k			
j	1	2	3	4
1	0	0	0	0
2	0	0	0	0
3	AB= × =	AB= × == =B= × A=	AB= × === =B= × A== === × AB=	AB= × ==== =B= × A=== === × AB==
4	0	0	0	0

By using this and the previous tables, we can simplify Equation (5.10) by removing all the null terms (and for brevity using the notation $p(.)$ to represent $p(., t)$) as follows:

$$\begin{aligned}\alpha(\texttt{AB=},t) = \quad & p(\texttt{AB=})p(\texttt{=}) \\ & + \tfrac{1}{2}\, p(\texttt{AB=})p(\texttt{==}) \quad + \tfrac{1}{2}p(\texttt{=B=})p(\texttt{A=}) \\ & + \tfrac{1}{3}\, p(\texttt{AB=})p(\texttt{===}) \;\; + \tfrac{1}{3}p(\texttt{=B=})p(\texttt{A==}) \;\; + \tfrac{1}{3}p(\texttt{===})p(\texttt{AB=}) \\ & + \tfrac{1}{3}\, p(\texttt{AB=})p(\texttt{====}) + \tfrac{1}{3}p(\texttt{=B=})p(\texttt{A===}) + \tfrac{1}{3}p(\texttt{===})p(\texttt{AB==})\end{aligned} \tag{5.12}$$

The complexity of this equation can be reduced by using hyperschemata to represent groups of schemata, to obtain:

$$\begin{aligned}\alpha(\texttt{AB=},t) = \quad & p(\texttt{AB=})p(\texttt{=}) \\ & + \tfrac{1}{2}\, p(\texttt{AB=})p(\texttt{==}) \;\; + \tfrac{1}{2}p(\texttt{=B=})p(\texttt{A=}) \\ & + \tfrac{1}{3}\, p(\texttt{AB=})p(\texttt{==\#}) + \tfrac{1}{3}p(\texttt{=B=})p(\texttt{A=\#}) + \tfrac{1}{3}p(\texttt{===})p(\texttt{AB\#})\end{aligned}$$

This is equivalent to reordering the terms by first size and shape of common region and then by crossover point.

Equation (5.12) is quite different from (5.11) obtained with the microscopic exact schema theorem. Thanks to the use of macroscopic quantities, it is general; i.e. independent of a particular population. Also, it clearly indicates how individuals sampling AB= can be assembled from individuals having different shapes and nodes. The schemata in this equation can be seen as the *building blocks* for AB=.

If we calculate the probabilities of selection of the schemata in the previous equation using Population 1, we obtain:

$$\begin{aligned}\alpha(\texttt{AB=},t) = \quad & \tfrac{4}{14} \times 0 \\ & + \tfrac{1}{2}\times\tfrac{4}{14}\times\tfrac{2}{14} + \tfrac{1}{2}\times\tfrac{4}{14}\times\tfrac{2}{14} \\ & + \tfrac{1}{3}\times\tfrac{4}{14}\times\tfrac{12}{14} + \tfrac{1}{3}\times\tfrac{4}{14}\times\tfrac{10}{14} + \tfrac{1}{3}\times\tfrac{6}{14}\times\tfrac{10}{14} \\ = \; & \frac{43}{147} \approx 0.2925\end{aligned}$$

This result is the same as the one in Equation (5.11) as expected. However, once the exact macroscopic formulation of the transmission probability of a schema is available, this is much easier to use in calculations than the corresponding microscopic description. Indeed, we can use it to calculate $\alpha(\texttt{AB=},t)$ for Population 2 with a simple pocket calculator, to obtain $\alpha(\texttt{AB=},t) \approx 0.1905$.

5.5.2 Comparison of Bounds by Different Schema Theorems

Let us consider a population containing 10 copies of each of the three programs in Figure 5.3, and let us focus on schema $H' =$(= = (= = =)). Since this schema has no defining nodes both its defining length, $\mathcal{L}(H')$, and the number of crossover points laying between defining nodes, $B(H')$, are zero. Therefore, Equation (5.9) gives $\Delta\alpha(H',t) = 0$. This means the GP schema theorem with schema creation correction (see Section 5.4.4) does not provide a better bound than the older one in [Poli and Langdon, 1997c], as summarised in Section 4.4.4.

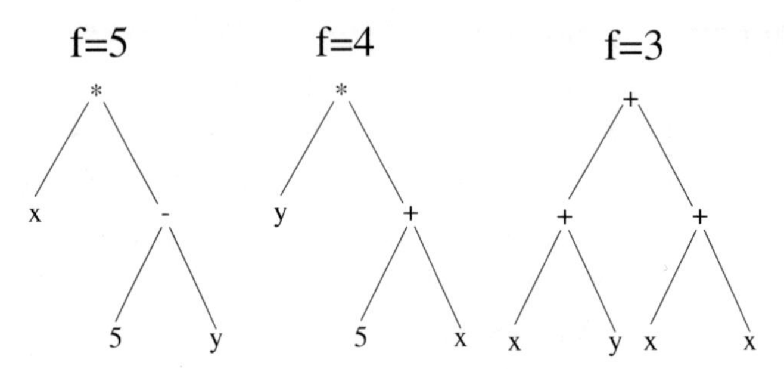

Fig. 5.3. Example population of programs with their fitnesses.

Table 5.1. Lower-bound and exact values for the expected number of programs matching `(* = (- 5 =))` in the next generation, $E\left[m\left(\texttt{(* = (- 5 =))}, t+1\right)\right]$, provided by different schema theorems for the population (size 30) in Figure 5.3 assuming 100% one-point crossover.

Equation	Method	Value
–	Selection only	12.5
(5.5) or (5.10)	Exact schema theorem	10.5
(5.8)	Schema theorem with schema creation correction	9.4
(4.15)	Old schema theorem	7.3

However, if one considers the schema $H'' =$`(* = (- 5 =))`, with simple calculations (assuming fitness proportionate selection) one obtains $\Delta\alpha(H'', t) \approx 0.07 \times p_{xo}$. This might look like a small difference, but if one multiplies this by the population size $M = 30$, one can see that the schema theorem with schema creation correction can provide a lower bound for $E[m(H'', t+1)]$ that is up to about 2.1 fold better than the old theorem (Equation 4.15) page 65. The maximum value being reached when $p_{xo} = 1$. This is a big difference considering that, as shown in Table 5.1, the correct value for $E[m(H'', t+1)]$ obtained from Equations (5.5) (microscopic) or (5.10) (macroscopic) is 10.5 (for reference: the value obtained if selection only were acting is 12.5), and the lower bound provided by the old schema theorem is only 7.3. So, in this example the "schema creation correction" improves the estimate by nearly 30%, providing a tight lower bound of 9.4 for $E[m(H'', t+1)]$.

We can also return to the two populations in the example in the previous section, and compare the bounds provided by different schema theorems. In that example the exact GP schema theorems gave $\alpha(\texttt{AB=}, t) \approx 0.2925$ for Population 1 and $\alpha(\texttt{AB=}, t) \approx 0.1905$ for Population 2. For comparison, for either population, the schema theorem with schema creation correction, Equation (5.8), gives the lower bound

$$\begin{aligned}\alpha(\texttt{AB=}, t) &\geq \frac{1}{3}p(\text{AB=})p(\text{===}) + \frac{1}{3}p(\text{=B=})p(\text{A==}) + \frac{1}{3}p(\text{===})p(\text{AB=}) \\ &\approx 0.1088\end{aligned}$$

which is nearly one-third of the correct value for Population 1 (0.2925) and half of the correct value for Population 2 (0.1905). This is because Equation (5.8) only takes account of schema creation events between programs of the same shape as AB=, i.e. $G(\texttt{AB=}) = \texttt{===}$. Incidentally, since there are no creation events of this type in the example in Section 5.5.1, the old GP schema theorem (Equation (4.15), page 65) gives exactly the same bound.

5.5.3 Example of Schema Equation for Binary Trees

Let us now consider a GP system using functions with two arguments, with programs of up to three levels. It is easy to write a schema theorem equation for schema (A B C). That is, the expected number of copies of the program composed of the binary function A with first argument B and second terminal C under 100% one-point crossover and no mutation is $\alpha(\texttt{(A B C)})$ times the number in the current generation, where $\alpha(\texttt{(A B C)})$ is given by:

$$\begin{aligned}
\alpha(\texttt{(A B C)}) = \; & \tfrac{1}{3}\, p(\texttt{(= (= = =) C)})\, p(\texttt{(A B (= = =))}) \\
+ & \tfrac{1}{3}\, p(\texttt{(= (= = =) C)})\, p(\texttt{(A B =)}) \\
+ & \tfrac{1}{3}\, p(\texttt{(= B (= = =))})\, p(\texttt{(A (= = =) C)}) \\
+ & \tfrac{1}{3}\, p(\texttt{(= B (= = =))})\, p(\texttt{(A = C)}) \\
+ & \tfrac{1}{3}\, p(\texttt{(A B C)}) \quad p(\texttt{(= (= = =) (= = =))}) \\
+ & \tfrac{1}{3}\, p(\texttt{(A B C)}) \quad p(\texttt{(= (= = =) =)}) \\
+ & \tfrac{1}{3}\, p(\texttt{(= B =)}) \quad p(\texttt{(A (= = =) C)}) \\
+ & \tfrac{1}{3}\, p(\texttt{(A B C)}) \quad p(\texttt{(= = (= = =))}) \\
+ & \tfrac{1}{3}\, p(\texttt{(= = C)}) \quad p(\texttt{(A B (= = =))}) \\
+ & \tfrac{1}{3}\, p(\texttt{(A B C)}) \quad p(\texttt{(= = =)}) \\
+ & \tfrac{1}{3}\, p(\texttt{(= B =)}) \quad p(\texttt{(A = C)}) \\
+ & \tfrac{1}{3}\, p(\texttt{(= = C)}) \quad p(\texttt{(A B =)}) \\
+ & \;\; p(\texttt{(A B C)}) \quad p(\texttt{=})
\end{aligned}$$

5.6 Exact Macroscopic Schema Theorem for GP with Standard Crossover

The theory presented so far applies to GP one-point crossover only. In recent work [Poli, 2001b] a general schema theory for GP with subtree-swapping crossover was presented. The theory is very general and includes, as special cases, the theory for one-point crossover just presented and the theory for standard crossover. In Sections 5.6.1 and 5.6.2 we will describe the specialisation of the theory to standard crossover with uniform selection of the crossover points. For brevity, we will omit the proofs of the theorems. The reader interested in the proofs and in the more general schema theory for subtree swapping crossover is referred to [Poli, 2001b].

The theory is based on an extension of the notion of hyperschema, the variable arity hyperschema and on the concept of Cartesian node reference systems. These are presented in the following sections.

5.6.1 Cartesian Node Reference Systems

A Cartesian node reference system can be defined by first considering the largest possible tree that can be created with nodes of arity $a_{\max}$, where $a_{\max} = \max_{f \in \mathcal{F}} \text{arity}(f)$. This maximal tree would include 1 node of arity $a_{\max}$ at depth 0, $a_{\max}$ nodes of arity $a_{\max}$ at depth 1, $a_{\max}^2$ nodes of arity $a_{\max}$ at depth 2, etc. Then one can organise the nodes in the tree into layers of increasing depth and assign an index to each node in a layer. We can then define a coordinate system based on the layer number d and the index i. This reference system can also be used to locate the nodes of non-maximal trees by using a subset of the nodes and links in the maximal tree. Figure 5.4 shows the expression `(A (B C D) (E F (G H)))` placed in a node reference system with $a_{max} = 3$. For example, the coordinates of `F` are (2,3). In this reference system it is possible to transform pairs of coordinates into integers by counting the nodes in breadth-first order (and vice versa). So, nodes `A`, `B`, `C`, `D`, `E`, `F` and `G` also have indices 0, 1, 4, 5, 2, 7, 8 and 25, respectively. Later we will use this property to simplify our notation.

Given a node reference system, it is possible to define functions and probability distributions over it. In [Poli, 2001b] these notions were used to model a large class of crossover operators including standard crossover. For brevity, we omit these models since they are not necessary when stating the theory that follows.

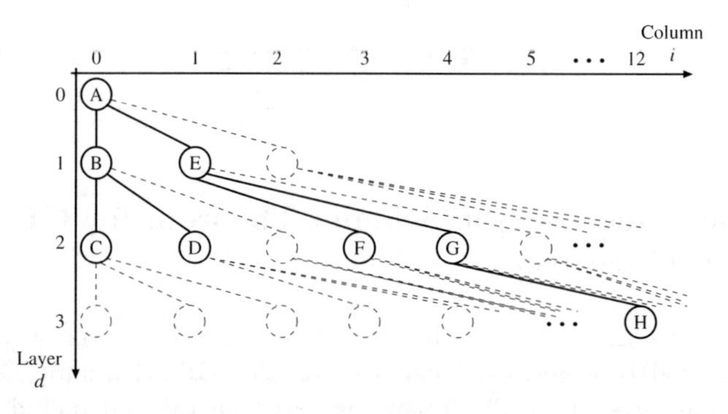

Fig. 5.4. The Cartesian node reference system. The root node, A, is always node (0,0), B is (1,0), C (2,0), D (2,1) and E (1,1). An individual tree (solid lines) need not fill the co-ordinate system. For example node B only has two arguments (rather than the maximum of three) so, for example, node (2,2) is empty. Missing nodes and links of the maximal tree are drawn with dashed lines. Only four layers are shown.

5.6.2 Variable Arity Hyperschema

In order to state a macroscopic schema theorem for standard crossovers, we need to introduce a different form of schemata.

Definition 5.6.1. *A* variable arity (VA) Hyperschema *is a rooted tree composed of internal nodes from the set $\mathcal{F} \cup \{=, \#\}$ and leaves from $\mathcal{T} \cup \{=, \#\}$, where $\mathcal{F}$ and $\mathcal{T}$ are the function and terminal sets* [Poli, 2001b]. *The operator* = *is a "don't care" symbol which stands for exactly one node, the terminal* # *stands for any valid subtree, while the function* # *stands for exactly one function of arity not smaller than the number of subtrees connected to it.*

For example, the VA hyperschema `(# x (+ = #))` represents all the programs with the following characteristics: (1) the root node is any function in the function set with arity 2 or higher, (2) the first argument of the root node is the variable `x`, (3) the second argument of the root node is `+`, (4) the first argument of the `+` is any terminal, and (5) the second argument of the `+` is any valid subtree. If the root node is matched by a function of arity greater than 2, the third, fourth, etc. arguments of such a function are left unspecified, i.e. they can be any valid subtree.

Once VA hyperschemata are defined, we can define two useful functions. Given a fixed-size-and-shape schema and the coordinates of one or two nodes, these functions return the VA hyperschemata that are used to build instances of the schema. For simplicity in these definitions we use a single index to identify nodes. We can do this because, as indicated previously, there is a one-to-one mapping between pairs of coordinates and natural numbers.

Definition 5.6.2. *The function $L(H, i, j)$ returns the variable arity hyperschema obtained by: (1) rooting at coordinate j in an empty reference system the subschema of the schema H below crossover point i, (2) labelling all the nodes on the path between node j and the root node with* # *function nodes, and (3) labelling the arguments of those nodes which are to the left of such a path with* # *terminal nodes.*

Definition 5.6.3. *The function $U(H, i)$ returns the VA hyperschema obtained by replacing the subtree below crossover point i with a* # *node.*[4]

The functions $L(H, i, j)$ and $U(H, i)$ are designed to return *exactly* the hyperschemata needed to create H using crossover. $U(H, i)$ is the hyperschema representing all the trees that match the *upper* portion of H (i.e. the parts of H not below crossover point i). $L(H, i, j)$ is the hyperschema representing all the trees that match the *lower* portion of H, but where the matching portion is at some arbitrary position j. The combined effect of these definitions is that if one crosses over *any* individual matching $U(H, i)$ at point i with *any*

[4] This function corresponds exactly to the function $U(H, i)$ defined in Section 5.4.3, since the variable arity hyperschema returned by the function does not include any # functions, and so is in fact a standard hyperschema.

individual matching $L(H,i,j)$ at point j, the resulting offspring is always an instance of H. Furthermore, this is the only way to construct an instance of H. ($L(H,i,j)$ and $U(H,i)$ are discussed in more detail in [Poli, 2001b]).

To better understand how $U(H,i)$ and $L(H,i,j)$ are constructed, let us consider an example; throughout this example we will use the 2-D coordinate system, so positions i and j are ordered pairs. Let us take our schema to be $H =$`(* = (+ x =))`, and our coordinates to be $i = (1,0)$ and $j = (1,1)$. Figure 5.5 illustrates how we construct $U(H,i)$. It shows the initial schema H, with the crossover point i marked, and the lower part of the schema shaded. The lower grid shows $U(H,i)$, which is obtained by simply replacing the shaded subtree (in this case just the terminal `=`) with a '#'.

The upper of the three coordinate grids for $L(H,i,j)$ in Figure 5.6 again illustrates the initial schema H with the crossover point i marked. Now, however, the shaded area (the part of H below i) needs to be translated to position j, as shown in the second coordinate grid. The third coordinate grid then shows the insertion of '#' symbols: (1) along the path from the root to j (in this case just $(0,0)$) and (2) in all argument positions to the left of '#' symbols (in this case just $(1,0)$). This placement of '#' symbols, combined with the fact that we allow '#'s to represent functions of varying arity, ensures that $L(H,i,j) =$`(# # =)` represents *all* the possible trees whose subtrees at position j match the lower part of H (i.e. the part below position i).

5.6.3 Macroscopic Exact Schema Theorem for GP with Standard Crossover

Once the functions $L(H,i,j)$ and $U(H,i)$ are available, it is easy to state the following:

Theorem 5.6.1 (Macroscopic Exact GP Schema Theorem for Standard Crossover). *The total transmission probability for a fixed-size-and-shape GP schema H under standard crossover with uniform selection of crossover points is*

$$\alpha(H,t) = (1 - p_{xo})p(H,t) + \tag{5.13}$$
$$p_{xo} \sum_{k,l} \frac{1}{N(G_k)N(G_l)} \sum_{i \in H \cap G_k} \sum_{j \in G_l} p(U(H,i) \cap G_k, t) p(L(H,i,j) \cap G_l, t)$$

where the first two summations range over all the possible program shapes (i.e. all the fixed-size-and-shape schemata including only `=` *symbols) G_1, G_2, $\cdots$, and $N(K)$ is the number of nodes in the schema K.*

As an example let us specialise this result to the case of linear structures.

In the case of function sets including only unary functions, programs and schemata can be represented as linear sequences of symbols like $h_1h_2...h_N$. In this case:

- The hyperschema $U(h_1h_2...h_N, i) = h_1h_2...h_i$#.

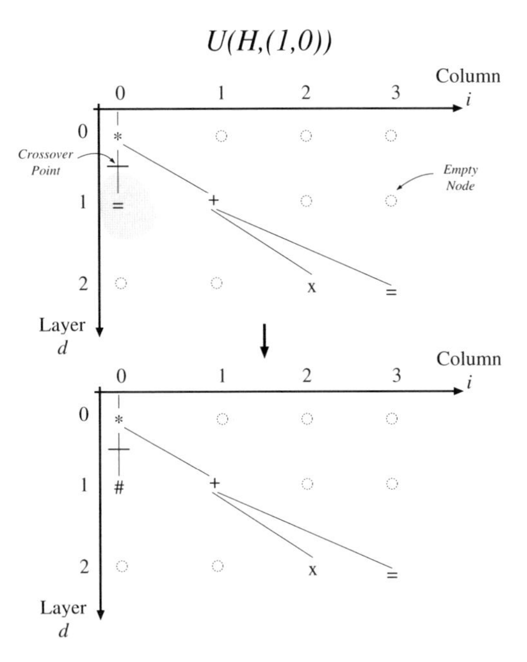

Fig. 5.5. Phases in the construction of the variable arity hyperschema building block $U(H,(1,0))$ of the schema H =(* = (+ x =)) within a binary tree node coordinate system, i.e. $a_{max} = 2$. The top panel shows the location of H and the crossover point. The lower panel shows the variable arity hyperschema, which is produced by replacing every thing below the crossover point (shown shaded in top panel) with a leaf # wild card symbol. See also Figure 5.6.

- The hyperschema $L(h_1h_2...h_N,i,j) = (\#)^j h_{i+1}...h_N$, where the notation $(x)^n$ means x repeated n times. This is equivalent to $(=)^j h_{i+1}...h_N$ since only linear structures are allowed.
- The set $U(h_1h_2...h_N,i) \cap G_k = \begin{cases} h_1h_2...h_i(=)^{k-i} & \text{for } i < k, \\ \emptyset & \text{otherwise.} \end{cases}$
- The set $L(h_1h_2...h_N,i,j) \cap G_l = \begin{cases} (=)^j h_{i+1}...h_N & \text{if } l = j-i+N, \\ \emptyset & \text{otherwise.} \end{cases}$

By substituting these quantities into Equation (5.13) and performing some simplifications one obtains the following result

$$\alpha(h_1...h_N,t) = (1-p_{xo})p(h_1...h_N,t) + \tag{5.14}$$
$$p_{xo}\sum_k \frac{1}{k} \sum_{i=0}^{\min(N,k)-1} p(h_1...h_i(=)^{k-i},t) \sum_j \frac{p((=)^j h_{i+1}...h_N,t)}{j-i+N}$$

which can be shown to be entirely equivalent to the schema theorem for linear structures reported in [Poli and McPhee, 2001b], where we provided a direct proof of it.

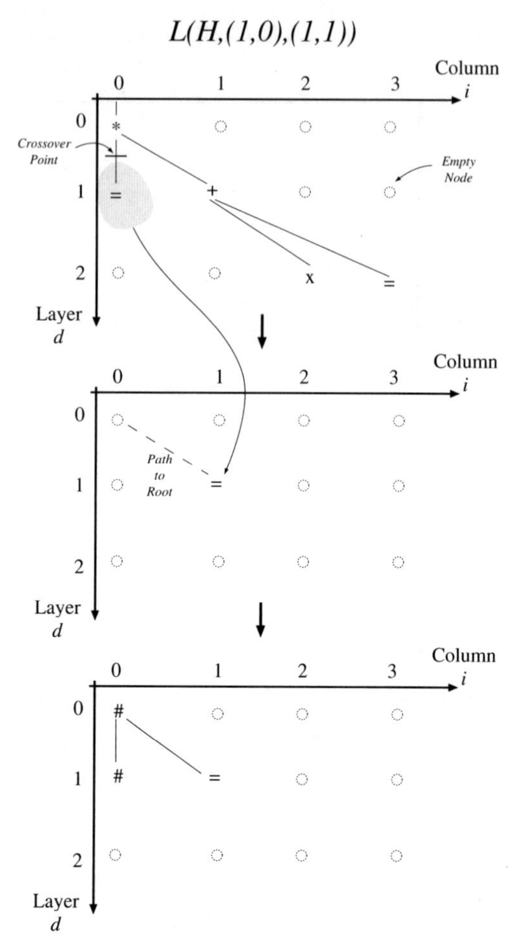

Fig. 5.6. Construction of variable arity hyperschema building block $L($`(* = (+ x =))`$,(0,1),(1,1))$ of the schema $H =$ `(* = (+ x =))` within a node coordinate system with $a_{max} = 2$. See also Figure 5.5. Top panel shows the schema H in the binary Cartesian coordinate system. The location of the crossover point $(0,1)$ is shown with a short line across the link above the shaded don't care symbol =. Node $(1,1)$ is occupied by a + node. The middle panel shows the path between this + node and the root. Note that the + node and all of its arguments have been replaced by the subtree taken from the crossover point (shaded in the top panel). The bottom picture shows the final step. All the nodes on the path are replaced with # and their left arguments are also replaced by #.

5.7 Summary

We have introduced exact schema theorems for GP for two different operators, one-point crossover (Section 5.4) and standard crossover (Section 5.6), and we have illustrated how these can be applied to specific schemata or specific classes of schemata (Section 5.5). In the next chapter we will analyse how exact schema theorems can be used to understand the internal mechanisms of genetic programming more deeply.

6. Lessons from the GP Schema Theory

This chapter considers different ways in which we can use the schema theory described in the previous chapters to better understand the dynamics of genetic programming populations. We will also illustrate the possible uses of the schema theory to clarify concepts such as effective fitness (next section), biases in GP (Section 6.2), building blocks (Section 6.3), problem hardness and deception (Section 6.5). We also discuss new ideas inspired by schema theories (Section 6.4).

6.1 Effective Fitness

The concept of effective fitness is a useful tool to better understand the effects of the operators on the search performed by a GA or a GP system.

Effective fitness is a measure of the reproductive success of an individual or group of individuals (schema). Fitness determines, on average, how many times an individual or group of individuals (schema) will be used as a parent (or parents). However, because of the action of genetic operators, fitness alone cannot tell us how many of its offspring will have the genetic makeup of the parent (or will belong to the schema). The concept of effective fitness is an attempt to rescale fitness values so as to incorporate the creative and disruptive effects of the operators.

This concept has been proposed three times independently in the literature, as described in the following sections.

6.1.1 Goldberg's Operator-Adjusted Fitness in GAs

Goldberg modelled the effects of crossover and mutation on the path followed by genetic algorithm populations by introducing the notion of *operator-adjusted fitness* (not to be confused with Koza's adjusted fitness [Koza, 1992]). [Goldberg, 1989a, page 155] defined the adjusted fitness of a schema as

$$f_{\mathrm{adj}}(H,t) = f(H,t)\Big(1 - p_{xo}\frac{\mathcal{L}(H)}{N-1} - p_m\mathcal{O}(H)\Big) \tag{6.1}$$

(As before $\mathcal{L}(H)$ is the defining length of schema H, $\mathcal{O}(H)$ is the number of non-don't care bits in it and N is the number of bits in the individual). In the case of low mutation rate (i.e. $p_m << 1$), fitness proportionate selection and assuming that only $N-1$ crossover points are available, this allowed him to reformulate his version of Holland's schema theorem[1] as

$$E[m(H,t+1)] \geq \frac{m(H,t)}{\bar{f}(t)} f_{\mathrm{adj}}(H,t) \tag{6.2}$$

This clearly indicates that an alternative way of interpreting the effects of crossover and mutation is to imagine a genetic algorithm in which selection only is used, but in which each individual (or schema) is given a fitness f_{adj} rather than the original fitness f. (Of course this is simplistic in that it ignores the genetic search being carried out by crossover and mutation).

A way of interpreting the operator-adjusted fitness is that a GA could be thought to be attracted towards the optima of f_{adj} rather than those of the fitness function f. If, for a particular problem, these optima do not coincide, then the problem can be said to be *deceptive* [Goldberg, 1989a].

The definitions of adjusted fitness and deception are only approximations since they are based on a version of Holland's schema theorem, which is itself only an approximation. So, Equation (6.1) can only provide a lower bound for the true adjusted fitness of a schema in a binary genetic algorithm.

6.1.2 Nordin and Banzhaf's Effective Fitness in GP

The concept of effective fitness in GP was introduced in [Nordin and Banzhaf, 1995, Nordin *et al.*, 1995, Nordin *et al.*, 1996] to explain the reasons for bloat and active-code compression in genetic programming. The effective fitness is defined by reformulating the approximate equation:

$$P_j^{t+1} \approx P_j^t \frac{f_j}{\bar{f}^t} \left(1 - p_{xo} \frac{C_j^e}{C_j^a} p_j^d\right) \tag{6.3}$$

which describes "the proliferation of individuals from one generation to the next" [Nordin *et al.*, 1995]. In the equation:

C_j^a = Number of nodes in program j.
C_j^e = Number of nodes in the active part (in contrast to the intron part) of program j.
p_{xo} = Crossover probability.
p_j^d = Probability that crossover in an active block of program j leads to worse fitness for the offspring of j.
f_j = Fitness of individual j.
$\bar{f}^t$ = Average population fitness at generation t.
P_j^t = Proportion of programs like j at generation t.
P_j^{t+1} = Average proportion of offspring of j which behave like j at generation $t+1$.

[1] Goldberg's version of Holland's schema theorem is applicable when both parents are selected with fitness proportionate selection. It is similar to Whitley's version (cf. page 33) but it provides a slightly weaker bound.

The approximately equal to sign, "≈", in the equation should really be "≥", but it was kept with the justification that the reconstruction of individuals with the same behaviour as j (due to crossover applied to individuals different from j) was a rare event. As noted in [Nordin and Banzhaf, 1995] this equation resembles a schema theorem. Indeed, it is similar to Rosca's schema theorem [Rosca, 1997a] (cf. Section 4.1).

Assuming fitness proportionate selection, the equation can be rewritten as

$$P_j^{t+1} \approx P_j^t \frac{f_j^e}{\bar{f}^t} \tag{6.4}$$

where

$$f_j^e = f_j \left(1 - p_{xo} \frac{C_j^e}{C_j^a} p_j^d\right) \tag{6.5}$$

is the *effective fitness* of program j. Equation (6.4) clearly indicates how Nordin and Banzhaf's notion of effective fitness and Goldberg's notion of operator-adjusted fitness are essentially the same idea, although specialised to different representations and operators.

6.1.3 Stevens and Waelbroeck's Exact Effective Fitness in GAs

Stephens and Waelbroeck [Stephens and Waelbroeck, 1997, Stephens and Waelbroeck, 1999] independently rediscovered the notion of effective fitness. Using our own notation, the *effective fitness of a schema* is implicitly defined through the equation

$$E\left[\frac{m(H,t+1)}{M}\right] = \frac{m(H,t)}{M} \cdot \frac{f_{\text{eff}}(H,t)}{\bar{f}(t)}$$

assuming that fitness proportionate selection is used. This has basically the same form as Equation (6.4). Indeed, the two equations represent nearly the same idea, although in different domains. Noticing that $E\left[\frac{m(H,t+1)}{M}\right] = \alpha(H,t)$ and that $\frac{m(H,t)}{M\bar{f}(t)} = \frac{p(H,t)}{f(H,t)}$, one obtains

$$f_{\text{eff}}(H,t) = \frac{\alpha(H,t)}{p(H,t)} f(H,t) \tag{6.6}$$

$$= f(H,t)\left[1 - p_{xo}\left(1 - \sum_i \frac{p(L(H,i),t)p(R(H,i),t)}{(N-1)p(H,t)}\right)\right] \tag{6.7}$$

where we used the value of $\alpha(H,t)$ given in Equation (5.3).

Equation (6.7) is similar to Equations (6.5) and (6.1), but there are important differences: f_j^e and $f_{\text{adj}}(H,t)$ are *approximations* (of unknown accuracy, being in fact lower bounds) of the true effective fitness of an *individual* in a standard GP system and of a schema in a GA, respectively, while $f_{\text{eff}}(H,t)$ is the *true* effective fitness for a *schema* in a binary GA. In addition, $f_{\text{eff}}(H,t)$ of a schema can be bigger than $f(H,t)$ if the building blocks for H are abundant and relatively fit. While the estimates/bounds given by f_j^e and $f_{\text{adj}}(H,t)$ ignore the constructive effects of crossover and therefore are always smaller than f_j and $f(H,t)$, respectively.

6.1.4 Exact Effective Fitness for GP

It is easy to extend to GP with one-point crossover the notion of effective fitness provided by [Stephens and Waelbroeck, 1997, Stephens and Waelbroeck, 1999]. By using the definition of effective fitness given in Section 6.1.3 and the expression for the total transmission probability $\alpha(H,t)$ of a schema under one-point crossover in Equation (5.10) (page 83) we obtain:

$$f_{\text{eff}}(H,t) = \frac{\alpha(H,t)}{p(H,t)} f(H,t)$$

$$= \quad f(H,t)\left[1 - p_{xo}\left(1 - \sum_j \sum_k \sum_{i \in C(G_j,G_k)} \frac{p(L(H,i) \cap G_j,t)p(U(H,i) \cap G_k,t)}{NC(G_j,G_k)p(H,t)}\right)\right]$$

It is important to stress that this equation gives the *true effective fitness for a GP schema* under one-point crossover: it is not an approximation or a lower bound. Also, like for the effective fitness of schemata representing strings, in GP $f_{\text{eff}}(H,t)$ can be bigger than $f(H,t)$ if the building blocks for H are abundant and relatively fit. This shows that crossover does not always have the destructive connotation often attributed to it in the GP literature.

It is possible to obtain a definition of effective fitness for standard crossover, too. Again for simplicity, we will consider the case in which crossover points are selected uniformly at random. In this case, from Equation (5.13) (page 92) we obtain that the effective fitness of a fixed-size-and-shape GP schema H under standard crossover with uniform selection of crossover points is

$$f_{\text{eff}}(H,t) =$$

$$f(H,t)\left[1 - p_{xo}\left(1 - \sum_{k,l} \sum_{i \in H \cap G_k} \sum_{j \in G_l} \frac{p(U(H,i) \cap G_k,t)p(L(H,i,j) \cap G_l,t)}{N(G_k)N(G_l)p(H,t)}\right)\right]$$

6.1.5 Understanding GP Phenomena with the Effective Fitness

Bloat and Code Compression. In [Nordin and Banzhaf, 1995, Nordin *et al.*, 1995, Nordin *et al.*, 1996], Equation (6.5) was used to explain the reasons for bloat and active-code compression in GP with standard crossover. The idea is that individuals for which the ratio between active code and total code, C_j^e/C_j^a, is smaller would appear to have a higher effective fitness (i.e. a higher chance of reproducing accurately). A small C_j^e/C_j^a ratio can be achieved in two ways: (1) by having large amounts of inactive code so as to have a large total length C_j^a with respect to the active code length, or (2) by having a very compact code, represented by a small C_j^e (or both). So individuals with lots of introns would propagate better and therefore the average length of the members of the population would grow, leading to the phenomenon known as *bloat.* At the same time, evolution would give higher chances of success to individuals which achieve a certain behaviour with a minimum number of instructions, leading to *active code compression.*

Since Equation (6.5) is only an approximation, it is still an open question whether these conclusions are applicable in general. Perhaps it will be possible to investigate issues such as bloat and code compression in a more rigorous way by using the exact schema theorems described in the previous chapter and the related exact definitions of effective fitness presented in the previous sections.

For example, schemata can easily represent classes of programs with the same active code (and, therefore, the same behaviour) but different amounts of inactive code. For example, see the three schemata in Figure 6.1. These schemata contain programs having exactly the same fitness. So the schemata have the same fitness too.

It seems possible to use the exact definition of effective fitness proposed in Section 6.1.4 to establish whether there is an effective fitness advantage in having large amounts of inactive code, and if so why and when. So, for example, it might be possible to show that the schemata in Figure 6.1 have different effective fitnesses despite the fact that their average fitnesses are the same. Studies of this nature may lead to a better understanding of bloat and to better antibloat measures.[2]

It is also possible to represent equivalent code fragments using schemata. For example, each of the three schemata in Figure 6.2 represents a different way of implementing the same behaviour using different amounts of active code but the same inactive code. For each program in each schema, it is possible for there to be a corresponding programs in any another schema with exactly the same behaviour. If the corresponding programs do exist in the population, then the three schemata will have the same fitness.

No exact study on this subject is available yet, but it seems possible to use the concept of effective fitness to establish whether there is an advantage in having more compact active code, and if so why and when. For instance, it might be possible to show that the effective fitness of the schema on the

[2] Studies on the evolution of size based on exact GP schema theories have very recently become available [McPhee and Poli, 2001]. In these studies the exact schema theory for standard crossover operating on linear structures has been used to better understand the changes in size distribution in GP, on the hypothesis of infinite populations. No study for tree-like structures is available, yet, although equations for the expected average size of tree-like programs in the next generation are available in [Poli, 2001b].

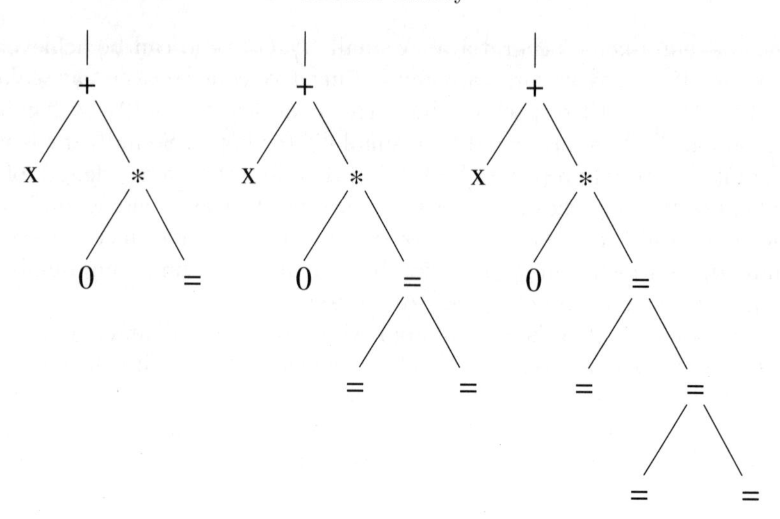

Fig. 6.1. Schemata representing programs with the same active code (and behaviour) but with different amounts of inactive code.

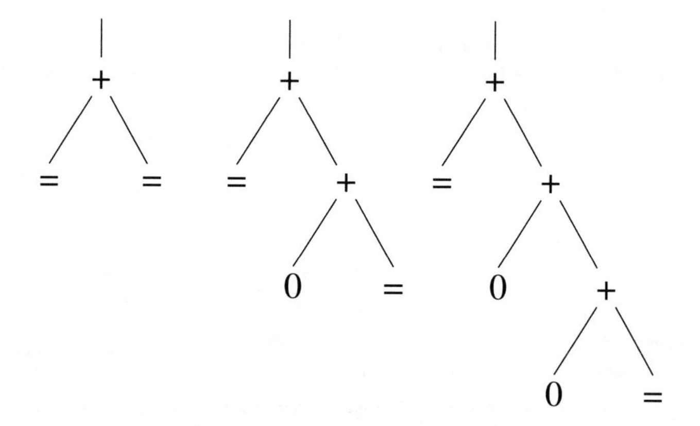

Fig. 6.2. Schemata representing different ways of implementing the potentially same behaviour using different amounts of active code but exactly the same inactive code.

left in Figure 6.2 is higher than that of the other two schemata. This kind of study may lead to a better understanding of code compression and to better generalisation-promotion measures.

Effective Fitness Landscapes. Effective fitness can be used to define a related concept of *effective fitness landscape* [Stephens, 1999, Stephens and Vargas, 2000, Stephens and Vargas, 2001]. The idea is simply that, at any given generation, one could imagine that a genetic algorithm is performing a sort of parallel hill-climbing on a landscape. The neighbourhood structure of the landscape is determined by the search operators, like for a normal fitness landscape. However, the height associated to the points in the search space is not given by the fitness function, but by the effective fitness of each point. Since the effective fitness changes dynamically generation after generation, the effective fitness landscape is a dynamic landscape. Evolution is forced to continue until the effective fitness landscape becomes completely flat.

The notion of effective fitness landscape is very useful to understand why there can be flows of individuals from parts of the search space to other parts even in the absence of an explicit selective pressure. This is because although the fitness landscape is flat the effective fitness landscape is not. This causes evolution to continue.

As an example of this phenomenon, let us consider a GP system with standard crossover (and uniform selection of the crossover points) in which only unary functions are used. So, the system operates on variable size but linear structures. Then, let us assume that the fitness function always returns 10 regardless of the program being examined, and that the population is initialised with equal proportions of programs of lengths between 1 and 39 nodes. The fitness landscape is flat but the initial distribution of lengths will undergo some changes. For example, we would expect that programs of length bigger than 39 be eventually generated. How can we explain this dynamics in the system? One very natural way to do that is to look at the effective fitness landscape explored by GP at different generations. Due to its high dimensionality ($\gg 2$) we cannot really plot this landscape but we can calculate and plot the effective fitness of programs of different lengths at different generations. This can be done either experimentally by performing GP runs and collecting the appropriate data, or by *iterating the schema equations*. By doing this one obtains the expected behaviour of a finite population, which is also equivalent to the exact behaviour of an infinite population. Figure 6.3 shows the effective fitness of programs of different lengths at generation 1, 2, 3 and 4 (solid lines) obtained by iterating the schema equations on the assumption of infinite populations along with the actual fitness of such programs (represented by the crosses). Let us explain what is happening in Figure 6.3.

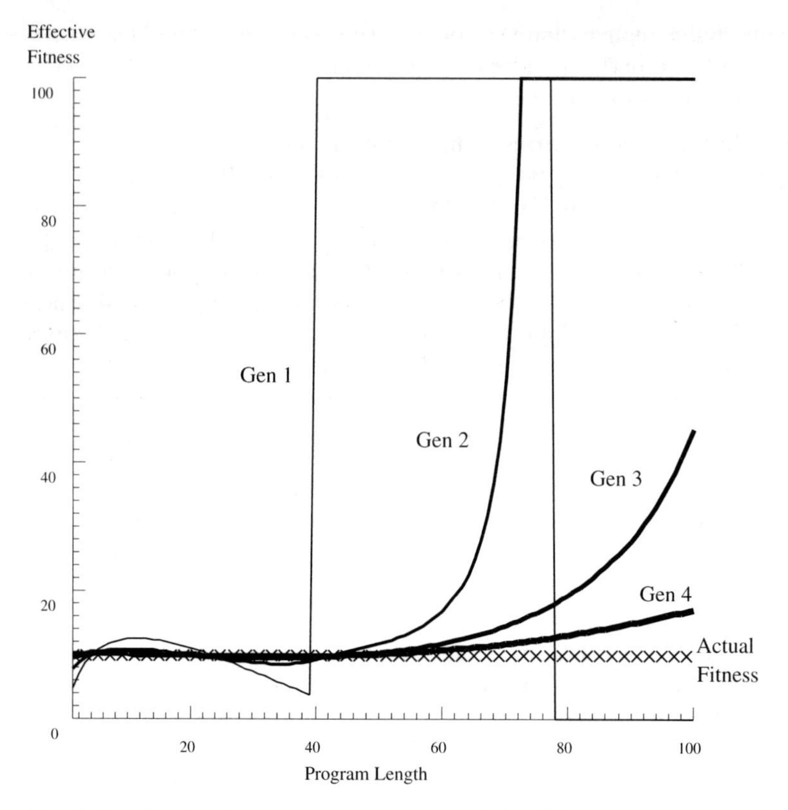

Fig. 6.3. Effective fitness vs. program length for different generations. In order to display infinite effective fitness values they are clipped at 100. Fitness values in the range $[100, \infty)$ are also clipped.

Let us consider an infinite population on our flat landscape. Equation (6.6) becomes

$$f_{\text{eff}}(H,t) = 10\frac{\alpha(H,t)}{\alpha(H,t-1)}$$

(by using Equation 5.2 (page 72) at time $t-1$ and noting $p(H,t) = \lim_{M\to\infty} E[m(H,t)]/M = \alpha(H,t-1)$). $\alpha(H,t)$ can be interpreted as the probability of transmitting programs matching H at generation t or, by virtue of the infinite population assumption, as the proportion of programs matching H at generation $t+1$. So, with a flat fitness landscape, the effective fitness of a schema is a measure of how much the proportion of individuals matching a given schema changes from one generation to the next.

Now, let us start by analysing the plot for generation 1, (i.e. the first generation created by standard crossover). The schemata we are interested in are very simple, they just match programs of a given length. ($H = =^i$

where i is the program length of interest). Since, at generation 1 programs of lengths between 40 and 77 can be created which did not exist in generation 0, the effective fitness of such programs is infinitely large. This explains the large "box" in the plot. The effective fitness of programs of length bigger than 77 is undefined because we had none at generation 0 and we still have none at generation 1. Conventionally we assigned a fitness of 0 to such programs to indicate that there appear to be no effective evolutionary pressure to sample such programs. The round shape of the plot between lengths 1 and 39 is due to the sampling bias induced by standard crossover [McPhee and Poli, 2001].

At generation 2, programs of lengths bigger that 77 are created, while there were none at the previous generations. So, their effective fitness is infinite. The effective fitness of shorter programs is flatter. This tendency is confirmed by the plots of generations 3 and 4. This indicates that the length distribution is approaching a stable state, so evolution is slowing down at least for relatively short programs. So, as predicted, the effective fitness landscape seems to be approaching a flat fitness landscape.

6.2 Operator Biases and Linkage Disequilibrium for Shapes

Knowing the biases introduced by the operators used in genetic programming is very important, since it allows an informed choice of operators and parameter settings for particular problems. Exact mathematical formulations of these biases might also allow the definition of optimum operators and parameters for different classes of problems.

The availability of exact schema models for different operators allows a formal study of the biases of those operators and a comparison of their performance, as recently shown in studies based on a version of the schema theorem for standard crossover applicable to linear structures [McPhee and Poli, 2001]. These studies show, in the absence of other biases, that standard crossover will tend to heavily sample the space of smaller-than-average programs and is unable to focus its search on programs of a particular size. This means that the crossover bias may counteract the tendency of selection to prefer fitter programs. This does not happen with one-point crossover, which is instead completely unbiased with respect to program length.

The quantity $\delta(h_1 \in L(H,i))\delta(h_2 \in U(H,i))$ (which we used to obtain the exact schema theorems for one-point crossover in the previous chapter), and a similar quantity for standard crossover, can be considered as measurement functions [Altenberg, 1995] for the distribution of parents and crossover points. By using different measurement functions one can obtain schema-theorem-like results that clarify the behaviour (and highlight other biases) of the operators and explain the influence of their parameters on the search. For example, in [Poli and Langdon, 1998a] we studied the amount of genetic material exchanged by different crossovers in GP. This revealed that (for trees) standard crossover is generally a local search operator. Therefore, standard GP searches the space of programs like a set of hill-climbers. Also, we showed that one-point crossover is slightly less local, while uniform crossover, an operator in which the offspring is formed by swapping nodes in the common region with uniform probability, is a global search operator.

Another interesting form of bias of crossover becomes apparent if one analyses the schema equations for schemata without defining symbols (i.e. program shapes). For example, let us consider a genetic programming system using two-input functions, with individuals of up to three levels. Figure 6.4 shows the shapes of some of the possible parent programs and of the children that could be created by one-point crossover. The schema evolution equation for the shape `(= = =)` (cf. Figure 6.4, top left) under one-point crossover becomes (after simplification):

$$\alpha(\texttt{(= = =)}) = p(\texttt{(= = =)}) + \frac{2}{3}\Delta p \tag{6.8}$$

where we used the fact that the sum of the selection probability of all shapes must always be 1 and

$$\begin{aligned} \Delta p = \quad & p(\texttt{(= (= = =) =)}) \times p(\texttt{(= = (= = =))}) \\ & - p(\texttt{(= = =)}) \qquad\quad \times p(\texttt{(= (= = =) (= = =))}) \end{aligned}$$

This result is quite surprising. It states that if $\Delta p = 0$, shape `(= = =)` will evolve (on average) as if selection only was acting on it, while if $\Delta p \neq 0$ evolution will tend to increase or decrease the proportion of individuals in `(= = =)` so as to make Δp approach 0. Why?

When one crosses over two parents belonging to any of the four shapes `(= = =)`, `(= (= = =) (= = =))`, `(= (= = =) =)` and `(= = (= = =))` (see Figure 6.4), only offspring having one of these shapes can be produced. For example, when one crosses over an instance of the schema `(= = =)` with an instance of the schema `(= (= = =) (= = =))`, there are four possible outcomes: an instance of `(= (= = =) =)`, an instance of `(= = (= = =))` , an instance of `(= = =)` (if the crossover point is above the root node and the parent donating the root is in `(= = =)`) or an instance of `(= (= = =) (= = =))` (if the crossover point is above the root node and the parent donating the root is in `(= (= = =) (= = =))`). Likewise, if one crosses over an instance of the schema `(= (= = =) =)` with an instance of the schema `(= = (= = =))` , there are exactly the same four possible outcomes. Some of the outcomes of crossovers between programs with three to seven nodes are illustrated in Figure 6.4. In all other possible combinations of parents (not shown in the figure), only offspring with the same shapes as one of their parents are produced. So, there is a constant migration of individuals from one shape to another. Until the probabilities of creating all these instances are exactly balanced, there will have to be a net flow of individuals among the four schemata `(= = =)`, `(= (= = =) (= = =))`, `(= (= = =) =)` and `(= = (= = =))`.

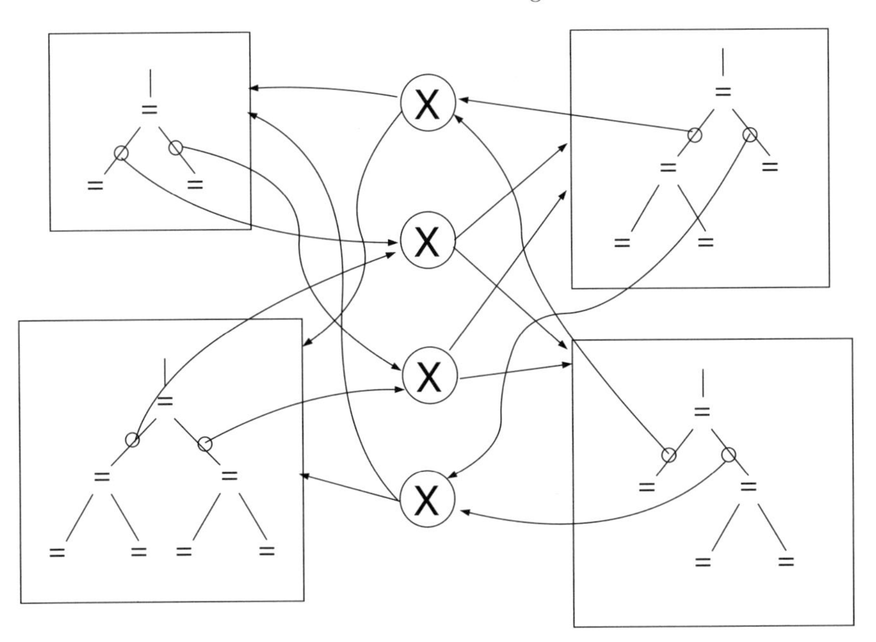

Fig. 6.4. Four binary trees with up to three levels. Large Xs indicate some of the possible one-point crossover operations. Small circles indicate crossover points (in the common region), while arrows pointing to boxes indicate the shape of the two children produced by each crossover.

It is easy to convince ourselves that equilibrium is reached only when $\Delta p = 0$ by comparing Equation (6.8) with the schema evolution equations for the schemata `(= (= = =) (= = =))`, `(= (= = =) =)` and `(= = (= = =))`. (All have the form $\alpha(H) = p(H) \pm \frac{2}{3}\Delta p$). The comparison shows that evolution tends to make the value of Δp approach 0. The condition $\Delta p = 0$ can be considered to be something like a *linkage equilibrium equation for shapes*. Until this form of equilibrium is reached, evolution will continue *even on a flat landscape*. This is due to the bias imposed by one-point crossover. Similar kinds of biases should be expected by other crossover operators. These biases might be the reason for the fact, reported in various experimental studies (e.g. [Langdon *et al.*, 1999]), that GP populations seem to tend towards programs with a particular level of "bushiness" (see Chapter 11).

6.3 Building Blocks in GAs and GP

The *building block hypothesis* [Goldberg, 1989b] is typically stated informally by saying that a genetic algorithm works by combining short, low-order schemata of above average fitness (building blocks) to form higher-order ones over and over again until it converges to an optimum or near-optimum solution.

The building block hypothesis and the related concept of deception have been strongly criticised in [Grefenstette, 1993], in the context of binary GAs, where it was suggested that these ideas can only be used to produce first-order approximations of what really happens in a GA. In [O'Reilly and Oppacher, 1995] the criticisms of the building block hypothesis have been extended to genetic programming with standard crossover, arguing that the hypothesis is even less applicable to GP because of the highly destructive effects of crossover. The analysis of the "survival-disruption GP schema theorem" (Inequality (4.15) page 65) reported in [Poli and Langdon, 1997c, Poli and Langdon, 1998b] and the experimental results in [Poli and Langdon, 1997a] suggested that the latter criticism might not apply to GP with one-point crossover. Therefore, we argued that the building block hypothesis might still be considered a reasonable first approximation of the behaviour of GP with one-point crossover.

Stephens and Waelbroeck analysed their schema theorem (cf. Equation (5.3) page 73) and claim that the presence of the terms (equivalent to) $p(L(H,i),t) \times p(R(H,i),t)$ indicates that GAs do build higher-order schemata (H) by juxtaposing lower-order ones (the $L(H,i)$'s and $R(H,i)$'s) [Stephens and Waelbroeck, 1997, Stephens and Waelbroeck, 1999]. Note here that the building blocks of schema H are well defined. They are the $L(H,i)$ and $R(H,i)$ pairs. There are up to $N-1$ pairs of building of building blocks for each high order schema. However, these building blocks are not necessarily all fitter than average, short or even of low-order. The GA will use them as building material provided there are enough of them. Of course fitness does have an important role, since (1) it is helps determine $p(L(H,i),t)$ and $p(R(H,i),t)$ and (2) the fitness of individuals in the previous generation will have played a part in deciding how many copies of each building block there are.

Since the macroscopic exact GP schema theorem (Section 5.4.5) is a generalisation of Stephens and Waelbroeck's schema theory, exactly the same can be said for GP with one-point crossover. Indeed, the presence of the terms $p(L(H,i) \cap G_j, t) \times p(U(H,i) \cap G_k, t)$ suggests that the schemata $L(H,i) \cap G_j$ and $U(H,i) \cap G_k$ (for all meaningful values of i, j and k) are a form of building blocks for fixed-size-and-shape schemata. So, it is arguable that GP with one-point crossover also builds higher-order schemata by juxtaposing lower-order ones. Again, the building blocks need not necessarily be all fitter than average, short or even of low-order. If there are many copies of a pair of building blocks $L(H,i) \cap G_j$ and $U(H,i) \cap G_k$, one-point crossover will tend to assemble them into the higher-order schema H even if they are of below-average fitness.

Exactly the same could be said for standard crossover. In Equation (5.13) (page 92) the schemata $U(H,i) \cap G_k$ and $L(H,i,j) \cap G_l$ play the role of building blocks for standard crossover. So, we could say that also GP with standard crossover builds higher-order schemata by juxtaposing lower-order ones. Again, for this to happen the building blocks need not necessarily be all fitter than average, short or even of low-order.

However, the term building block may be slightly misleading when used for GP. The term seems to suggest that a GA/GP system will reliably and reproducibly put together step-by-step building blocks of higher and higher order until a solution is found. In the creation of a particular schema H in a binary GA with one-point crossover, the number of pairs of building blocks that can possibly create H is one less than to the number of defining bits in H (i.e. at most $2\mathcal{O}(H) - 2$ schemata are involved). So, for GAs, at least for low order schemata, there are very few building blocks pairs each of which has a relatively high chance of being selected. In GP with standard crossover (and to a lesser extent GP with one-point crossover) a huge number of schemata can lead to the creation of instances of a particular schema. Obviously, in any finite population there are only a finite number of schemata, and so only a finite number of schemata can contribute to the creation of a particular schema. However, the collection of building blocks that GP can use to build instances of a schema still remains very large and so, the probability that any *pair* of them is chosen to create a particular schema in a particular selection/crossover process can only be very very small. So, although the schema equations indicate exactly the ways in which a particular schema can be assembled from lower order schemata (building blocks), they also suggest that each such event will happen only very rarely. As a consequence, it is entirely possible that none of the instances of a particular schema created in a certain generation will have been created using the same two building blocks. This suggests that the schemata $U(H,i) \cap G_k$ and $L(H,i,j) \cap G_l$ should be interpreted as the building material used by crossover to create instances of H, but the process of choosing between pairs of such schemata should considered largely non-reproducible and random.

6.4 Practical Ideas Inspired by Schema Theories

Most people see GA/GP theory as a way of modelling existing algorithms, representations and operators. However, theory can also indicate new directions. For example, the introduction of one-point crossover for GP in [Poli and Langdon, 1997c] was not an act of creative innovation. One-point crossover was suggested as a natural operator to act on fixed-size-and-shape schemata by the theory itself: in developing the theory it became clear that without an operator that swapped subtrees in exactly the same position in the two parents, the mathematics initially involved in obtaining a schema theorem was much too complex. In turn, this operator suggested other "homologous" operators based on the concept of a common region, such as GP uniform crossover [Poli and Langdon, 1998a], which later led to the notion of smooth operators [Page *et al.*, 1999, Poli and Page, 2000].

Many initialisation strategies have been proposed in the GP literature [Koza, 1992, Banzhaf *et al.*, 1998a, Bohm and Geyer-Schulz, 1996, Iba, 1996, Luke, 2000b, Ratle and Sebag, 2000], some of which have a theoretical motivation [Langdon, 1999, Langdon, 2000]. Other theoretically sound initialisation strategies could be derived thanks to the knowledge of the biases of the operators obtained from exact schema theories. In general-purpose evolutionary algorithms, a good strategy might be to initialise the population so as to minimise the biases of the operators in the first generation – a particularly important stage in a run. If instead one is interested in optimising/specialising an evolutionary algorithm for a particular class of applications, a good initialisation strategy might be to maximally bias the search operators towards areas of the search space with the highest density of solutions.

6.5 Convergence, Population Sizing, GP Hardness and Deception

The schema theorems presented so far predict, at best, the expected or average number of instances of a schema in the next generation. In fact, in a finite population the actual number will be distributed around this estimate. In [Poli, 1999b], Chebyshev's inequality [Spiegel, 1975] was used to transform the exact statement of the mean and variance of $m(H, t+1)$ into a probabilistic statement about the interval with which $m(H, t+1)$ will vary (see Section 5.2). By using this in conjunction with Stephens and Waelbroeck's schema theory, [Poli, 2000c] derived a recursive conditional schema theorem for genetic algorithms which allows one to predict the probability of obtaining a solution for a problem in a fixed number of generations assuming that the fitness of the building blocks and of the population are known. This in turn led to the formulation of conditional population sizing equations.

Although these results are very conservative, nothing prevents one from using the macroscopic exact genetic programming schema theorems (presented in Sections 5.4.5 and 5.6.3) to extend this conditional-convergence theory to GP. In the future, this might allow us to identify rigorous strategies to size the population and therefore to calculate the computational effort required to solve a given problem using GP.

The availability of rigorous computational effort equations would open the way to a precise definition of "GP-friendly" ("GP-easy") fitness functions. Such functions would simply be those for which the number of fitness evaluations necessary to find a solution with say 99% probability in multiple runs is (much) smaller than 99% of the effort required by exhaustive search or random search without resampling. An alternative approach to define GP-hardness would be to use the exact definition of effective fitness to correct and extend the ideas introduced in [Goldberg, 1989a] to obtain a rigorous definition of *deception* for GP (a topic that was first studied in [Langdon and Poli, 1998c]; (see also Chapters 9 and 10).

6.6 Summary

Schema theories are characterisations of the way that genetic operators affect the syntactic components of a representation and how they may bring these components together. Various schema theories were described in Chapters 3–5. This chapter considered different ways in which we can use them to better understand the dynamics of genetic programming populations.

Schema theories make no assumptions about the semantics of the syntactic structures manipulated by an evolutionary algorithm and the performance of the individuals that the evolutionary algorithm searches: they only require that a semantics and a performance measure exist.

The next chapter is explicitly concerned with programs, their semantics and how fitness is distributed in program spaces.

7. The Genetic Programming Search Space

We shall investigate how the space of all possible programs, i.e. the space which genetic programming searches, scales. Particularly how it changes with respect to the size of programs. We will show, in general, that above some problem dependent threshold, considering all programs, their fitness shows little variation with their size. The distribution of fitness levels, particularly the distribution of solutions, gives us directly the performance of random search. We can use this as a benchmark against which to compare GP and other techniques. These results are demonstrated in this chapter using a combination of enumeration and Monte Carlo sampling. For the interested reader, formal proofs are given in Chapter 8. Informal arguments are presented in Section 7.5 to extend our results to modular GP, memory and Turing complete GP. Section 7.6 considers the relationship between tree size and depth for various types of program. We finish this chapter with a discussion of these results and their implications.

7.1 Experimental Exploration of GP Search Spaces

For the very shortest programs, it is feasible to generate and test every program of a given size. (The size of a tree program is the number of internal nodes in the tree plus the number of leaves in the tree). However, as Figure 7.1 makes clear, the number of possible programs grows very quickly with their size and so we must fall back on randomly sampling programs. We use the random tree method given in [Alonso and Schott, 1995] to sample uniformly all the programs of a specific size. Typically we sample 10,000,000 programs of each size. C++ code to generate random programs is available via `ftp://ftp.cs.bham.ac.uk/pub/authors/W.B.Langdon/gp-code/rand_tree.cc`. (`ntrees.cc` contains C++ code to calculate how many programs there are in the search space; cf. Figures 7.1 and 7.15).

The ramped half-and-half method [Koza, 1992, page 93] is commonly used to generate the initial population in genetic programming (GP). Half the random programs generated by it are full (i.e. every leaf is the same distance from the root). Therefore, we also explicitly consider the subspace of full trees. In some cases this subspace is radically different from the whole space.

7.2 Boolean Program Spaces

The Boolean functions have often been used as benchmark problems. The program trees we will consider are composed of n terminals (D0, D1, ...) which are the Boolean inputs to the program and four sets of the Boolean logic functions {NAND}, {XOR}, {AND, OR, NAND, NOR} and {AND, OR, NAND, NOR, XOR}. There are 2^{2^n} Boolean logic functions of n inputs. NAND by itself is sufficient to construct any of them, and therefore so are the last two sets. XOR by itself can only generate 2^n of them, but as we shall see that adding it to the function set can dramatically affect the whole search space. The fitness of each tree is given by evaluating it as a logical expression for each of the 2^n possible combinations of inputs. Its fitness is the number of fitness cases where its output agrees with that of the target Boolean function [Koza, 1992].

There are $n^{(l+1)/2}|F|^{(l-1)/2} \times \frac{(l-1)!}{((l+1)/2)!((l-1)/2)!}$ different trees of size l [Koza, 1992, Alonso and Schott, 1995, page 213]. $|F|$ is one, four or five, depending which of the four function sets is used. (Note this formula is simple as each function (internal node) has two arguments). The number of programs rises rapidly (approximately exponentially) with increasing program size l (see Figure 7.1). Of course, if no bounds are placed on the size or depth of programs then the number of them is unbounded, i.e. the search space is infinite.

7.2.1 NAND Program Spaces

Because of the ease of manufacture of NAND gates in integrated semiconductors and because any Boolean function can be constructed from a network of NAND gates, NAND gates are the principal active component in digital electronics. It is therefore an interesting function to study, and in this section we give the number of each Boolean logic function that can be constructed from program trees composed only of NAND of various sizes for two, three and four inputs.

Two-Input NAND Program Spaces. There are $2^{2^2} = 16$ Boolean functions of two inputs. The proportion of NAND trees that evaluate to each is plotted in Figure 7.2.

Not surprisingly, there are two peaks at size 1, both of height 0.5 which correspond to the functions D0 and D1, and no other functions are possible. Likewise there is a peak of the same height for size 3 which is NAND itself, and two smaller peaks for ND0 (not D0) and ND1 (not D1). Three functions can be constructed from trees of size 5 (i.e. two NAND gates and three inputs). Seven functions of size 7, and so on. Its not until tree size 15 (7 NAND gates) that all of the 16 possible functions can be constructed from a tree of one size. (Although all of them can be constructed from trees of size 13 or less).

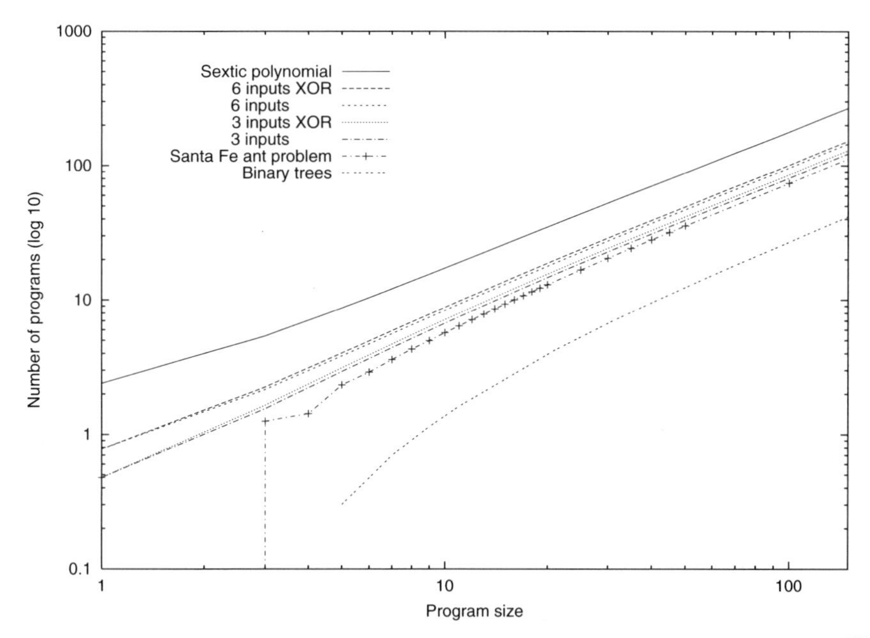

Fig. 7.1. Size of various search spaces (note log log scale). E.g. there are 1.556 10^{176} Sextic polynomial programs of size 99 (i.e. made of 44 functions and 45 inputs or constants)

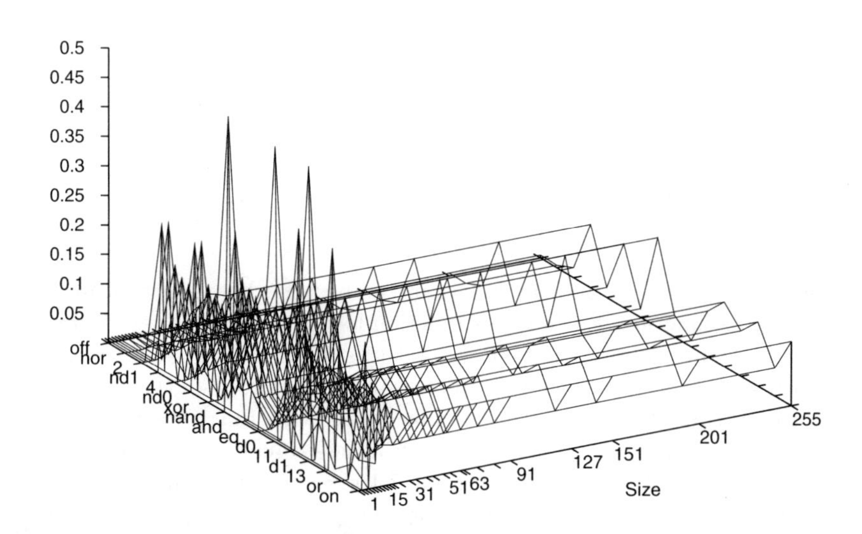

Fig. 7.2. Proportion of NAND trees that yield each two-input logic function.

Note XOR can be fabricated from 5 NAND gates and 6 terminals in a nearly full tree with 11 nodes and a height of 4.

From Figure 7.2 we can also see that each of the functions quickly converges towards some limiting proportion of the NAND trees of a given size. Using this convergence property, we can give the proportion of the search space occupied by each of the sixteen functions; see Table 7.1.

Table 7.1. Proportion of two-input NAND programs, and their fitness values (for the always on and XOR problems). Column 1 contains each program's truth table, expressed as a decimal integer. Lower table gives proportion by fitness value.

Rule		Proportion	Fitness	
			Always on	Odd-2-parity
0	off	0.00490	0	2
1	nor	0.00415	1	1
2		0.01689	1	3
3	nd1	0.10710	2	2
4		0.01696	1	3
5	nd0	0.10745	2	2
6	xor	0.01430	2	4
7	nand	0.15121	3	3
8	and	0.03695	1	1
9	eq	0.01088	2	0
10	d0	0.07727	2	2
11		0.10920	3	1
12	d1	0.07702	2	2
13		0.10898	3	1
14	or	0.04753	3	3
15	on	0.10922	4	2
			Proportion	
		Fitness 0	0.00495	0.01088
		1	0.07495	0.25928
		2	0.39403	0.48296
		3	0.41691	0.23259
		4	0.10921	0.01430

Each of the $2^{2^n} = 16$ Boolean functions can be regarded as a problem with its own fitness function. Each fitness function has $2^n + 1 = 5$ values. A simple problem is to find a tree that always returns 1 (the always-on problem). This corresponds to Function 15. Function 15 has, of course, the maximum fitness value (equal to 4). The fitness value of the other functions are given in column 4 of Table 7.1. (Column 5 gives their fitness values for the odd-2 parity (XOR) problem). The lower part of Table 7.1 gives the fraction of the search space with a particular fitness value for the two example problems. Note the relationship between function and fitness (on a given problem) is fixed. Therefore, since the proportion of the search space occupied by each function does not change with respect to size above the threshold, the distribution of fitness values is independent of size above the same threshold. This is true for all possible (two-input) problems.

Functions 10 and 12 (D0 and D1) are equivalent to each other in the sense that the other is produced by exchanging inputs. Function 10 and 12 are equally common in the search space. There are three other pairs of equivalent functions, 2 and 4, 3 and 5 (ND1 and ND0), and 11 and 13. It is also apparent that lexical ordering of functions is not the most convenient form for graphical presentation. In other plots they will be presented in order of decreasing frequency, and only data for one function of an equivalent set will be plotted.

Three-Input NAND Program Spaces. There are $2^{2^3} = 256$ Boolean functions of three inputs. In Figure 7.3 we plot the proportion of NAND trees which implement each of these functions. Fortunately many of the functions are equivalent to each other. Taking this into account they reduce to 80 classes. (In Figure 7.3 we exploit this by giving just the plot for one example function from each equivalence class. We use the class ordering given in [Koza, 1992, Table 9.2]).

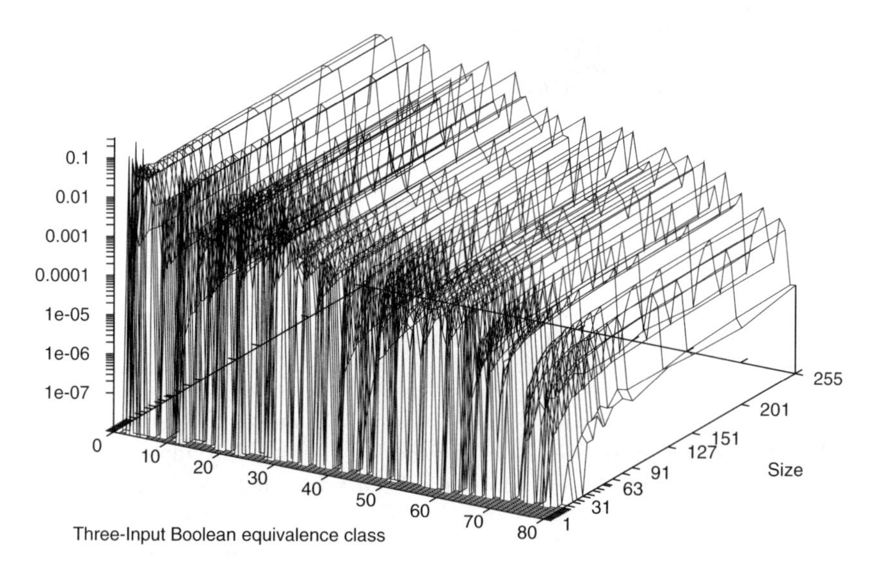

Fig. 7.3. Proportion of NAND trees that yield each three-input equivalence class.

Comparing Figures 7.2 and 7.3 we see they have several features in common. In particular the proportion of each function changes initially as we look at programs of increasing size but once some threshold has been exceeded, it changes scarcely at all with size. As might be expected the variation between each function is far greater than when using just two inputs.

Four-Input NAND Program Spaces. There are $2^{2^4} = 65536$ Boolean functions of four inputs. Again many of these are equivalent to each other and so these reduce to 4176 classes. In Figure 7.4 we plot the proportion of NAND trees that evaluate to each of these classes. The ordering of the classes is given by their measured frequency in trees of size 255.

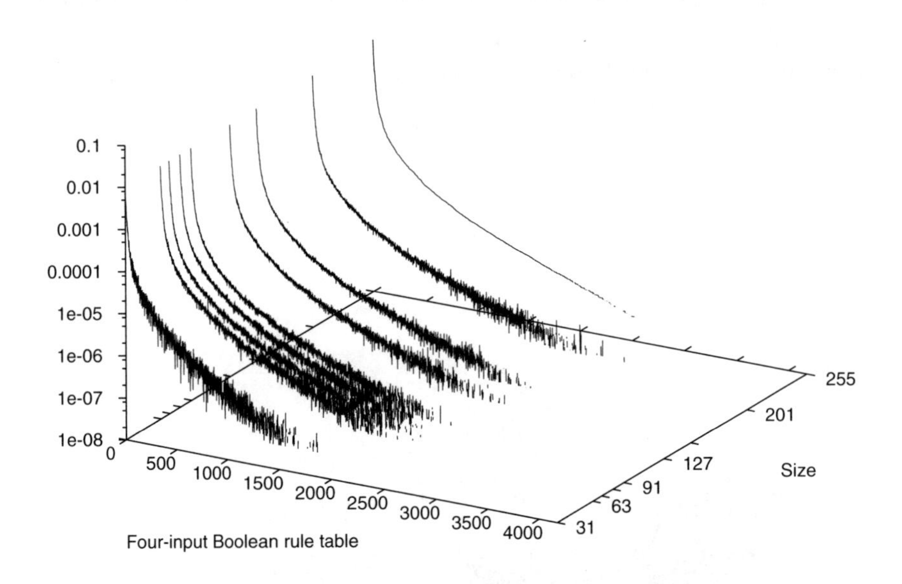

Fig. 7.4. Proportion of NAND trees that yield each four-input equivalence class (data with a signal-to-noise ratio of less than 3.0 are excluded)

Comparing Figure 7.4 with Figures 7.2 and 7.3; we see the same common features. While the proportion of each function changes initially (for clarity, data for short programs are not plotted in Figure 7.4), if we look at programs containing 15 or more NAND gates (i.e. size 31 or more) the proportion scarcely changes with size. This appears to be true for all 65536 functions but, as the data is based on Monte Carlo sampling, the data for the rarer functions is correspondingly noisy.

Again the variation between each function is far greater than when using just two or three inputs. Indeed none of the functions in the 842 rarest equivalence classes were discovered in Monte Carlo sampling of 10,000,000 programs of size 255. This includes both the odd and even parity functions with four inputs.

7.2.2 Three-Input Boolean Program Spaces

In this section we consider all the Boolean functions for $n = 3$ when using the larger function sets (i.e. {AND, OR, NAND, NOR} and {AND, OR, NAND, NOR, XOR}). As we said in Section 7.2.1, there are 256 of them but they can be split into 80 equivalence classes.

Comparing Figure 7.5 with Figure 7.3, we see the bigger search space shares many characteristics with that produced by NAND on its own. In particular, it shows that a certain minimum size is required before the problem can be solved, and that the minimum size depends on the difficulty of the problem. Then the proportion of programs that belong to the equivalence class grows rapidly with program size to a stable value which appears to be more-or-less independent of program size. Figure 7.6 shows these characteristics are retained if we extend the function set to include XOR. Note that adding the XOR function radically changes the program space. In particular, as might be expected, the two parity functions (equivalence classes 79 and 80) are much more prevalent. Also the range of frequencies is much reduced. For example, 68 of the 80 equivalence classes have frequencies between 0.1/256 and 10/256, rather than 28 as with the standard function set.

While Figures 7.5 and 7.6 can be used to estimate the fitness space of each three-input Boolean function across the whole space, there are some interesting parts of these spaces where certain functions are more concentrated than elsewhere. There are far more parity functions amongst the full trees than there are on average. When XOR is added to the function set, there are again a higher proportion of parity functions but the difference between the full trees and the rest of the search space is less dramatic.

7.2.3 Six-Input Boolean Program Spaces

In this section we investigate the distribution of six-input Boolean functions using the two larger function sets (i.e. {AND, OR, NAND, NOR} and {AND, OR, NAND, NOR, XOR}). It is difficult to analyse all the Boolean functions with more than four inputs. Instead, we have concentrated on what are generally considered to be the easiest and hardest Boolean functions of six inputs. Namely the always-on-6 function and the even-6 parity function. Figures 7.7 and 7.8 show the proportion of programs of various sizes with each of the possible fitness scores (hits). Figure 7.9 shows the same when XOR is added to the function set. Always-on-6 and even-6 parity, both with and without XOR, have the same property of the fitness being nearly independent of size.

The fitness distribution of the even-6 parity problem is much tighter than that of the binomial distribution that would be produced by selecting Boolean functions uniformly at random from the 2^{2^n} that are available [Rosca, 1997c, page 62]. I.e. centred on $2^n/2 = 2^{n-1}$ with standard deviation of $\sqrt{2^n/4} = 2^{n/2-1}$. The measured standard deviation is only 0.3 rather than 4. (Chapter 8 shows that while in big random trees the distribution of outputs may be

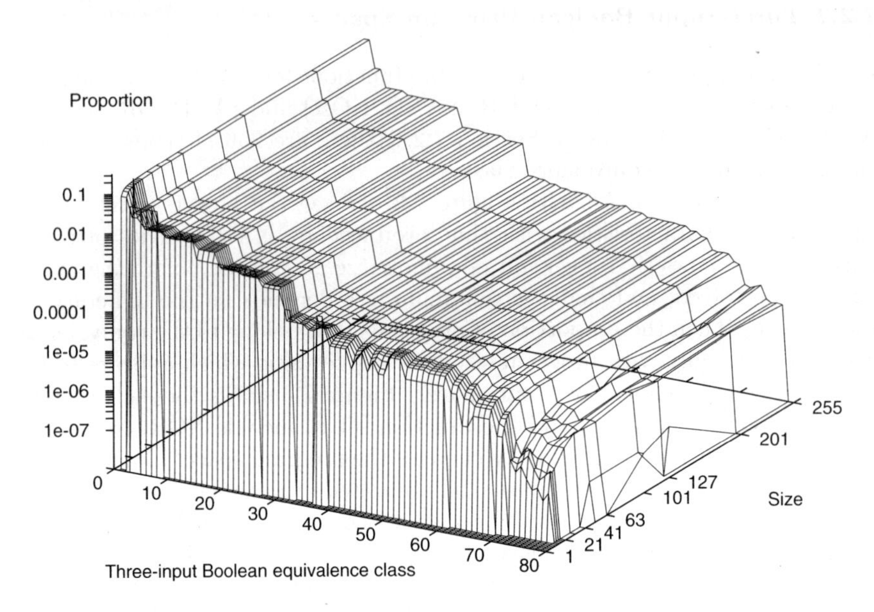

Fig. 7.5. How the search space of all possible programs (of various sizes) that can be created using the function set {AND, OR, NAND, NOR} and three inputs is divided between the 256 possible functions. (One function from each equivalence class is plotted)

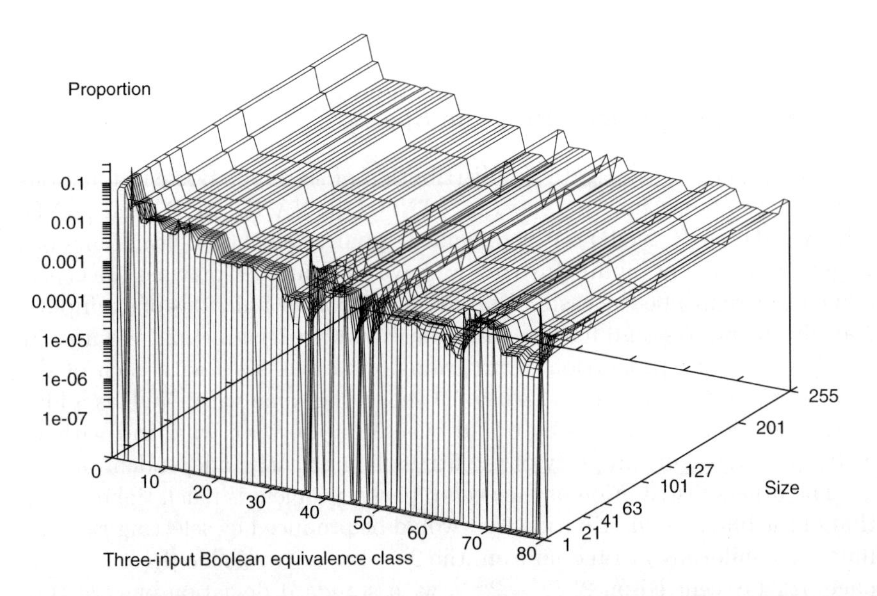

Fig. 7.6. Proportion of possible three-input Boolean programs with function set {AND, OR, NAND, NOR, XOR}, of various sizes, that implement each function.

random there are good reasons for expecting the distribution of functionality not to be). Such a tight fitness distribution and in particular the absence of a high fitness tail suggests that the problem will be hard for adaptive algorithms. When discussing the evolution of evolvability [Altenberg, 1994b] assumes that high fitness tails exist and can be found by evolutionary search algorithms. Of course we would hope that they would be better at finding and exploiting such tails than random search.

As expected, adding XOR to the function set greatly increases the width of the even-6 parity fitness distribution, and it retains its near independence of program size (see Figure 7.9). The standard deviation is now 0.92 rather than 0.34. Adding XOR means it becomes feasible to solve the six-input parity problem using random search. Our Monte Carlo simulations show solutions (i.e. programs scoring 64 hits) occupy about $2\ 10^{-7}$ of the whole search space.

Figure 7.7 shows the distribution of number of trues returned is a sawtooth curve. Again, convergence to a limiting distribution is rapid. The distribution of number of trues returned when XOR is added to the function set is a little changed, but retains its sawtooth appearance and near independence of program size.

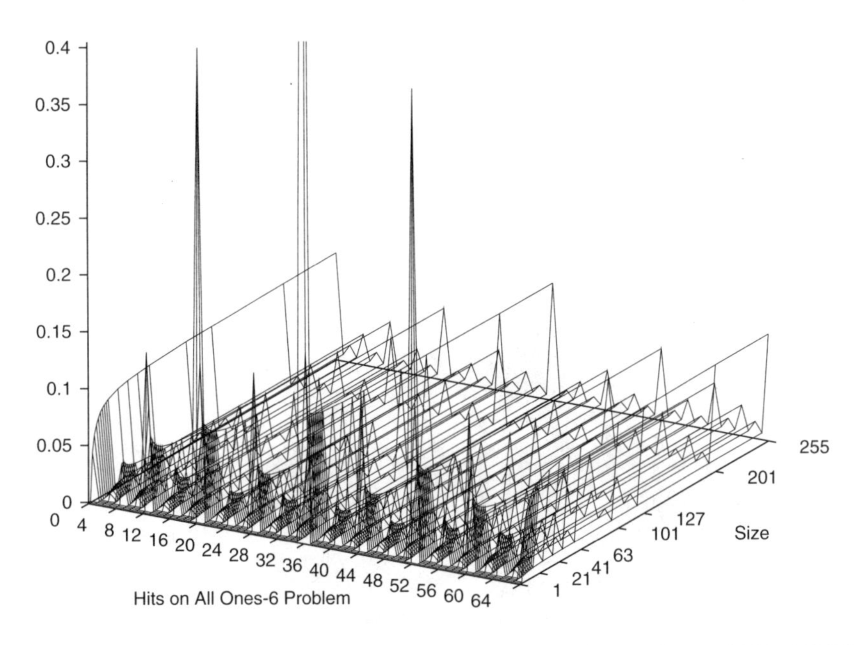

Fig. 7.7. Proportion of six-input Boolean functions {AND, OR, NAND, NOR} by number of ones returned by them (note the linear vertical scale)

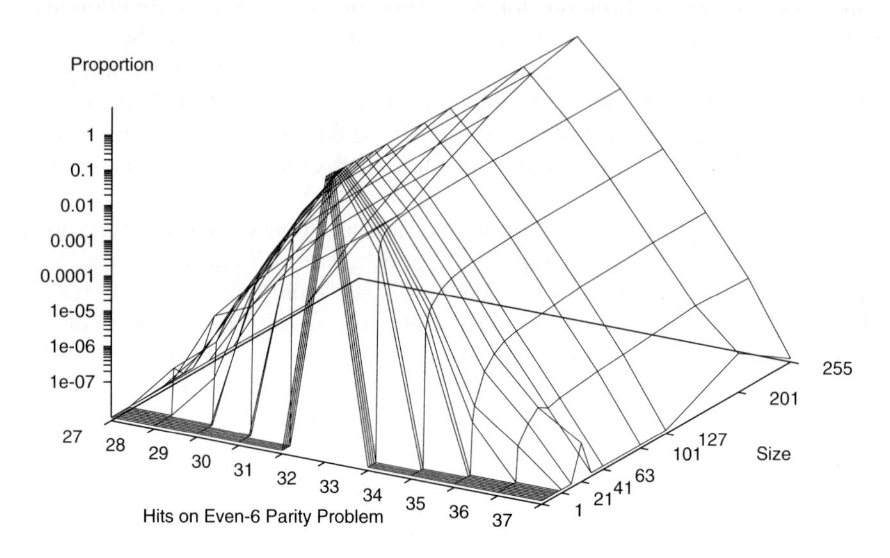

Fig. 7.8. Even-6 parity program space when using function set {AND, OR, NAND, NOR} and six inputs. Based on Monte Carlo sampling. No program scored less than 27 or more than 37 hits.

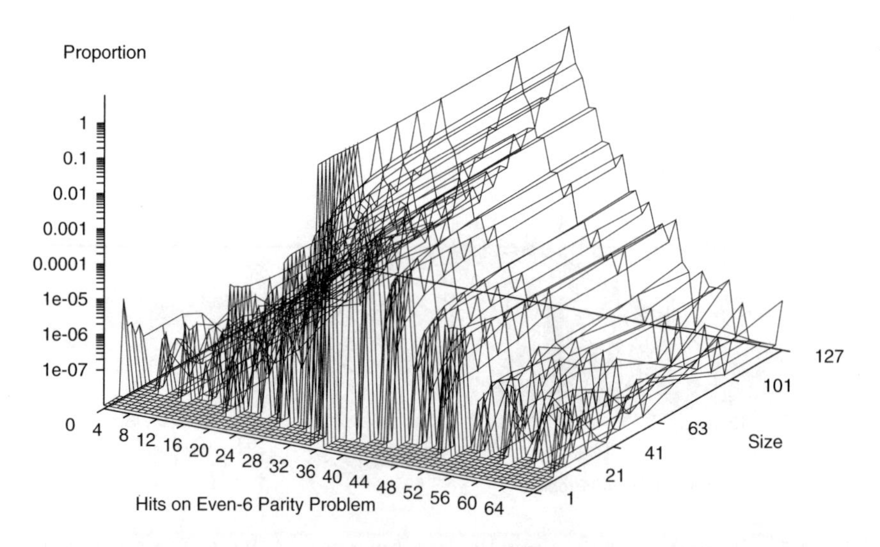

Fig. 7.9. Even-6 parity program space when using function set {AND, OR, NAND, NOR, XOR}. Note the noise due to Monte Carlo sampling.

7.2.4 Full Trees

Restricting our search to just the full trees yields a fitness distribution that is similar to that for the even-6 parity problem, see Figure 7.10. In particular, we have the convergence of fitness distribution once the tree size exceeds a threshold. There is a small variation with size but it does appear to decrease as we consider bigger trees. The distribution of fitness values observed is considerably wider with a range of 25–38 (twice that for the whole search space; see Figure 7.8) and a standard deviation of 0.68. Adding XOR to the function set further widens the distribution (the standard deviation becomes 1.8).

Searching just the full trees yields a similar fitness distribution for the always-on-6 problem as for the whole search space.

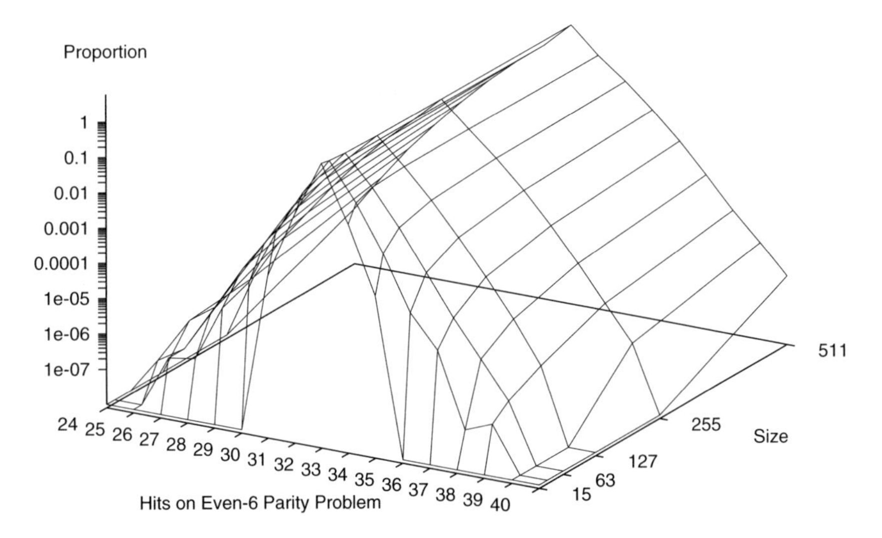

Fig. 7.10. Even-6 parity full tree program space {AND, OR, NAND, NOR}. Based on Monte Carlo sampling. No program scored less than 24 or more than 40 hits.

7.3 Symbolic Regression

Symbolic regression is the problem of finding in a functional form a model of some dataset. Note that symbolic regression means finding both the coefficients and the form of the function itself. Often such models are used to predict values for unknown data points. Here we investigate a benchmark symbolic regression problem.

We use the sextic polynomial regression problem [Koza, 1994, pages 110–122]. The sextic polynomial is $x^6 - 2x^4 + x^2$, which can be rewritten as the product of three squares: $x^2(x-1)^2(x+1)^2$. The problem for GP is to match it over the range $-1.0 \ldots 1.0$ using $+$, $-$, $\times$, protected division, the input x and random constants. We used 250 random constants, chosen from the 2001 numbers between -1 and $+1$ with a granularity of 0.001. (By chance, no constant was repeated and none of the three special values of -1, 0 or 1 were included). The 50 test points we used in our experiments were chosen uniformly at random from within the range -1 to 1. No granularity was imposed. Again by chance, none of the three special values of -1, 0 or 1 were included and no value was repeated. Apart from limiting ourselves to 250 constants, this is as described in [Koza, 1994, pages 110–122].

7.3.1 Sextic Polynomial Fitness Function

The fitness of each program is given by its absolute error over all the test cases [Langdon *et al.*, 1999]. This is as described by [Koza, 1994], except that we divide by the number of test cases (50) to yield the average discrepancy between the value it returns and the target value. All calculations were performed in standard floating point representations.

7.3.2 Sextic Polynomial Fitness Distribution

The distribution of fitness is given in Figure 7.11. It is apparent that symbolic regression shares many of the characteristics of the more difficult Boolean problems. The proportion of good programs is very small (as expected), but again we see that above a small threshold the proportion of good programs in a given fitness range converges to a value which is independent of their size. Figure 7.11 shows that the proportion of very bad programs does show variation w.r.t. size. However, it appears to reach a stable value but the threshold size is bigger. Each test case where a floating point exception occurs is given a penalty of about 2,000 [Langdon *et al.*, 1999]. Figure 7.11 suggests bigger programs are more likely to cause floating point exceptions, but when programs are big enough the proportion of these programs also converges to a limit. In practice such programs have little effect as they are never selected to be parents of the next generation.

7.4 Side Effects, Iteration, Mixed Arity: Artificial Ant

Our last example is in some ways the most complex. We report the distribution of fitness in a benchmark GP problem, which combines side effects and iteration, and by including functions with three inputs (arity-3 functions) investigates non-binary trees too. The problem is the artificial ant following the

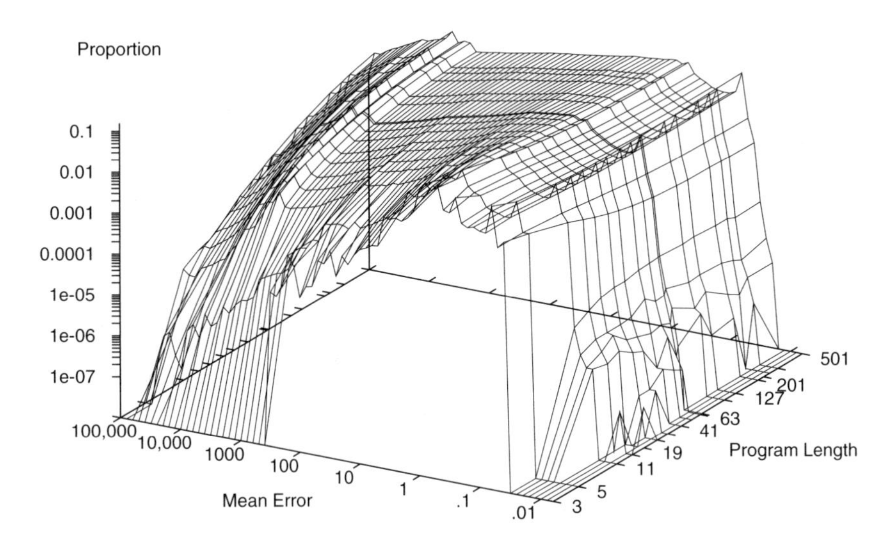

Fig. 7.11. Sextic polynomial fitness distribution (constants and input are sampled equally)

Santa Fe trail. (This problem is described in more detail, including a schema analysis, in Chapter 9. Here we concentrate upon variation w.r.t. size). In this problem the program tree is repeatedly executed. The program controls an artificial ant using the side effects of special leaves (move forward, turn left and turn right). The idea is to move the ant along a twisting trail (hence the name "Santa Fe trail"); see Figure 7.12. The trail is made of food pellets and the program can sense when there is food immediately in front of the ant. The ant eats each pellet it reaches. A program achieves maximum fitness if the ant eats all the food within 600 time steps.

The problem is made more difficult by making the ant's world toroidal, so rather than running into the edge of its world, on reaching an edge it reappears on the far side of its world (because of this Figures 7.12 and 9.1, page 153, are equivalent).

Figure 7.13 again suggests that, provided programs exceed some small fitness dependent threshold, the distribution of program fitnesses is roughly independent of their size. (Much of the fluctuation seen at the higher fitness levels is due to sampling noise inherent in Monte Carlo measurements).

In Figure 7.14 (solid line) we present the data for just one fitness value (value 89, the solutions). There are no solutions with less than ten nodes. From 11 to 18 nodes the proportion of solutions in the search space rises rapidly (but not monotonically) to a peak from which it falls. For programs with more than 30 nodes, the concentration of solutions appears to change only slowly with program size. It appears to eventually fall to near zero.

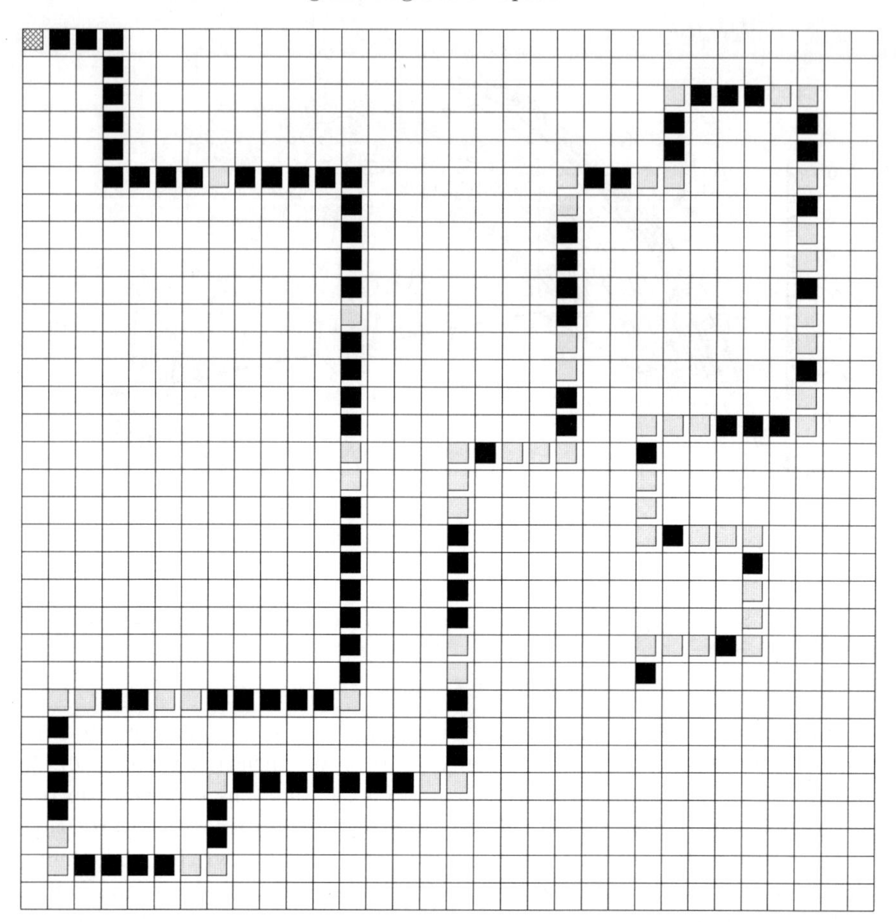

Fig. 7.12. The Santa Fe trail. Solid squares are food items. The ant starts at the top left (shown with cross-hatching) facing to the right. For convenience, gaps in the trail where there is not a food pellet are shown by empty boxes. They have no significance to the execution of the programs or their fitness.

In the standard Santa Fe problem, the function set includes both Prog2 and Prog3; however it is not necessary to have both to solve the problem. The two dashed lines in Figure 7.14 show the density of solutions when only one of them is included. While the data are subject to sampling noise, it appears that both subspaces formed by excluding one or other of the Prog functions are richer in solutions than the original one. Again in both subspaces, it appears there is only very slow variation in the density of the maximum fitness value with respect to size above some threshold (about 50 or 100 nodes).

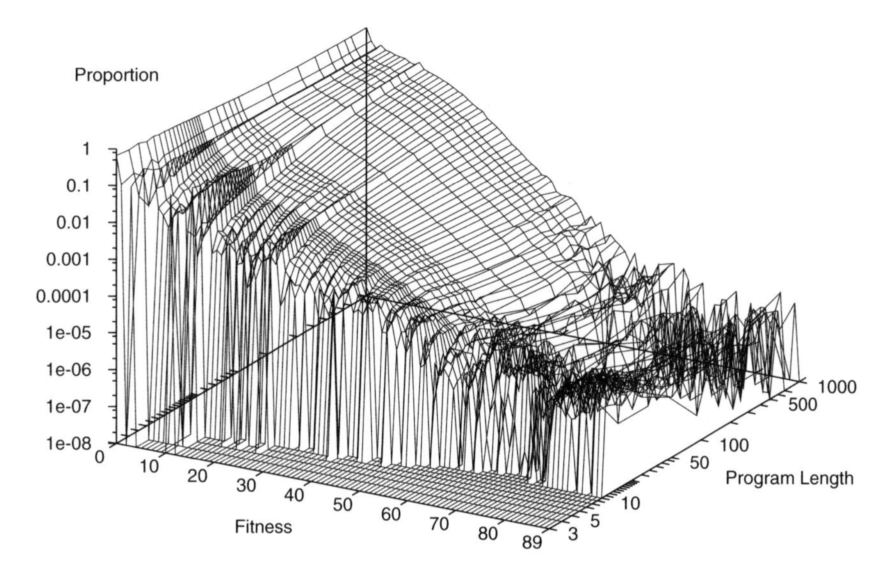

Fig. 7.13. Proportion of Artificial Ant programs of a given size according to their fitness. Values for sizes 15 and above are based on Monte Carlo sampling and so are subject to noise.

7.5 Less Formal Extensions

In the previous sections we have shown 66344 examples from diverse problems where the distribution of fitness in programs of a certain size does indeed tend to a limiting distribution that is independent of the size. Chapter 8 proves this for linear programs and standard GP. This section extends this result to Automatically Defined Functions (ADFs), memory and finally Turing complete programs.

7.5.1 Automatically Defined Function

We can extend our argument to cover programs evolved using ADFs [Koza, 1994]. Each ADF can be viewed as a tree in its own right and so, if the ADF exceeds the threshold size, the distribution of possible functions that the ADF can implement will also converge to a limiting distribution as the ADF gets bigger. For each function, its value is determined by its input(s). The value of each of its inputs is given by a subtree in the calling ADF or main program. When the subtree exceeds the threshold size, the distribution of values used when calling the ADF will also converge, and thus the distribution of values returned by the ADFs will also converge and so, finally, will the distribution of values returned by the program as a whole.

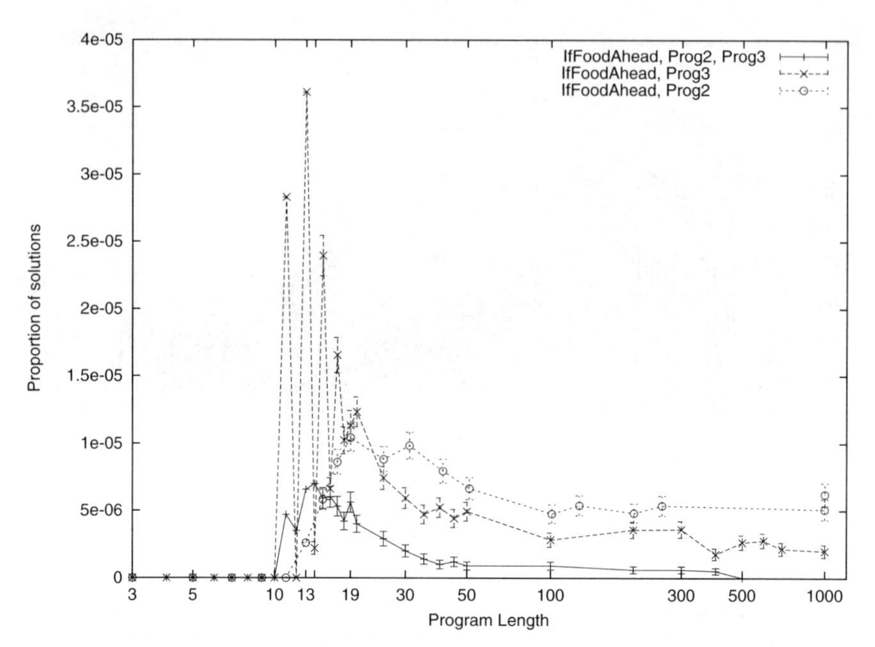

Fig. 7.14. Proportion of Artificial Ant programs of a given size that solve the Santa Fe trail problem, using three different function sets. Error bars indicate the standard error. With the two function sets (those with two functions) there is little variation with size (note log scale) for larger programs. There may be some variation above 400 when using the function set with all three functions (solid line). Cf. Table 9.1 page 152

7.5.2 Memory

It should be possible to extend this argument to covers tree programs that have additional state (i.e. memory) outside the tree. We have not measured the distribution of programs with external memory. It may be that memory radically changes the threshold size. However, we suggest that our claim will hold for non-recursive programs that include memory and subroutines.

7.5.3 Turing-Complete Programs

Finally, to cover all programs we need to consider the addition of either recursion or loops. (In principle, adding either makes the computer programming language Turing-complete, i.e. able to calculate everything that a computer can). The proof contained in Section 8.1 for long linear programs can be extended to Turing-complete programs, provided they halt. First, we note that Section 8.1 requires the programs to complete in l instructions. But if l is big enough, programs longer than l have the same distribution. So exactly when the program halts is not crucial, since the distribution will be the same.

Secondly, it is assumed that each instruction is independent of the previous one. But remember that the distribution is being calculated for all possible programs, that is for programs constructed from instructions chosen at random from those available. In program loops, the instructions are executed in an ordered sequence. However, if each loop is small compared to the program it is reasonable to treat a repeated sequence of random instructions as if it was simply a random sequence.

It seems reasonable that a similar result will also apply to big trees including iteration (e.g. for loops) or recursion and memory. However, in both cases we anticipate the greater complexity available may radically affect the threshold size. Perhaps to such an extent that it is not obvious that the fitness distribution for large Turing-complete programs will tend to a limiting distribution that is independent of size. Or if they do, the size threshold may be too large.

7.6 Tree Depth

So far in this chapter we have not considered the shape of program trees. For simplicity we will just consider binary trees, but distributions for trees with larger or a variable number of branches at each (internal) node are similar. The size and maximum depth of programs are of course related. A tree of a given size cannot exceed a certain maximum depth (that of a tree of the chosen size but composed of only one long chain of functions, with all side branches terminating immediately in leaves). Similarly its depth cannot be less than a minimum. (The minimum depth is that of a (nearly) full tree where every branch is continued and where leaves only occur at the maximum depth (or one level closer to the root)). In the case of binary programs (i.e. those composed only of two input functions), the maximum and minimum depths are given by $(l+1)/2$ and $\lceil \log_2(l+1) \rceil$. In fact most programs lie between these two extremes, see Figure 7.15.

In binary trees the number of programs of a given size and depth is $|T|^{(l+1)/2}|F|^{(l-1)/2}$ times the number of possible tree shapes of that size and depth. (Where $|T|$ is the number of different terminals and $|F|$ is the number of different functions from which programs can be constructed). Note that this depends on tree size but not tree depth. Figure 7.15 was calculated using the number of unlabelled binary trees of each size and depth. Since the ratio of the number of programs to the number of trees does not depend on tree depth, Figure 7.15 also applies to the distribution of programs. (Similar formulae apply for non-binary trees if their internal nodes all have the same branching factor).

The overwhelming number of trees (programs) are neither full nor minimal, but are randomly shaped. That is, if plotted in Figure 7.15 they would lie close to the mean curve. In these large binary random trees, about half the functions have one or more terminals as their arguments [Sedgewick and

Flajolet, 1996, page 241]. That is, not only are they relatively sparse (their average height is $2\sqrt{\pi(l-1)/2} + O(l^{1/4+\epsilon})$ [Sedgewick and Flajolet, 1996, page 256], which is greater than the height of full binary trees $\lceil \log_2 l + 1 \rceil$) but so too are the subtrees within them. That is, the whole search space contains a lower proportion of full or nearly full subtrees than do full trees.

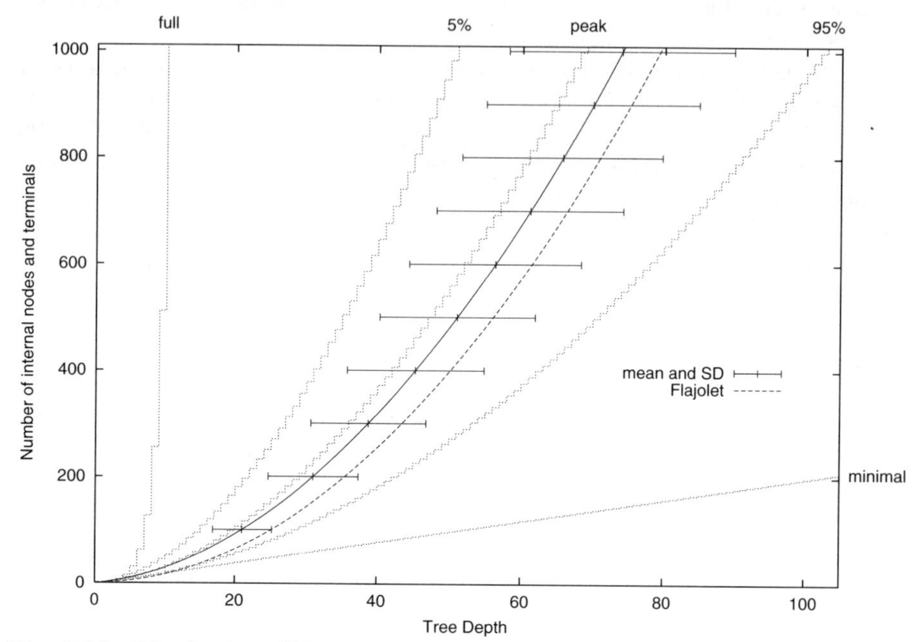

Fig. 7.15. Distribution of binary trees by size and maximum depth. Solid line and error bars indicate the mean and standard deviation of the depth for trees of a give size. The dashed line is the large tree limit for the mean, i.e. $2\sqrt{\pi(\text{internal nodes})}$ (ignoring terms $O(N^{1/4})$). The full tree and minimal tree limits are shown with dotted lines, as are the most likely shape (peak) and the 5% and 95% limits (which enclose 90% of all programs of a given size).

7.7 Discussion

7.7.1 Random Trees

On average half the random trees sampled using the ramped half-and-half method [Koza, 1992, page 93] are full. Therefore, particularly if the depth parameter is increased beyond the usual six (equivalent to a maximum size of 63), the chances of finding at random both the even-3 and the odd-3 parity functions are considerably higher when using it than when using uniform search. In contrast, the ramped half-and-half method is less likely to find solutions to the Santa Fe ant trail problem than uniform search, (see Table 9.3, page 158). This suggests that the best method to use to create the initial random population is problem dependent.

Since nearly full subtrees can be used to form XOR from NAND gates, this may be a partial explanation for why the parity functions are so rare in the whole search space but are comparatively more frequent in full trees, cf. Section 7.2.4. (Section 11.5 discusses the importance of tree shape in the context of bloat).

7.7.2 Genetic Programming and Random Search

We have discussed the number of programs that implement each function as a fraction of the total number of programs, particularly the proportion which are solutions. This corresponds directly to the difficulty of the problem for random search and establishes a benchmark with which to compare GP and other techniques. In [Koza, 1992, Chapter 9] GP performance is shown not to be the same as random search. Indeed, in the case of all but a few of the simplest problems, which both GP and random search easily solve, GP performance is shown to be superior to random search. It is commonly assumed that problems that are harder for random search will also be harder for any practicable search technique. There is a little evidence to support this. For example, [Koza, 1992, Figure 9.2] shows a strong correlation in the three-input Boolean problems between difficulty for random search and difficulty for GP. Thus the distribution of solutions in the search space can give an indication of problem difficulty for GP.

7.7.3 Searching Large Programs

As the density of solutions changes little with program size, there is no intrinsic advantage in searching programs that are bigger than the threshold. It may be that some search techniques perform better with bigger programs, perhaps because together they encourage the formation of smoother, more correlated or easier to search fitness landscapes [Poli and Langdon, 1998a]. However, in practice searching bigger programs is liable to be more expensive both in terms of memory and time (since commonly the CPU time to perform each fitness evaluation rises in proportion to program size). Chapter 11 deals with these points in detail.

At present we do not know in advance where the threshold is. A line of research would be to devise a means of predicting it. This could be of practical value by replacing existing adhoc measures to preset the upper bound on the size of programs using a more principled approach.

7.7.4 Implications for GP

We should not get too downhearted about these results, since they refer to the space which genetic programming and other techniques search, rather than to GP itself. For example, we know that GP does indeed find solutions in a reasonable time and that these may be general [Langdon, 1998c]. Nevertheless, we suggest randomly produced solutions are unlikely to generalise. This lack of generality is important and warrants investigation. Does the apparent performance of GP stem from the use of trees rather than linear programs, or because the programs it uses are small? The success of linear GP argues against the shape of the programs being fundamental. However, commercial linear GP systems are not necessarily free of "tree-like" bias. For example, Discipulus[1] write-protects its program's inputs. This means even in long random programs, there is a reasonable chance of an input being separated from an output by only a small number of program steps. This may introduce a bias which makes some functions much more likely than others. This may, in turn, be more important than the choice of tree or linear representation. In principle, the importance of biases in the search space to the success of GP, like the importance of biases in its search operators, could be investigated experimentally.

It is worth noting that we have only considered the case where all components of the programs are of the same type. So there are no restrictions on how functions work with each other. Performance gains for GP have been reported by using strong typing [Montana, 1995] or grammars [Whigham, 1996a] which control the interactions between the program components. In the case of the Boolean problems, examples include [Janikow, 1996] who gainfully employed typed inputs, while Yu structures her inputs as a list [Yu and Clack, 1998]. As we saw in Section 6.4, work is continuing into understanding the role of different function sets and search operators on the Boolean problems [Page *et al.*, 1999, Poli and Page, 2000].

7.8 Conclusions

In the first four sections of this chapter we presented experimental data from a number of common benchmark problems, which indicate that after some problem dependent size threshold, the distribution of program functionality across the whole search space tends to a limit. (The next chapter treats this formally). Informal extensions to Automatically Defined Functions (ADFs), and programs containing side effects, particularly memory and loops or recursion were given in Section 7.5.

Section 7.6 considered the relationship between tree size and depth for various types of programs, while Section 7.7 discussed some of the implications for genetic programming.

In general the number of programs of a given size grows approximately exponentially with that size. Thus, the number of programs with a particular fitness score or level of performance also grows exponentially; in particular the number of solutions also grows exponentially.

[1] RML Technologies, Inc. Littleton, CO, USA 80127.

8. The GP Search Space: Theoretical Analysis

We start this chapter by formally proving results given in Chapter 7 for linear and tree based GP. The rest of the chapter gives proofs for other properties of program search spaces. The implications of these results have been previously discussed in Section 7.7. For the reader who does not wish to study the proofs in detail, the main results are summarised next.

Summary

The distribution of program functionality (and hence fitness) in conventional linear programs tends to a limit as the programs get longer (Section 8.1). Furthermore convergence in the limit is exponentially fast. The threshold size is $O(1/\log(\lambda_2))$, where λ_2 is the second largest eigenvalue of the computer's Markov state transition matrix.

Where the function set is symmetric, each function generated by long random linear programs is equally likely, and so the random chance of finding a long solution is $2^{-|\text{test set}|}$.

Section 8.2 proves that, provided certain assumptions hold, there is a limiting distribution for functionality implemented by large binary tree programs (cf. Sections 7.1–7.3).

Section 8.3 proves in large trees composed only of XOR functions that the density of order n parity solutions is $1/(2^{n-1})$, provided the tree is of the right size. Also, the fitness landscape is like a needle-in-a-haystack, so any adaptive search approach will have difficulties.

8.1 Long Random Linear Programs

In this section we will prove, provided certain assumptions hold, that each output generated by long random linear programs is equally likely, and that this is true regardless of the program's inputs and its length. Consequently, the chance of finding at random a solution reduces exponentially with the size of the test set.

The state of a computer is determined by the contents of its memory. For our purposes, registers, condition flags, etc. within its CPU and input and output registers are regarded as part of its memory, but we exclude the program counter (PC). If the computer has N bits of memory it can be in up to 2^N states. Program execution starts with all memory initialised. Execution of each program instruction moves the computer from one state to another. Usually the next state will be different but it need not be. For example, an instruction which sets a register to zero will not change the state if the register is already zero. We will assume there is at least one such state and instruction.

We will assume the designer of the computer (or GP experiment) has ensured that it is possible to reach every state. This is done since: (1) inaccessible states correspond to unusable memory, i.e. to inefficient use of hardware; and (2) it makes it possible to transform any input to any output.

Further results can be obtained where the instruction set is symmetric. By symmetric we mean that if there is one instruction that moves the computer from one state to another there is also an instruction that moves it in the other direction.

Consider a program as a sequence of instructions each of which transforms the computer's state. In particular, consider the case where the program contains l instructions chosen at random. The computer starts from an initial state (given by the program's inputs) and terminates l states later. The program's output is then the state of the computer's outputs. Note the program itself need not be linear, it can contain branches, loops, function calls, etc. provided it executes state changing instructions at random and terminates after l of them. We exclude the program counter from the machine's state so its contents and thus the address of the next instruction need not be random. Program termination could be forced by an external event or by fixing a path l instructions long for the program counter; cf. linear GP [Nordin, 1997].

We can represent the computer by a probability vector of length 2^N. When executing a fixed program, the computer will be in exactly one state at a time. That is, its probability vector will contain one element of 1.0 and the rest will be zero. We can view each instruction as an $2^N \times 2^N$ matrix which, when multiplied by the current probability vector (state), yields the next probability vector (next state). The numbers in the matrix (its elements) are either zero or one, and there is exactly one "1" in each row. Therefore it is a stochastic matrix (i.e. its elements can be treated as probabilities which sum to unity [Feller, 1970, page 375]).

If we consider all possible programs of length l, we can define the average state at a particular time (i.e. after a certain number of instructions) as the mean probability vector u at that time. In a particular program the next state is given by the current state vector multiplied by a particular matrix. When considering all programs, we can say the average next state is given by the mean probability vector when multiplied by the average instruction matrix. Note the next state is given by the current state alone. It does not depend upon earlier events. Thus, on average, the state of the computer can be represented by a Markov process. The Markov transition probability matrix M is the mean of all the instruction state matrices.

Since M is the mean of stochastic matrices, it too will be stochastic. At least one of the elements on its diagonal will be greater than zero. The period of this state will therefore be 1, i.e. it will be aperiodic [Feller, 1970, page 387]. Therefore the greatest common divisor of all the states is 1. Since any state can be reached, M is irreducible. Thus M corresponds to an irreducible ergodic Markov chain so that u will tend to a limit u_∞ that is independent of the starting state (i.e. the program's inputs) as the length l of the program is increased [Feller, 1970, page 393].

If the instruction set is symmetric, M will be symmetric and therefore not only its rows but its columns will each sum to 1.0. I.e. M is doubly stochastic. Therefore, in the limit all states are equally probable [Feller, 1970, page 399]. I.e. there is a limiting probability distribution for the states of the computer and at the limit each state is equally likely. (If the instruction set is asymmetric, there is still a limit but the states are no longer equally likely).

8.1.1 An Illustrative Example

Consider a system with two Boolean registers R_0 and R_1. At the start of program execution each is loaded with an input. When the program terminates its answer is given by R_0. There are eight instructions:

$$\begin{array}{llll}
R_0 \leftarrow \text{AND} & (R_0, R_1) & R_1 \leftarrow \text{AND} & (R_0, R_1) \\
R_0 \leftarrow \text{NAND} & (R_0, R_1) & R_1 \leftarrow \text{NAND} & (R_0, R_1) \\
R_0 \leftarrow \text{OR} & (R_0, R_1) & R_1 \leftarrow \text{OR} & (R_0, R_1) \\
R_0 \leftarrow \text{NOR} & (R_0, R_1) & R_1 \leftarrow \text{NOR} & (R_0, R_1)
\end{array}$$

There are $2^2 = 4$ states ($R_1 R_0 = 00, 01, 10, 11$). The instruction matrices are:

$$\begin{array}{cccc}
R_0 \leftarrow \text{AND} & R_1 \leftarrow \text{AND} & R_0 \leftarrow \text{NAND} & R_1 \leftarrow \text{NAND} \\
1\,0\,0\,0 & 1\,0\,0\,0 & 0\,1\,0\,0 & 0\,0\,1\,0 \\
1\,0\,0\,0 & 0\,1\,0\,0 & 0\,1\,0\,0 & 0\,0\,0\,1 \\
0\,0\,1\,0 & 1\,0\,0\,0 & 0\,0\,0\,1 & 0\,0\,1\,0 \\
0\,0\,0\,1 & 0\,0\,0\,1 & 0\,0\,1\,0 & 0\,1\,0\,0
\end{array}$$

$$\begin{array}{cccc}
R_0 \leftarrow \text{OR} & R_1 \leftarrow \text{OR} & R_0 \leftarrow \text{NOR} & R_1 \leftarrow \text{NOR} \\
1\,0\,0\,0 & 1\,0\,0\,0 & 0\,1\,0\,0 & 0\,0\,1\,0 \\
0\,1\,0\,0 & 0\,0\,0\,1 & 1\,0\,0\,0 & 0\,1\,0\,0 \\
0\,0\,0\,1 & 0\,0\,1\,0 & 0\,0\,1\,0 & 1\,0\,0\,0 \\
0\,0\,0\,1 & 0\,0\,0\,1 & 0\,0\,1\,0 & 0\,1\,0\,0
\end{array}$$

As an example consider, the first instruction, $R_0 \leftarrow$ AND. This performs the logic AND operation on R_0 and R_1 and then writes the result into R_0. Suppose initially $R_1 = 1$ and $R_0 = 0$. This means that the probability vector of the inputs is $u_0 = (0\ 0\ 1\ 0)$. We can calculate the probability vector after performing $R_0 \leftarrow$ AND by multiplying u_0 and M:

$$u_1 = u_0 M = (0\ 0\ 1\ 0) \times \begin{pmatrix} 1\,0\,0\,0 \\ 1\,0\,0\,0 \\ 0\,0\,1\,0 \\ 0\,0\,0\,1 \end{pmatrix} = \begin{pmatrix} 0 \\ 0 \\ 1 \\ 0 \end{pmatrix}$$

Decoding u_1 we see $R_1 = 1, R_0 = 0$. We can check that this is correct by performing the AND operation and storing its result directly. That is AND(0,1) $= 0$, so R_0 is set to 0 while R_1 is unchanged.

If we use each of the instructions with equal probability, the Markov transition matrix is the average of all 8, i.e.

$$M = 1/8 \begin{pmatrix} 4\,2\,2\,0 \\ 2\,4\,0\,2 \\ 2\,0\,4\,2 \\ 0\,2\,2\,4 \end{pmatrix}$$

We can see that $u_\infty = 1/4(1,1,1,1)$ satisfies $u_\infty M = u_\infty$, so it is the limiting probability distribution. Alternatively, we can prove this by noting that M is symmetric and hence doubly stochastic; it has at least one non-zero diagonal term thus the theorem from [Feller, 1970, page 399] holds and so in the limit all states are equally probable.

The eigenvalues λ and corresponding eigenvectors of M are

$$\begin{array}{ll} \lambda_{00}{=}1/2 & (\ 0\ {-1}\ \ 1\ \ 0\ ) \\ \lambda_{01}{=}1/2 & ({-1}\ \ 0\ \ 0\ \ 1\ ) \\ \lambda_{10}{=}1 & (\ 1\ \ 1\ \ 1\ \ 1\ ) \\ \lambda_{11}{=}0 & (\ 1\ {-1}\ {-1}\ \ 1\ ) \end{array}$$

Note since M is symmetric, all the eigenvalues are real.

8.1.2 Rate of Convergence and the Threshold

The eigenvectors form an orthonormal set. This means any vector can be expressed as a linear combination w of them. In particular, this is true of any probability vector u, so $u = wE$. Where E is the $n \times n$ matrix formed by the n eigenvectors of M and E^{-1} is its inverse (w can be calculated since $w = uE^{-1}$). Since E is composed of the eigenvectors of M, $EM = \Lambda E$, where Λ is the $n \times n$ diagonal matrix formed from the eigenvalues of M. (I.e. element i,i of Λ is equal to the ith eigenvalue of M, and all off-diagonal elements are zero). Thus the probability vector of the next state is

$$\begin{aligned} u_1 &= u_0 M \\ & w_0 E M \\ & w_0 \Lambda E \\ & w_1 E \end{aligned}$$

Where $w_1 = w_0 \Lambda$. The new probability vector u_1 can also be re-expressed as a linear combination of the eigenvectors of M. It is actually w_1. Note that if we express u_0 and u_1 in terms of the eigenvectors of M, we can see that each of the components of u_0 has been shrunk by a factor given in Λ. I.e. by the corresponding eigenvalue. (As M is stochastic, none of its eigenvalues will exceed unity. So the components of u_1 can only be the same or shrink, they cannot grow).

The probability distribution of the second state is $u_2 = u_1 M = w_0 \Lambda E M = w_0 \Lambda^2 E$, and so that of the ith state is $u_i = w_0 \Lambda^i E$.

As i increases, only the components with the largest eigenvalues will survive, other components will vanish exponentially quickly. I.e. the probability distribution will converge to the limit. The slowest terms to be removed that are not part of the limiting distribution are given by the eigenvectors corresponding to the second largest eigenvalue λ_2. The number of steps, i.e. the threshold size, is dominated by the magnitude of this eigenvalue.

We can calculate approximately how may steps h will be needed before the largest transient term is less than some $\epsilon > 0$. Ignoring the smaller eigenvalues:

$$\begin{aligned} (\lambda_2)^h &< \epsilon \\ h \log(\lambda_2) &< \log(\epsilon) \\ h &> \log(\epsilon)/\log(\lambda_2) \\ \text{Threshold size} &= \mathrm{O}(1/\log(\lambda_2)) \end{aligned}$$

Returning to our example, suppose both inputs are 0. Then $u_0 = (1, 0, 0, 0)$. From M we can calculate its eigenvalues (i.e. the diagonal elements of Λ), E, and E^{-1}:

$$\Lambda = 1/2 \begin{pmatrix} 1\,0\,0\,0 \\ 0\,1\,0\,0 \\ 0\,0\,2\,0 \\ 0\,0\,0\,0 \end{pmatrix}, E = \begin{pmatrix} 0 & -1 & 1 & 0 \\ -1 & 0 & 0 & 1 \\ 1 & 1 & 1 & 1 \\ 1 & -1 & -1 & 1 \end{pmatrix}, E^{-1} = 1/4 \begin{pmatrix} 0 & -2 & 1 & 1 \\ -2 & 0 & 1 & -1 \\ 2 & 0 & 1 & -1 \\ 0 & 2 & 1 & 1 \end{pmatrix}$$

$$\begin{aligned} w_0 &= u_0 E^{-1} &&= (\ \ 0, -1/2, 1/4, 1/4) \\ w_1 &= w_0 \Lambda &&= (\ \ 0, -1/4, 1/4, \ \ 0) \\ u_1 &= w_1 E &&= (1/2, \ \ 1/4, 1/4, \ \ 0) \\ u_2 &= w \Lambda^2 E &&= (\ \ 0, -1/8, 1/4, \ \ 0) E \\ & &&= (3/8, \ \ 1/4, 1/4, 1/8) \end{aligned}$$

Note all elements of u_2 are already within 50% of their limit values (remember $u_\infty = (1/4, 1/4, 1/4, 1/4)$). In this example the other eigenvalues are not close to 1.0 and so convergence is rapid. Indeed, $-1/\log(\lambda_2) = -1/\log(1/2) = 1.442\,695$.

8.1.3 Random Functions

So far in this section we have talked about the distribution of outputs produced by averaging across all programs of a given length. We can readily extend the above arguments to the functions implemented by the computing machine. We replace the current state of the machine after executing l instructions by the function it has implemented. For a finite machine, there are a finite number of functions it can implement. We give each a number. This number is the equivalent of the state of the machine in the previous discussion.

For a machine with N bits, there are $(2^N)^{2^N}$ functions. Unless the input and output registers both occupy all N bits, many of these will be equivalent when we consider functions implemented by programs. However, since memory that is not in the input or output register may by used by the programs, we need to consider all of them initially. So our probability vector is now 2^{N2^N} elements long and each instruction is represented by a $2^{N2^N} \times 2^{N2^N}$ square matrix. As before, at any one time the probability vector has exactly one element that has the value "1" and all the others are zero. Similarly, there is exactly one "1" in each row of each instruction's transition matrix and all other elements are zero. Again the average probability vector is the mean of the vector across all programs after a given number of steps, and the average function transition matrix is the mean of the function transition matrices in the instruction set.

To reuse the Markov proof, we make assumptions about the computer that are equivalent to those we made in Section 8.1. As before, for simplicity, we assume there is some instruction that does not change the state of the machine. However, we will strengthen this to insist that it makes no change regardless of the inputs to the program. Thus the function implemented will be unchanged. As before, this ensures at least one diagonal element of the average transition matrix is non-zero.

Once again we will assume that the computer has been designed such that, all functions from the input register to the output register are possible. If the input register is m bits wide and the output register consists of n bits, there are $(2^n)^{2^m}$ such functions. These are the externally visible functions that the computer can actually implement. Each external function can be implemented by any of $2^{N2^N}/(2^n)^{2^m} = 2^{(N2^N - n2^m)}$ internal functions. If we use this equivalence we can reduce the average instruction matrix from $2^{N2^N} \times 2^{N2^N}$ to $2^{n2^m} \times 2^{n2^m}$. The new matrix remains stochastic and is fully connected (irreducible).

We now have all the conditions needed and Markov results can be applied. That is, the distribution of external functions implemented by long linear random programs tends to a limit[1]. Convergence is exponentially fast, being given by the magnitude of the second largest eigenvalue. Finally, if the external function transition matrix is symmetric then, in the limit of long programs, each function is equally likely.

8.1.4 The Chance of Finding a Solution

A program is a solution if it passes all the test conditions (c.f. fitness cases). Suppose there are T non-overlapping tests (i.e. it is possible, in principle, for a program to pass them all). Each test specifies a number of input bits and a target output pattern of n bits. After executing a long random program, each final state is equally likely. In particular, each combination of bits in the output register is equally likely. Thus the chance of generating exactly the target bit pattern is 2^{-n}. There are many functions that implement this transformation. If the average external function transition matrix, as well as the state transition matrix, is symmetric, there is equal chance of selecting at random any of them. In particular, if the test cases do not overlap, the chance of having selected one of them which also passes the second test case is also 2^{-n}. Therefore the chance of finding a program that passes both the first and the second test case is 2^{-2n}. Generalising, the chance of finding a program that passes all T tests (i.e. of solving the problem) is 2^{-nT}. This can be viewed as the chance of finding a solution via random search is given by the information content of the test set (nT bits).

8.2 Big Random Tree Programs

In this section we will prove that, provided certain assumptions hold, there is a limiting distribution for functionality implemented by large binary tree programs. Initially we allow functions with any number of inputs but later subsections simplify this by only considering binary functions.

8.2.1 Setting up the Proof for Trees

We wish to establish a similar result for trees that we have already shown for linear programs. We will start by assuming that all the state information is held within the tree (i.e. the functions have no side effects) and that the output of the program is returned only via its root. It should be possible to combine this result with that for linear programs to include trees with external memory. We start with an example tree.

[1] The limit does not depend upon the initial conditions. However, this is unimportant since each program always implements the same function, the identity function, before it starts.

Consider the expression $(x-y/3)\times(10+2)$. It has four functions ($-$, $/$, $\times$ and $+$) and five leaves ($x, y, 3, 10$ and 2). When we represent it as a tree (Figure 8.1) it would usually be interpreted depth first, i.e. by evaluating x, then $y/3$, then $(x-y/3)$, then $10+2$ and finally multiplying these results together. However, as there are no side effects the expression can be evaluated in a variety of different orders, all of which yield the same answer. In particular, the tree can be evaluated from the deepest node upwards, e.g. evaluate y first, then $y/3$. As we reach each new function node in the tree, we have to stop and save the value we have calculated until the values of the function's other arguments have also been calculated. When one execution thread is blocked, we create a new one from the next unprocessed leaf. In this case 10 and then 2, so we can calculate $10+2$ but when this new thread reaches the $\times$ node it also has to stop. So we start a new thread from x. When it reaches the $-$ node, all its arguments are known and so we can restart the deepest path thread. It calculates $x-y/3$ and moves up the tree. On reaching the root ($\times$), all its arguments are now known so we can perform the final multiplication using the current value and the previously stored result of $10+2$. This is the value of the tree and evaluation halts.

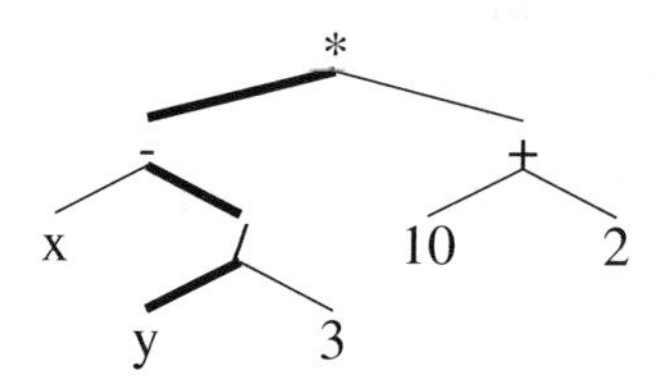

Fig. 8.1. The expression $(x-y/3)\times(10+2)$ represented as a tree. The longest path is shown by the thick lines.

We will now repeat the analysis used for linear programs. Instead of having a linear execution path, the program is a tree but we will concentrate on the longest path within the tree (shown with thick lines in Figure 8.1).

The state of the program at each point in this path is determined by the current value at that point in the path (remember we are excluding additional memory). Its initial value is given by the leaf we start from. Each function along the way to the root will potentially change it, and then the new value will be propagated towards the root. If the function has more than one argument then, in general, before we can determine the transformation that the function will make, we will have to evaluate all its other arguments.

In our example the first function we reach is divide. Divide has two arguments. However, once we have calculated the value of its other argument (it is 3 in our example), we can treat divide as a function with one argument. That is, this divide node transforms the current value by dividing it by 3. Note in general the transformation at a given step will change each time the program is executed with different inputs.

This framework is similar to the linear case, in that in random trees we can view each function along the longest path (in conjunction with its other inputs) as causing a random transformation of the current value. However, the process is not, in general, a Markov chain because even for random trees the transformation matrices change as we get further from the leaf.

Let u be a vector whose elements correspond to each of the possible values. For example, if we are dealing with an integer problem then there are 2^{32} possible values ($n = 2^{32}$) and u has 2^{32} elements. u is the probability density vector of the current value. For a given program at a given time, one element of u will be 1.0 and all the others will be zero.

For each function of arity a, there is n^{a+1} hypercube transformation matrix. When all the inputs to the function are known they are converted to a probability vectors such as u. By multiplying the transformation matrix by each of them in turn we obtain the output probability vector. The output of the function is given by the non-zero element in the output probability vector.

Since we wish to treat the current path separately from the function's other inputs, we split its n^{a+1} hypercube transformation matrix into a n^{a+1} transformation matrices, one for each argument. First we determine which argument of the function the current path is. We then choose the corresponding transformation matrix and multiply it by each input in turn, but exclude the current path. This yields an $n \times n$ stochastic matrix which when multiplied by the current probability vector yields the next probability vector. As in Section 8.1, exactly one element in each row is 1.0 and the rest are zero; e.g. each binary function has two $n \times n \times n$ transformation matrices. If the current path is its first argument, we use the first, otherwise we use the second. If the function is symmetric then the two matrices will be the same. If $n = 2^{32}$ then each matrix is $2^{32} \times 2^{32} \times 2^{32}$ and 2^{64} of its 2^{96} elements will be 1.0. All the others are 0.

We now start the analysis of random programs. Starting from a leaf, the non-zero element of u will correspond to one of the inputs to the program. (For simplicity we will treat constant values as being inputs to the program). We define the average value of u as being the mean of its values across all possible programs. Thus, initially all the elements of u which correspond to one of the input values will be non-zero and all other elements will be zero; we call this u_0.

We then come to a function (of arity a) as its ith input. The probability distribution u_1 after the first function is given by $u_0 M_1$, where M_1 is the transformation matrix corresponding to the ith input of the first function. Since this is the deepest function, the other branches must also be random leaves and so their probability distributions will also be u_0. Thus $M_1 = u_0^{a-1} N_{afi}$, where N_{afi} is the n^{a+1} hypercube transformation matrix for input i of function f (a indicates its arity). On average, the new probability density function will be the mean of all functions of arity a. Also on average, the path we have chosen is equally likely to reach any of the inputs of f. So we can also average across all values of i. Let $N_a = (1/f_a)\sum_{j=1}^{j=f_a}(1/a)\sum_{k=1}^{k=a} N_{ajk}$, where f_a is the number of functions of arity a in the function set. So, on average, $M_1 = u_0^{a-1}N_a$ and $u_1 = u_0 M_1$.

8.2.2 Large Binary Trees

To avoid the complexity associated with considering multiple function arities, we will assume that all internal nodes are binary. We define $N = N_2$, so $M_1 = u_0 N$.

The probability vector u_1 is now propagated up the tree to the next function. Its other argument is determined, and N_{2jk} is multiplied by it to yield M_2. Again we have to average across all programs, so we use the mean N transformation matrix. The functions' arguments may be either leaves or trees of height one. If they are leaves, their probability vector will be identical to u_0 (since they are also random). If they are functions, the probability vector is the same as in the longest path because the same functions are equally probable in each branch. So $u_1' = u_1 = u_0 M_1$. Let p_{20}' be the average number of functions which are children of the function at distance 2 from the leaf along the longest path (excluding those on the longest path). Also, $p_{00}' = 1 - p_{20}'$ is the average number of children which are leaves. Thus $M_2 = p_{00}' u_0 N + p_{20}' u_1 N$. Note $M_2 \neq M_1$ and so the process is not Markovian.

$$
\begin{aligned}
M_1 &= u_0 N \\
u_1 &= u_0 u_0 N \\
M_2 &= p_{00}' u_0 N + p_{20}' u_1 N \\
M_2 &= p_{00}' u_0 N + p_{20}' u_0 u_0 N N
\end{aligned}
$$

At the third level

$$M_3 = p_{000}' u_0 N + p_{200}' u_0^2 N N + p_{220}'(u_0^2 N)(u_0 N)N + p_{222}'(u_0^2 N)(u_0^2 N)N$$

Where p_{ijk}' refer to the proportion of children of the fourth node along the longest path (but excluding those on the longest path). p_{000}' is the proportion of programs where there is only one child (so it is a leaf). p_{200}' is the fraction where the child is a function but both its children are leaves. p_{220}' is the fraction where the child has one child which is a function and p_{222}' is the remainder. That is, p_{222}' is the fraction of programs where the other child of the fourth node is a full subtree as deep as the child on the longest path (Figure 8.2).

The $n \times n$ submatrices of N are stochastic. Thus $u_0 N$ for arbitrary u_0 will also be stochastic. The product of stochastic matrices is itself stochastic.

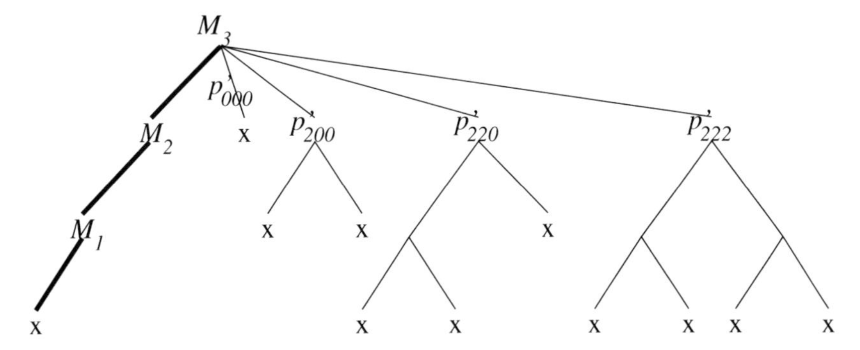

Fig. 8.2. Schematic showing shape of trees which corresponding to terms in M_3 (i.e. p'_{000}, p'_{200}, p'_{220} and p'_{222}). The bold link indicates the longest path, which is used in the analysis of random tree programs. x is a leaf.

Thus, no matter how high we go in the tree, each of the matrices in M_i will be bounded.

As we go higher in the tree the number of terms in M_i is 2^{i-1}. However, in random trees the number of branches we encounter that are the same size as the longest one falls rapidly. That is the higher coefficients p' are not only bounded (they are non-negative but sum to 1.0), but the higher-order ones vanish. In fact, in large random binary trees the chance that the other subtree is a terminal tends to a constant value of approximately 0.5 [Sedgewick and Flajolet, 1996, page 241]. That is, $p'_{0\ldots0}$ tends to 0.5 as we get further from the leaf (i.e. as i increases, M_i is dominated by the first few terms). Hence, while $M_i \neq M_1$, $M_{i+1} \approx M_i$, and so the current value along the longest path will become Markovian. Therefore, the probability distribution of the output of large random trees does not depend upon their size. From the eigenvalues and eigenvectors of M_∞, we can calculate the probability distribution of the output of large random trees and how big the threshold size is. Note that, unlike linear programs, these may depend upon the program's inputs.

8.2.3 An Illustrative Example

We take the same four Boolean functions as before (i.e. AND, NAND, OR and NOR) and apply them to a Boolean regression problem of n-bits. Each is a binary function of one bit input and so has two $2 \times 2 \times 2$ transition matrices. Since each function is symmetric, in each case the two matrices are the same.

As for each function, we also include the function with the opposite output (e.g. AND and NAND), and the mean matrix has the same value at every element. This means M_1 is independent of u_0 with every element being the same; i.e. M_1 is irreducible, doubly stochastic, with non-zero diagonal elements. Thus the output of the first random function is independent of its inputs, and is equally likely to be 0 as 1. Since the coefficients p' sum to 1, this is true for every M_i. Thus the output of the whole tree is independent of its inputs and is equally likely to be 0 as 1 regardless of the size of the tree. (The threshold is zero. Figure 7.2 shows a non-zero threshold as it refers to functions rather than program outputs).

8.2.4 The Chance of Finding a Solution

Because random trees have some inputs near their root, they are more likely to implement program functionality that needs few operations on the inputs than any of the possible functions chosen at random. We can repeat the analysis in Sections 8.2.1–8.2.2 but replacing the current value by the current functionality (cf. Section 8.1.3). That is, instead of considering the value output at each node in the tree, we use an index number of the function implemented by the subtree rooted at that node. For example, in an i-bit-input m-bit-output problem, there are 2^{m2^i} possible functions, each of which can be given an $m2^i$ bit index number; so the number of possible values $n = 2^{m2^i}$. The n^{a+1} hypercube transformation matrices N_{ajk} now operate on function index values rather than on actual values, but we can define average behaviour, etc., as before and the analysis follows through. That is, the functionality of large random trees tends to a distribution that is independent of the tree size. The distribution and the threshold size are again given by the eigenvalues and eigenvectors of the limit value of M.

8.2.5 A Second Illustrative Example

Returning to the example in Section 8.2.3, the transformation matrices now depend upon the order of the problem. If we take the case where there are two inputs, then there are a $2^{1^{2^2}} = 16$ possible functions. So each function has two $16 \times 16 \times 16$ transition matrices. The functions are still symmetric, so again we need only consider one of each pair.

However, this does not mean each function is equally likely. When we consider the average function transition matrix, it is irreducible, stochastic, with non-zero diagonal elements, but it is not symmetric. Thus, there is a limiting distribution independent of the inputs, but it is not uniform. Also, the matrix has several non-zero eigenvalues, so while convergence is rapid it is not instantaneous. Thus M_∞ depends on the distribution of p', i.e. of subtree sizes. To construct a simple model for p', we assume that all internal nodes in the trees are directly connected to at least one leaf node. Figure 8.3 plots the function distribution produced by this model and the measured distribution. The distribution for full trees is shown for comparison.

It is clear that the simple model is approximately correct. A full model would need to consider the distribution of subtree sizes (i.e. p') more carefully. In the case of trees containing only XOR, the output of a tree does not depend upon its shape and we can give an exact theoretical result for the limiting distribution (cf. Section 8.3).

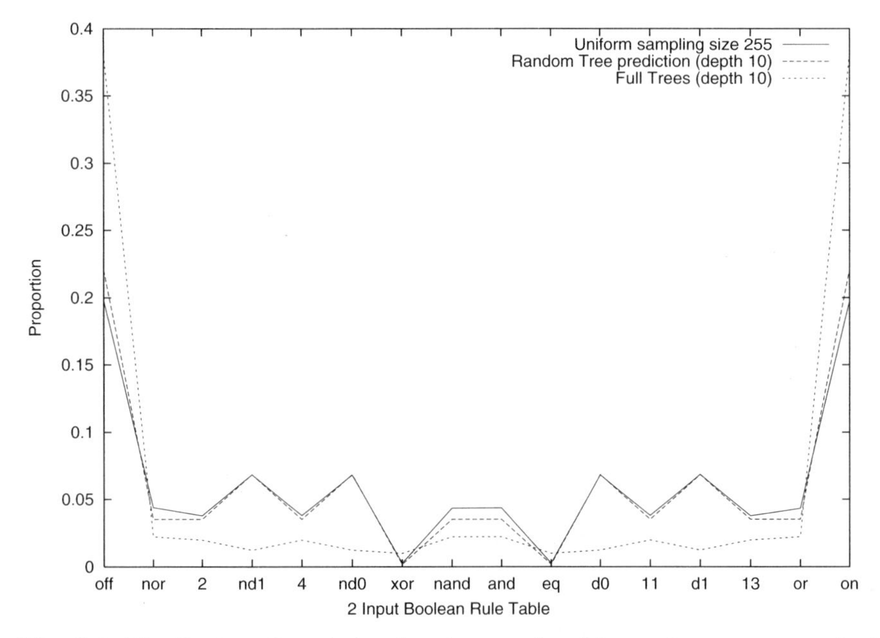

Fig. 8.3. Distribution of two-input functions produced by random trees comprised of AND, NAND, OR and NOR. The solid line gives the measured values, cf. Figure 7.2, page 115. Dotted lines are numerical predictions based on simplified models of random trees.

8.3 XOR Program Spaces

In this section we give a theoretical analysis that shows in the limit as program size grows the density of parity solutions when using EQ or XOR is independent of size, but falls exponentially with number of inputs (Section 8.3.2). But first we start with the functions that can be created using EQ and XOR, and in particular the form of the solutions to the parity problems.

8.3.1 Parity Program Spaces

Even-parity solutions, where n is even, are of the form (EQ D0 (EQ D1 (EQ D2 ... (EQ D_{n-2} D_{n-1})...))). However, given the symmetry of the EQ building block, the inputs to the program can occur in any order. Further, (EQ D_x D_x) is true so (EQ D_y (EQ D_x D_x)) = D_y. Therefore any pair of repeated inputs can be removed from the program without changing its output.

Odd-parity solutions, where n is odd, are of the form (XOR D0 (XOR D1 (XOR D2 ... (XOR D_{n-2} D_{n-1})...))). Again given the symmetry of the XOR building block, the inputs to the program can occur in any order. Further, (XOR D_x D_x) is false but (XOR D_x false) = D_x, so (XOR D_y (XOR

$D_x\ D_x)) = D_y$. Therefore, again any pair of repeated inputs can be removed from the program without changing its output.

A program will be a solution to the parity problem provided it contains an odd number of all n terminals. Thus all solutions contain $t = n+2i$ terminals, where $i = 0, 1, 2, \ldots$ Both EQ and XOR are binary functions, so programs are of odd length $l = 2t - 1$ and solutions are of length $l = 2t - 1 = 2n + 4i - 1$.

Any program with an even number (including zero) of one or more inputs effectively ignores those inputs. Given the nature of the parity function, such a program will pass exactly half the fitness cases.

Note while XOR or EQ may more readily create solutions to the parity problems, they are considerably more limited than NAND and can only generate 2^n of the possible 2^{2^n} functions (NAND can generate them all, thus the results of Section 8.1 do not directly apply). Unlike NAND, they show long-range periodicity by generating 2^{n-1} functions in trees of size $l = 2n + 4i - 1$ (for large i) and the other 2^{n-1} functions in trees of size $l = 2n + 4i + 1$.

8.3.2 The Number of Parity Solutions

If a program's length does not obey $l = 2n + 4i - 1$, then it cannot be a solution to the order-n parity problem and will score exactly half marks on the parity problems. If $l = 2n + 4i - 1$ is true, there is a chance that a randomly generated program will contain an odd number of each input and so be a solution. When calculating the fraction of programs of a given length that are solutions, we can ignore the number of different tree shapes. This is because the output of an XOR tree is determined only by how we label its leaves and not by its shape.

In the following Markov analysis, we consider the current state to be given by the oddness or evenness of the number of each of $n-1$ inputs. (Given this, and the fact that we are only considering programs which obey $l = 2n+4i-1$, the oddness or evenness of the number of times the remaining input appears in the tree is fixed. Therefore we need only consider $n-1$ rather than n inputs). The distribution of solutions to the parity problem is given by the chance of selecting a solution at random, i.e. of all $n - 1$ inputs having the correct parity. Suppose we create long random programs by adding two randomly chosen inputs i, j at a time. That is, the number of inputs of types i and j will both increase by one and so will change from an odd to and even number or vice versa. (If both inputs are of the same type ($i = j$), then it will swap back and there is no change of state). The chance of moving from one state to another does not depend upon how we got to that state; i.e. the process is Markovian and can be described by a stochastic state transition matrix. The chance of moving from one state to another is equal to the chance of moving in the opposite direction, i.e. the state transition matrix is symmetric. There is a $1/(n - 1) > 0$ chance of remaining in the same state after selecting the next pair of inputs. Thus there is an acyclic limiting distribution, and in it each of the states is equally likely [Feller, 1970, page 399]. There are 2^{n-1}

states, one of which corresponds to a solution to the parity problem. That is, in the limit of large programs the chance of finding a solution to the parity problem of order n in a tree composed of XOR (or EQ) is $1/(2^{n-1})$, provided the tree is the right size (and zero otherwise). In fact the fraction of each of the possible functions converges to the same limit. Thus (cf. page 114) the number of solutions of size l to the order-n parity problem using only the appropriate XOR or EQ building block is $2^{1-n}n^{(l+1)/2} \times \frac{(l-1)!}{((l+1)/2)!((l-1)/2)!}$ if $l = 2n+4i-1$, and zero otherwise. The reminder of this section experimentally confirms these results.

Figure 8.4 shows the fraction of 100,000 random programs which are solutions to the Even-6 parity or always-on-6 problems, for a variety of lengths. Figure 8.4 shows that the measurement closely approaches the theoretical large program limit for $l \geq 31$, i.e. more than 16 leaves. This is to be expected as the second largest eigenvalue of the Markov transition matrix is 0.36, which is far from the first. This means transient terms decrease by about $0.36 \approx e^{-1}$ every time the program increases by 2 leaves. Programs of size 31 have 5×2 leaves more than absolutely required, and $(0.36)^5 < 1\%$.

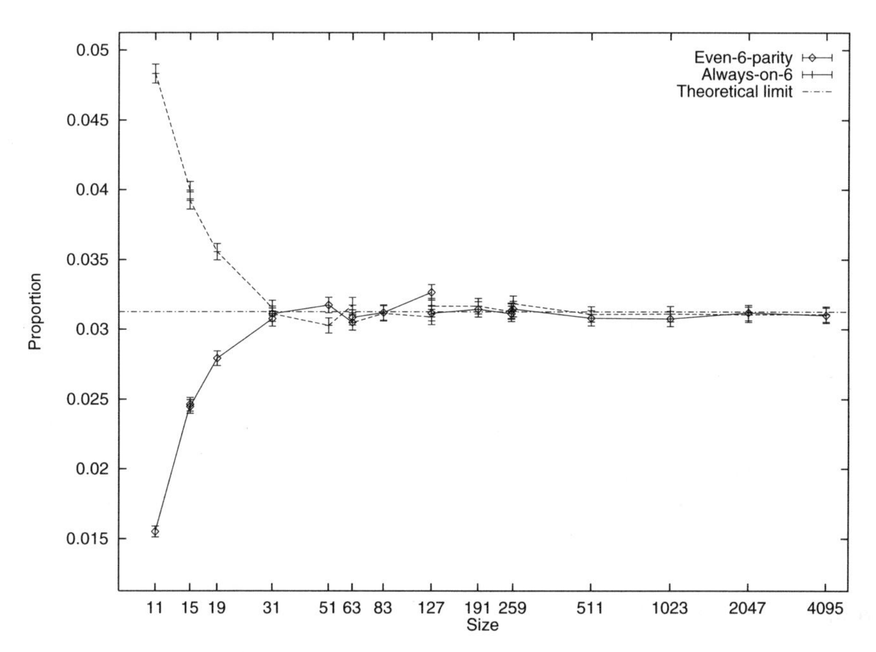

Fig. 8.4. Fraction of programs that solve Even-6 parity or Always-on-6 problems, using EQ (note log scale). Error bars show standard error.

Figure 8.5 shows the fraction of random programs which are solutions to the Even or Odd parity problems (100,000 or a million random trials). The sizes of the programs were chosen so that $l = 2n+4i-1$ is true and the number of leaves exceeds $6n/16$. $6n/16$ was chosen since a linear variation of threshold with number of inputs was assumed, and $2.33n$ was sufficient for $n = 6$. While the second eigenvalue changes only slowly with the number of inputs in the problem, the threshold appears to rise linearly with it. The threshold size $\approx n - 2$. Figure 8.5 again shows agreement between measurement and the theoretical large-program limit.

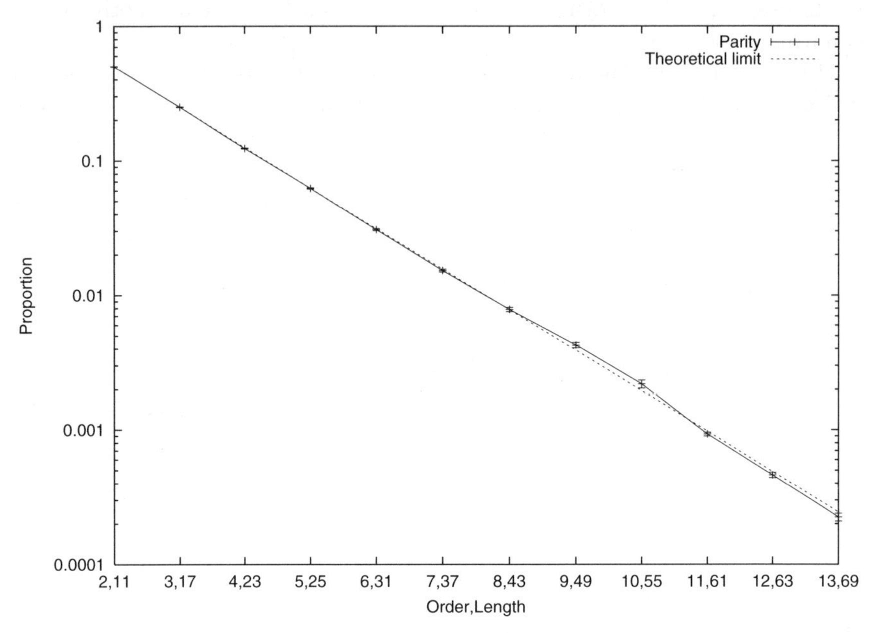

Fig. 8.5. Fraction of programs that solve parity problems given an appropriate building block. The size of programs increases with number of inputs to ensure leaves $\geq 2\frac{2}{3}$ inputs. Measurement and theory agree within two standard errors (error bars show standard error).

As the Markov analysis predicts, the fraction of the search space occupied by each of the possible four input functions composed of XOR converges rapidly to the same proportion as the trees get bigger (Figure 8.6).

8.3.3 Parity Problems Landscapes and Building Blocks

Section 8.3 has described the program space of the parity functions when given the appropriate functional building block (i.e. XOR or EQ). For all problems and at all program lengths, the fitness space is dominated by a central spike indicating almost all programs score exactly half marks. However, a small fraction of programs (actually 2^{1-n} of large programs, provided the tree is the right size, page 147) do solve the parity problems. For modest numbers

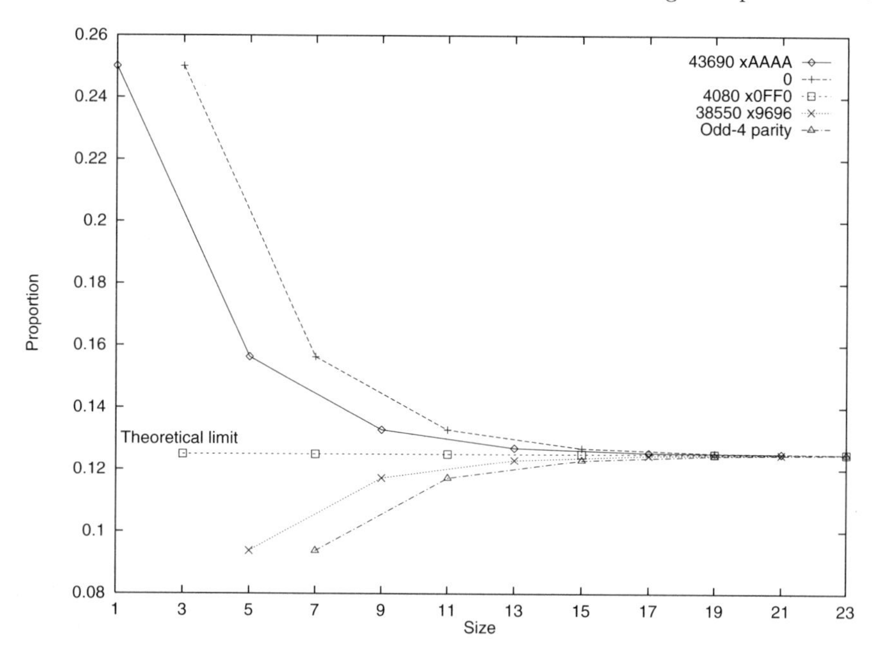

Fig. 8.6. Fraction of functions implemented by XOR trees with four inputs. Note the rapid convergence to the same limiting value. (Each member an equivalence class occurs the same number of times, so only one from each class is shown).

of inputs, modern computers are fast enough to make finding solutions by random search feasible.

In general, to derive a fitness landscape we would also need to consider how genetic or other operators move between points in the search space, as well as the fitness of those points (cf. Chapter 2). However, this is unnecessary in this case because either we have found a solution or the point in the search space scores half marks. That is, the program space contains no information which can guide the search towards a solution. Search techniques, such as GP, that rely on this information will be unable to out-perform random search on such a landscape. Indeed we might anticipate population based search without mutation performing worse than random search as genetic drift in a small population means the population may loose one or more primitives. If this happened then it becomes impossible to construct a solution. [Langdon, 1998c, page 135 and Chapter 8].

This suggests techniques which seek to solve the parity problems by evolving the appropriate building blocks, starting from AND, OR, NAND, NOR, etc., are unlikely to find minimal solutions directly. Such evolutionary techniques will probably have found programs with fitness well above half marks and so will reject any partial solution only composed of the discovered building blocks since these will score less than the partial solutions it has already discovered. It is possible that impure solutions to the problem may be found which subsequent evolution, perhaps under the influence of parsimony (i.e. giving a fitness reward to smaller programs) or beauty pressures, may evolve into a solution comprised only of building blocks.

8.4 Conclusions

We have proved that the distribution of functions implemented by linear programs converges in the limit of long programs exponentially fast. In the special case of symmetric instruction sets, in the limit each function is equally likely. Therefore, the chance of finding a long program on a symmetric machine which solves the problem is $2^{-|\text{test set}|}$.

We have also given the corresponding proof that the distribution of performance of large random binary trees (without side effects) also tends to a limit. Most trees are asymmetric and even if the function set is symmetric the limiting distribution is asymmetric. I.e. there is a much higher number of some functions than others. Functions that can be produced by small trees are frequent, not only in short programs, but also in very big ones. This suggests, in some cases, that there may be much more variation in big trees than in long linear programs. I.e. some functions are much more common and some much rarer.

We have shown that trees composed of only XOR or EQ functions can be treated as special cases of linear programs. This yields a limiting proportion of solutions to the n-input parity problems of $1/2^{n-1}$. This was confirmed by experiment. An empirical measure for the rate of convergence is also given. Together with the fitness function, these give the complete fitness landscape.

9. Example I: The Artificial Ant

In genetic algorithms, deception (i.e. false peaks in the fitness landscape) has long been recognised as an important barrier to performance however in genetic programming it has not received the recognition it deserves. In this and the next chapter we study two benchmark problems using the techniques described in the previous chapters. These show that deception is one of the reasons why these problems appear to be hard.

9.1 The Artificial Ant Problem

The artificial ant problem [Koza, 1992, pages 147–155] is a well-studied problem often used as a GP benchmark [Jefferson *et al.*, 1992, (John Muir trail)]; [Lee and Wong, 1995, Chellapilla, 1997, Harries and Smith, 1997, Luke and Spector, 1997, Ito *et al.*, 1998, Kuscu, 1998] [1]. Briefly, the problem is to devise a program that can successfully navigate an artificial ant along a twisting trail on a 32×32 toroidal grid (Figure 9.1). The program can use three operations, Move, Right and Left, to move the ant forward one square, turn to the right or turn to the left. Each of these operations takes one time unit. The sensing function IfFoodAhead looks into the square that the ant is currently facing and then executes one of its two arguments depending upon whether that square contains food or is empty. Two other functions, Prog2 and Prog3, are provided. These take two and three arguments, respectively, which are executed in sequence.

The artificial ant must follow the "Santa Fe trail", which consists of 144 squares with 21 turns. There are 89 food units distributed non-uniformly along it. Each time the ant enters a square containing food, the ant eats it. The amount of food eaten is used as the fitness measure of the control program. Table 9.1 gives the fitness function, function and terminal sets etc.; those parameters not shown are as in [Koza, 1994, page 655]. Table 9.1 also gives details relevant only to a GP search rather than the problem itself. These are identical to those used in [Langdon and Poli, 1997b].

---

[1] Unfortunately there is a tendency to introduce slight "improvements" which may subtly change the fitness search space. The version we use is as used by [Koza, 1992]. Note we use a time limit of 600, since the value documented in [Koza, 1992] may be in error.

Table 9.1. GP Parameters for the Ant Problem

Objective	Find an ant that follows the "Santa Fe trail"
Terminal set	Left, Right, Move
Functions set	IfFoodAhead, Prog2, Prog3
Fitness cases	The Santa Fe trail
Fitness	Food eaten
Selection	Tournament group size of seven, non-elitist, generational
Wrapper	Program repeatedly executed for 600 time steps.
Population size	500
Max program size	32,767
Initial population:	Created using "ramped half-and-half" with a maximum depth of six
Parameters	90% mutation, 10% reproduction
Termination	Maximum number of generations, G = 50

Initially it was believed that this was a hard problem and that genetic programming did well on it. However, we show that the problem is easy in the sense that it can be solved by enumeration and random search. In fact genetic programming, simulated annealing and hill climbing can all solve it, but their performance is only slightly better than random search.

Using the techniques of the previous chapters, we can see that the usual assumption that we are looking for a *single* solution is totally false. In fact, there are in excess of 10^{200} solutions to this simple problem (Figure 9.2). The (relatively) high density of solutions explains why random search is feasible on *this* problem. Also, since a few of the solutions are very short, enumeration can solve it.

Enumeration of a small fraction of the total search space and random sampling characterise it as rugged with multiple plateaus split by deep valleys and many local and global optima. This suggests it is difficult for hill climbing algorithms.

Analysis of the program search space in terms of fixed-size schemata suggests it is highly deceptive, and that for the simplest solutions large building blocks must be assembled before they have above-average fitness. In some cases we show solutions cannot be assembled using a fixed representation from small building blocks of above-average fitness. This suggests that the Ant problem is difficult for genetic algorithms.

As we saw in Section 7.4, random sampling of the program search space suggests on average that the density of global optima changes only slowly with program size. In fact it was this experiment that lead eventually to the general results given in Chapter 7. Note that the density of neutral networks linking points of the same fitness grows approximately linearly with program size. This is part of the cause of bloat.

Section 9.2 presents a systematic exploration of its program space. In Section 9.3 we calculate the number of fitness evaluations required by two

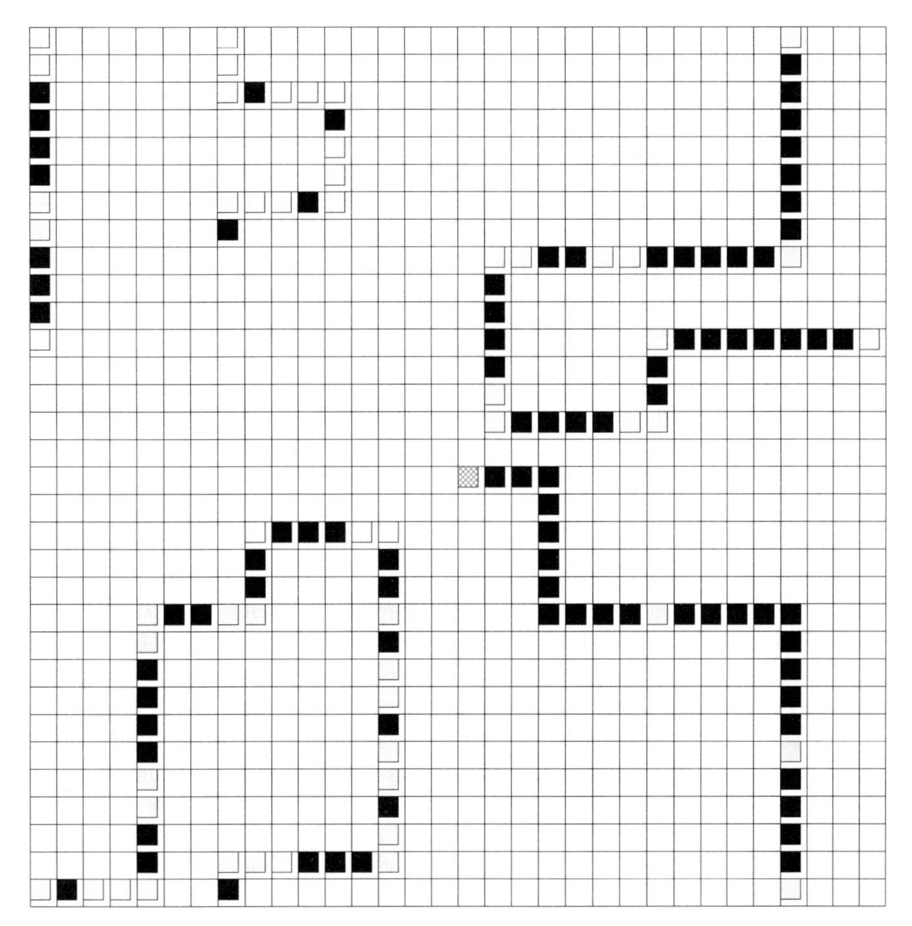

Fig. 9.1. The Santa Fe trail. Solid squares are food items. Note in this figure the ant starts at the centre (shown with cross-hatching) rather than at the top left corner (as the trail is conventionally drawn, cf. Figure 7.12 page 126). Due to the toroidal nature of the Ant's world, this has no effect but serves to highlight the fact than if the Ant goes the wrong way (e.g. up or left) it may still eat a lot of food. For convenience gaps in the trail where there is no food pellet are shown by empty boxes. These have no significance to the execution of the programs or their fitness.

types of random search to reliably solve the problem and then compare this with results for genetic programming and other search techniques. This shows that most of these techniques have broadly similar performance, both to each other and to the best performance of totally random search.

This prompts us to consider the fitness landscape (Section 9.4) and schema fitness and building blocks (Section 9.5) with a view to explaining why these techniques perform badly and to find improvements to them. In Section 9.6 we describe the simpler solutions. Their various symmetries and redundancies essentially mean that the same solution can be represented in an unexpectedly large number of different programs. In Section 9.7 we consider why the problem is important and how we can exploit what we have learnt.

In the final part of the chapter (Section 9.8) we use some of what we have learnt about the problem to reduce the deception in the problem and so make it easier to solve. We redefine the problem so that the ant is obliged to traverse the trail in approximately the correct order. We have also investigated including the ant's speed in the fitness function, either as a linear addition or as a second objective in a multiobjective fitness function, and GP one-point crossover [Langdon, 1998a]. A simple genetic programming system, with no size or depth restriction, is shown to perform approximately three times better with the improved training function. Finally, in Section 9.9 we give our conclusions.

9.2 Size of Program and Solution Space

The number of different programs in the ant problem is plotted against their size in Figure 9.2 (and is tabulated in the "Total" row at the bottom of Table 9.2). As expected, the number of programs grows approximately exponentially with the size of the programs. (The C++ code used to calculate the number of programs is available via anonymous ftp from `ftp.cs.bham.ac.uk` in `pub/authors/W.B.Langdon/gp-code/ntrees.cc`. File `antsol.tar.gz` contains 3916 solutions to this problem).

For the shorter programs it is feasible to explore the program space exhaustively. Table 9.2 summarises the program space for programs of up to size 14. Table 9.2 shows that the program space is highly asymmetric, with almost all programs having very low scores and the proportion with higher scores falling rapidly (but not monotonically) to a low point near 72. Above 72 it rises slightly. There is some dependence upon program size and, as expected, programs must be above a minimum size to achieve modest scores. However, above the minimum size the number of programs with a given score rises rapidly, being a roughly constant proportion of the total number of programs. There are an unexpectedly high number of solutions (albeit a tiny fraction of the total), and their number similarly grows with program size.

For longer programs, exhaustive search is not feasible and instead we sampled the program space randomly in a series of Monte Carlo trials for a number of program sizes. For each such size ten million programs were generated and tested. In the Ant problem there are usually multiple combinations of two and three argument functions which give a tree of a given size. Each corresponds to a different number of programs. One combination was chosen at random in proportion to this number, and then a tree with this combination of branching factors was created using the bijective random tree creation algorithm described in [Alonso and Schott, 1995, Chapter 4]. Each tree was converted to a program by labelling its nodes with a function or terminal of the correct arity chosen uniformly at random. This ensures that every program of the specified size has the same chance of being chosen.

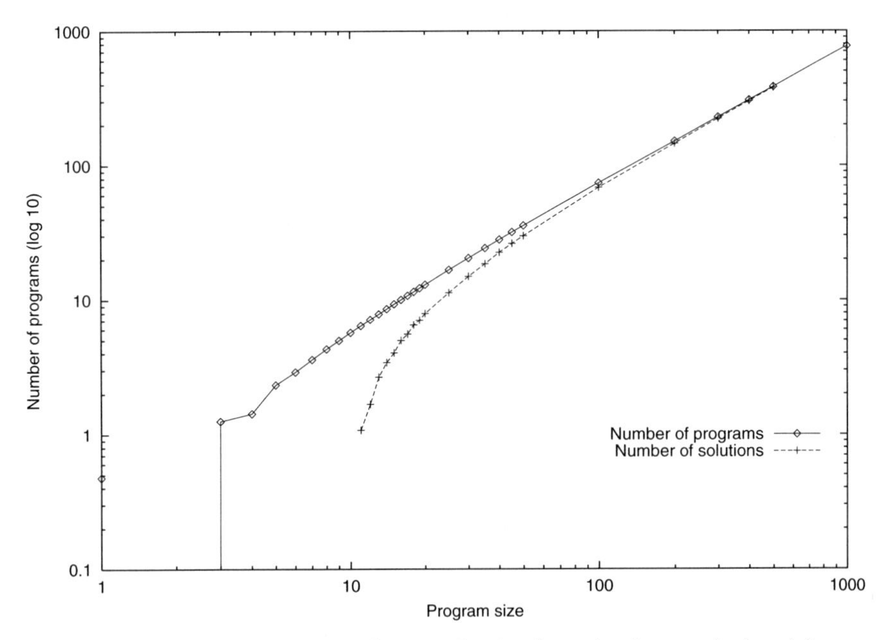

Fig. 9.2. Number of programs of a specific size (note log log vertical scale)

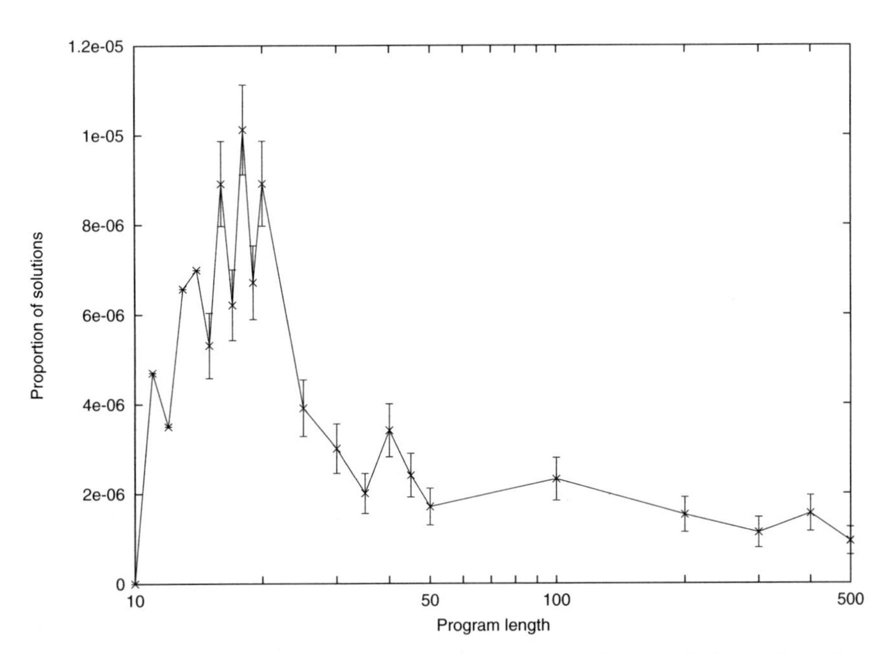

Fig. 9.3. Proportion of programs of a given size which are solutions. Error bars indicate standard error on Monte Carlo estimates.

Table 9.2. Number of trees and distribution of fitness for small Ant programs

Score	Program size													
	1	2	3	4	5	6	7	8	9	10	11	12	13	14
0	2	0	12	16	136	423	2262	10452	49088	252803	1227679	6443754	32595908	171997308
1	0	0	2	4	18	150	449	3058	13806	77269	380979	2070276	10954364	58002558
2	0	0	0	1	3	25	112	909	3429	21902	123174	646831	3587432	19987018
3	1	0	2	6	31	96	530	2779	11736	63996	318817	1656409	8501211	45339974
4	0	0	0	0	3	14	72	527	2526	16250	86999	501521	2820383	15927690
5	0	0	0	0	2	10	58	417	1844	13047	67895	390963	2213475	12466189
6	0	0	0	0	1	8	25	177	1155	6826	33174	216479	1248818	6766377
7	0	0	0	0	2	13	35	266	1601	8076	39428	240187	1324912	6872615
8	0	0	0	0	3	10	68	412	1818	10785	56857	303276	1580134	8846059
9	0	0	0	0	0	0	2	28	183	1392	8218	57485	348331	2053040
10	0	0	0	0	1	2	18	76	461	2758	12465	75079	406998	2276683
11	0	0	2	0	16	51	297	907	5876	27403	120960	659392	3245735	16642082
12	0	0	0	0	0	0	3	52	190	1326	8296	45293	258390	1525769
13	0	0	0	0	0	0	4	40	154	1203	8011	44681	266859	1548594
14	0	0	0	0	0	0	3	24	105	770	5437	27113	169161	1041738
15	0	0	0	0	0	0	9	75	150	1313	11513	41711	226363	1528861
16	0	0	0	0	0	0	0	8	34	350	2584	15053	104018	654943
17	0	0	0	0	0	2	4	35	167	1108	4962	33175	197400	1078896
18	0	0	0	0	0	0	0	2	13	190	1764	11192	83119	570147
19	0	0	0	0	0	0	0	10	66	593	3028	18180	101133	660977
20	0	0	0	0	0	2	4	20	105	763	3985	25601	179522	938185
21	0	0	0	0	0	0	1	13	56	448	2501	16089	107227	581413
22	0	0	0	0	0	0	0	16	71	639	2825	21205	129354	692285
23	0	0	0	0	0	0	0	20	47	489	2372	15321	83764	509240
24	0	0	0	0	0	4	12	47	293	1723	6688	39676	194484	1048249
25	0	0	0	0	0	0	0	1	36	315	1586	10698	65746	359170
26	0	0	0	0	0	0	0	2	23	240	1785	11356	76602	379975
27	0	0	0	0	0	0	0	7	16	222	1262	8820	54491	306671
28	0	0	0	0	0	0	0	2	18	153	1049	7208	51024	258757
29	0	0	0	0	0	0	0	0	11	118	605	4705	30333	184386
30	0	0	0	0	0	0	0	3	23	203	922	6483	40544	215169
31	0	0	0	0	0	0	0	0	12	120	624	5167	37145	191740
32	0	0	0	0	0	0	0	0	2	48	387	2859	21765	129411
33	0	0	0	0	0	0	0	5	13	132	898	5315	31898	158423
34	0	0	0	0	0	0	0	2	12	99	519	3648	23694	125950
35	0	0	0	0	0	0	0	1	5	50	298	2490	20548	105190
36	0	0	0	0	0	0	0	5	12	106	349	2905	16828	98260
37	0	0	0	0	0	0	1	0	18	104	595	3189	20800	99885
38	0	0	0	0	0	0	0	2	21	183	637	4054	17473	106100
39	0	0	0	0	0	0	0	0	6	79	226	1782	8274	54753
40	0	0	0	0	0	0	0	0	8	79	267	1888	9477	56608
41	0	0	0	0	0	0	0	0	6	57	245	1682	8812	49873
42	0	0	0	0	0	0	0	5	11	92	263	1821	9083	51056
43	0	0	0	0	0	0	0	6	8	78	190	1688	7681	47354
44	0	0	0	0	0	0	0	0	0	11	109	808	4988	27119
45	0	0	0	0	0	0	0	0	0	11	84	756	3929	24649
46	0	0	0	0	0	0	0	0	6	53	184	1165	5234	30441
47	0	0	0	0	0	0	0	1	5	42	133	1068	6063	30738
48	0	0	0	0	0	0	0	0	1	14	62	532	2978	16510
49	0	0	0	0	0	0	0	0	0	4	30	293	1917	11133
50	0	0	0	0	0	0	0	0	4	41	142	970	3467	20849
51	0	0	0	0	0	0	0	0	5	38	122	748	2731	16387
52	0	0	0	0	0	0	0	0	0	2	10	160	774	6193
53	0	0	0	0	0	0	0	0	0	13	22	307	1287	8526
54	0	0	0	0	0	0	0	0	0	0	1	26	128	1568
55	0	0	0	0	0	0	0	0	0	3	6	105	340	3437
56	0	0	0	0	0	0	0	0	0	0	10	73	289	2082
57	0	0	0	0	0	0	0	0	0	0	6	58	617	2212
58	0	0	0	0	0	0	0	0	0	0	3	24	146	949
59	0	0	0	0	0	0	0	0	0	0	0	0	20	316
60	0	0	0	0	0	0	0	0	0	0	8	46	229	1790
61	0	0	0	0	0	0	0	0	0	0	4	30	223	1435
62	0	0	0	0	0	0	0	0	0	0	0	2	29	1113
63	0	0	0	0	0	0	0	0	0	0	5	85	285	4538
64	0	0	0	0	0	0	0	0	0	0	0	1	23	337
65	0	0	0	0	0	0	0	0	0	0	0	0	53	1610
66	0	0	0	0	0	0	0	0	0	0	0	0	3	66
67	0	0	0	0	0	0	0	0	0	0	0	15	31	435
68	0	0	0	0	0	0	0	0	0	0	0	0	3	90
69	0	0	0	0	0	0	0	0	0	0	0	0	17	2394
70	0	0	0	0	0	0	0	0	0	0	0	0	0	1063
71	0	0	0	0	0	0	0	0	0	0	0	0	3	2344
72	0	0	0	0	0	0	0	0	0	0	0	0	1	18
73	0	0	0	0	0	0	0	0	0	0	0	0	7	525
74	0	0	0	0	0	0	0	0	0	0	6	42	119	868
75	0	0	0	0	0	0	0	0	0	0	0	0	0	146
76	0	0	0	0	0	0	0	0	0	0	0	0	14	174
77	0	0	0	0	0	0	0	0	0	0	4	26	113	733
78	0	0	0	0	0	0	0	0	0	0	6	34	158	991
79	0	0	0	0	0	0	0	0	0	0	4	16	137	755
80	0	0	0	0	0	0	0	0	0	0	12	104	499	3530
81	0	0	0	0	0	0	0	0	0	0	3	64	157	2126
82	0	0	0	0	0	0	0	0	0	0	2	10	60	363
83	0	0	0	0	0	0	0	0	0	0	0	60	76	1367
84	0	0	0	0	0	0	0	0	0	0	21	188	747	5559
85	0	0	0	0	0	0	0	0	0	0	3	223	459	5734
86	0	0	0	0	0	0	0	0	0	0	0	110	173	3103
87	0	0	0	0	0	0	0	0	0	0	27	194	563	3420
88	0	0	0	0	0	0	0	0	0	0	57	399	1188	6951
89	0	0	0	0	0	0	0	0	0	0	12	48	470	2676
Total	3	0	18	27	216	810	3969	20412	95256	516132	2554416	13712490	71521461	382794984
Mean	1	0	1.7	0.9	1.7	1.9	2.0	2.06903	2.24659	2.42444	2.44969	2.59006	2.70357	2.76907
SD	3.7	0	4.4	3.8	4.4	4.4	4.4	4.45867	4.57353	4.79021	4.76974	4.94338	4.98586	5.04057
Max	3	0	11	3	11	24	37	47	51	55	89	89	89	89

Figure 7.13 (page 127) shows that the proportion of programs with a given score is approximately constant for a wide range of program sizes. Since the total number of programs rises rapidly, this means that the number of programs with a given score also rises rapidly with their size, as described in the previous chapter.

With any Monte Carlo technique there will be some stochastic error in the estimates. In the case of rare events (such as finding a solution to the ant problem) this could be large. The stochastic error was kept reasonable by using a large number of trials, so a modest number of solutions were found at each size (between 9 and 101, and on average 39). An estimate of the stochastic error is shown in Figure 9.3.

9.3 Solution of the Ant Problem

Using the probability P of finding a solution, we can calculate the number of program evaluations needed to ensure that we find a solution (with probability of at least $1-\epsilon$). This is known as the "Effort" required [Koza, 1992, page 194]:

$$E = \frac{\log \epsilon}{\log(1-P)}$$

$$E \approx -\frac{\log \epsilon}{P}$$

Taking ϵ as 1%, we can calculate the number of fitness evaluations E required to find at least one solution (with at least 99% probability).

9.3.1 Uniform Random Search

Using uniform random search and taking the maximum value for P gives us a minimum figure of 450,000 for programs of size 18. However, as shown in Table 9.3, if we allow longer programs, P falls to produce a corresponding rise in E to 1,200,000 with programs of size 25, 2,700,000 with programs of size 50 and to 4,900,000 for size of 500. (Longer random programs also require more CPU time to evaluate).

9.3.2 Ramped Half-and-Half Random Search

Using the ramped half-and-half method with a depth limit of six, we created twenty million random programs of between 3 and 242 nodes. Six solutions to the Ant problem were found. This gives us an estimated Effort of 15,000,000. This is higher than the corresponding figures for uniform random search, indicating in the Ant problem that the bias in ramped half-and-half leads it to search less favoured regions of the program space. For example 51%

of the programs it generated contained ten or fewer nodes, and thus could not be solutions to the Ant problem. Another disadvantage of the ramped half-and-half method is that it will sample some programs repeatedly. (In this example one of the six solutions found was found twice).

9.3.3 Comparison with Other Methods

Table 9.3 gives E values for various methods of solving the Ant problem. It is clear that there are many techniques capable of finding solutions to the Ant problem, and although these have different performance the best methods typically only do marginally better than the best performance that could be obtained with random search.

Table 9.3. Effort required to solve the Santa Fe trail problem

Method		Effort/1000
Random (length = 18)		450
Random (length = 25)		1,200
Random (length = 50)		2,700
Random (length = 500)		4,900
Ramped half-and-half		15,000
Koza GP	[Koza, 1992, page 202]	450
Size limited evolutionary programming	[Chellapilla, 1997]	136
GP	[Langdon and Poli, 1997b]	450
Subtree mutation	[Langdon and Poli, 1998a]	426
Simulated annealing	50%–150%	748
	Subtree-sized	435
Hill climbing	50%–150%	955
	Subtree-sized	1,671
Strict hill climbing	50%–150%	186
	Subtree-sized	738
Population (data for best)	50%–150%	266
	Subtree-sized [Langdon, 1998b]	390
PDGP	Section 1.2.4	336

In the following sections we investigate the Ant problem fitness landscape to explain the disappointingly poor performance of these search techniques.

9.4 Fitness Landscape

As explained in Chapter 2, for a fitness landscape, it is important to consider the neighbourhood relationship, as well as the fitness function. We consider two programs in the program space to be neighbours if they have the same shape and if one can be obtained from the other just by changing one node. I.e. they are neighbours if making a point mutation to one program produces the other. This is the simplest neighbour relationship, which means we can avoid the complications inherent in crossover operations such as subtree crossover.

In the case of small programs (i.e. sizes 11, 12 and 13) we investigated the neighbourhoods of all the fitter programs, i.e. those with scores above 24. (In [Langdon, 1998b], in almost all runs the best individual found had a score better than 24). As expected, this showed many neighbours to be worse or much worse (i.e. score less than 24). It also showed that many individuals with fitness between 24 and 88 are local optima, in that none of their neighbours are fitter than them. With short programs, only a few neighbours have identical fitness.

The neighbourhoods of solutions are composed of low fitness programs. For programs of size 11 or 12, apart from programs which score 24–27 or 36, all neighbours of the solutions score < 24. Therefore, if a hill climber searching programs of size 11 or 12 finds a program scoring more than 36, we know (unless it restarts) it will never find a solution. (Figure 9.4 shows 50 runs of a variable size representation hill climber [Langdon, 1998b] most of which became trapped at suboptimal peaks. Similar behaviour is also seen with other search techniques such as GP). There are many more solutions of size 13, and they are structurally and operationally more diverse. So their neighbourhoods are also much bigger and more diverse and include programs with scores of 24–46, 52, 54, 63, 85, 87 and 88. However, five times as many of them have scores below 24.

For longer programs exhaustive enumeration of the landscape is not feasible, and instead we used Monte Carlo sampling. As before programs of a chosen size were sampled uniformly at random. Due to the rarity of high scoring programs only a small number (up to 19) with scores 24–89 were chosen and all their neighbours were created and tested. (In the Ant problem a program of size l has approximately $3l/2$ neighbours).

Figure 9.5 shows that for most program sizes and most fitness values there are a large number of programs that do not have any fitter neighbours. In contrast, Figure 9.6 shows that the average number of neighbours with the same fitness grows with program size. These same fitness neighbours displace those that are worse, and for the longest sizes almost all programs of intermediate fitness have a large number of neighbours with the same score.

9.5 Fixed Schema Analysis

Unlike conventional schema analysis, we define a schema's fitness as the mean score for *all* programs matching the schema. Our analysis shows that typically there is a large variation of program scores within a schema, with the standard deviation of scores being about the same as, or larger than, the schema's fitness. Thus a finite sample (such as a GA population) can only reliably be used to estimate the fitness of a schema if it contains multiple independent samples from the schema. If a GA is to reliably choose between schemata

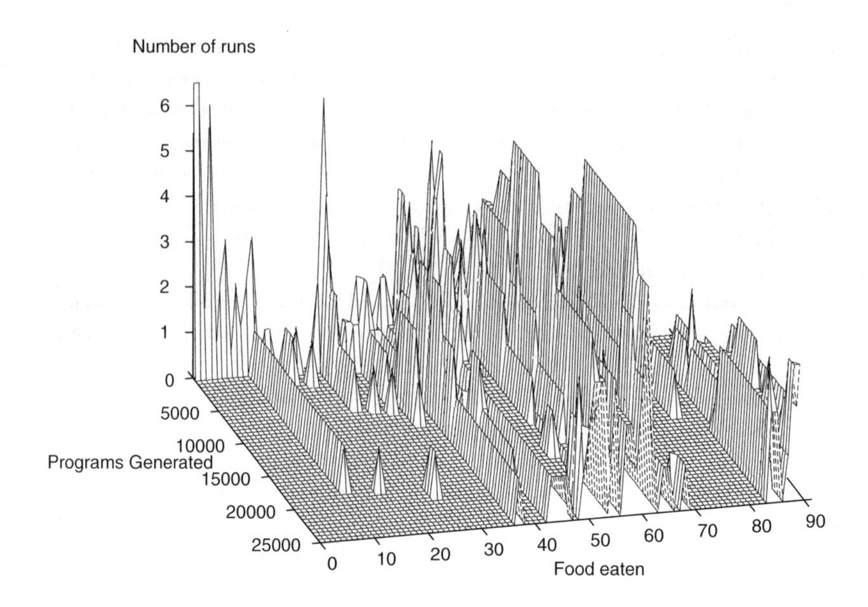

Fig. 9.4. Evolution of best fitness in 50 hill climbing runs using 50%–150% tree mutation. Hill climbers are likely to be stuck at local optima for long periods.

based on its estimate of their fitness, the number of samples (i.e. programs) must be even bigger where the schemata have similar fitness. (Of course the assumption that programs are independent is not justified after selection etc.).

9.5.1 Competition Between Programs of Different Sizes

The competition between schemata can be viewed as a hierarchy of competitions. The outermost competition being between schemata of different sizes. The next is between schemata with different shapes but of the same size. The final competition is between different schemata of the same size and shape. Figure 9.7 shows that the Ant problem is difficult at the outermost level. The region containing the highest concentration of solutions (size=18, cf. Figure 9.3) has a fitness of 2.9, but longer programs are on average fitter than this. While the standard deviation is large compared to the mean, a typical initial GP population is likely to be large enough to be able to reliably prefer solutions bigger than 18 over those containing 18 nodes.

Note Figure 9.7 actually suggests that the distribution of fitness values in the Ant program space does vary a little with respect to program size, even for very large programs. The effect is small but unexplained. Chapter 7 would

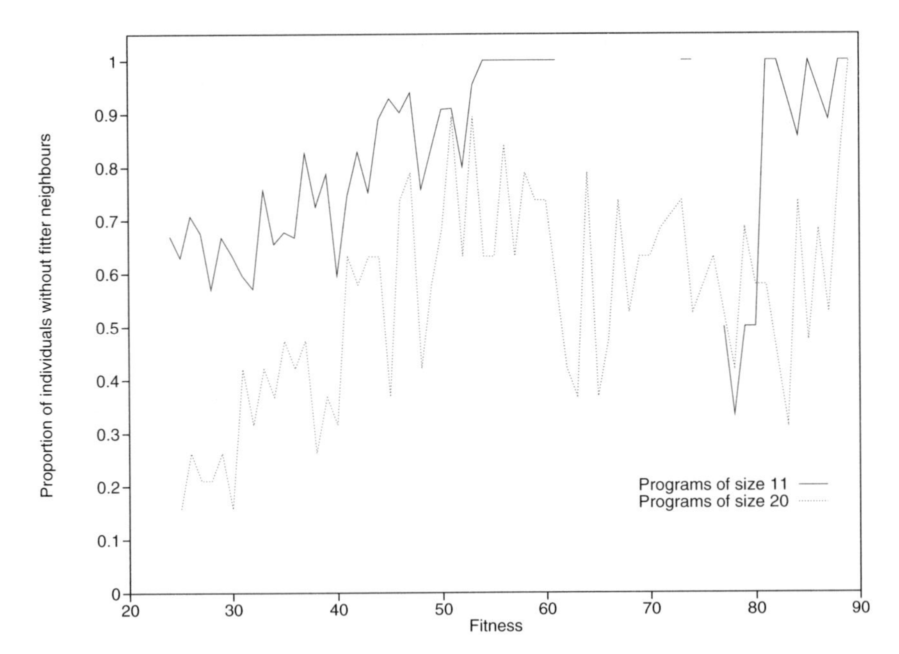

Fig. 9.5. Proportion of programs without fitter neighbours.

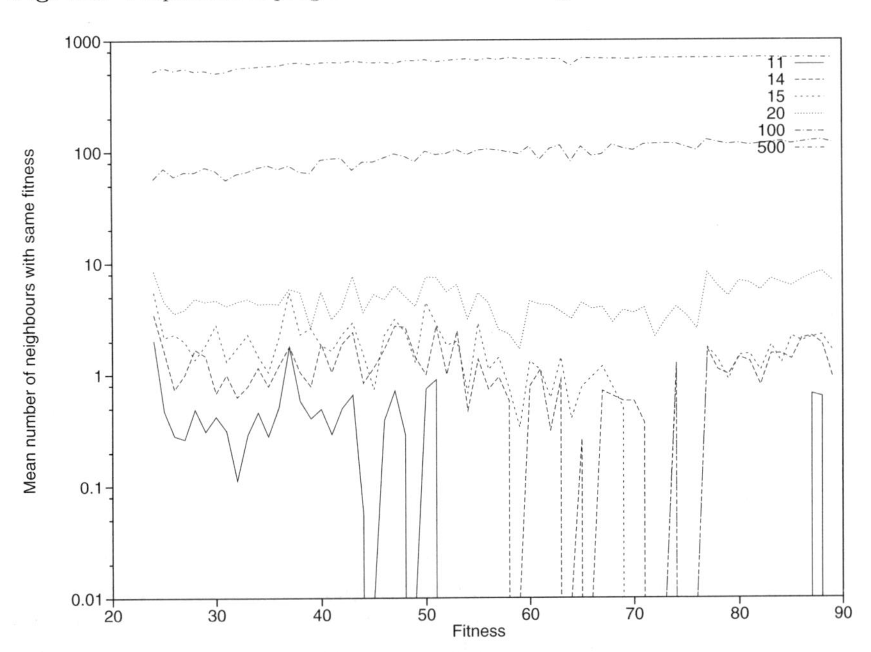

Fig. 9.6. Mean number of neighbours with the same score for various program sizes.

suggest that there should be essentially no such variation (above a threshold size). Note however that the formal proofs in Chapter 8 require trees without side effects, and so do not directly apply. (The effect does not directly explain the bloat, cf. Chapter 11, that is often seen in runs of the Ant problem. The fitness of average programs, to which Figure 9.7 refers, is significantly below that found during search. This means that average programs have no effect on the search and so cannot be directly responsible for bloat).

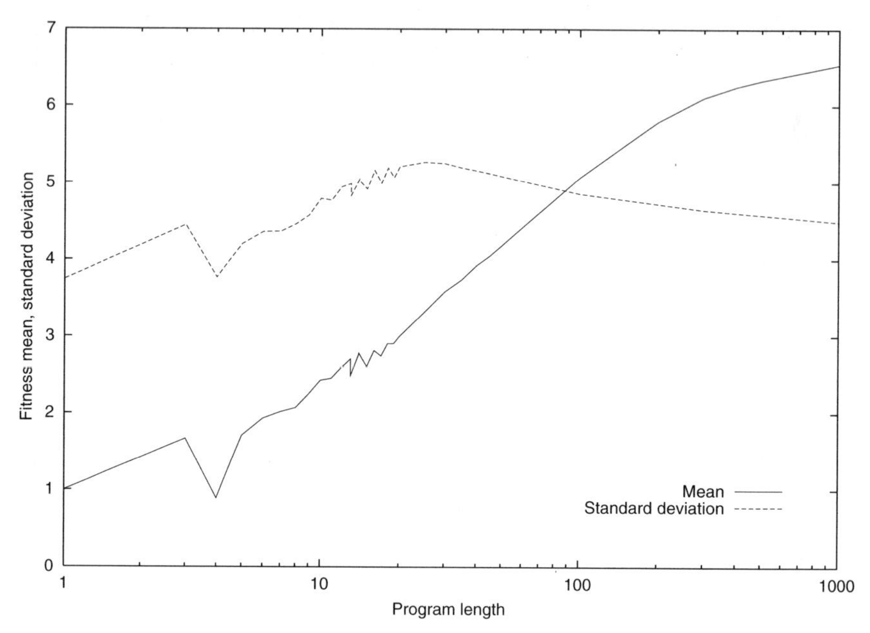

Fig. 9.7. Mean and standard deviation of all Ant program scores vs. their size. Note on average that trees bigger than 18 nodes (where solutions are clustered most tightly) have higher fitness. This means, at the very crudest schema level (program size), that the search space is "deceptive" in that it tends to guide GAs away from where solutions are most densely clustered (see Figure 9.3).

9.5.2 Competition Between Programs of Size 11

Only in the case of the smaller programs has it been possible to extensively study the whole of a fraction of the search space belonging to programs of one size. We concentrate upon those schemata that contain solutions and how their fitness compares with other schemata. (Section 9.6 will describe the solutions of size 11–13 to the Ant problem).

Competition Between Programs Shapes. Figure 9.8 shows the distribution of schema fitness for one of the hyperspaces containing solutions of size 11. (The other two hyperspaces are similar; a hyperspace is a schema with no defining nodes, cf. Definition 4.2.5 page 55). Looking at the order-zero schema, i.e. the hyperspace, we see that it has a fitness above the average for programs of the same size (broken horizontal line); however there are other hyperspaces that are fitter (dashed line, order zero). That is, on average programs of size 11 with the same shape as one of the solutions are of above average fitness, but there are other shapes with still higher average fitness. That is, the search space is deceptive at the hyperspace level (for size 11 hyperspaces). [2]

Competition Between Schemata of the Same Order. When comparing schemata of the same order we see, apart from those of order 11 (i.e. programs), that there are always schemata outside the hyperspace with higher fitness. However within the hyperspace, for a given order (except order 8), the fittest schema is always one containing a solution. This means the problem is deceptive at each schema order (except 11), since for every schema of size 11 containing a solution there are others of the same order with higher fitness which do not contain solutions.

It was feasible to consider all schemata of order 4 or less. (I.e. including schemata that do not contain solutions). As Figure 9.8 shows, there are many schemata that do not contain a solution (□) which are fitter than many of the same order that do (×). There are also small components of solutions with below average fitness. It is not until more than five components have been assembled that all schemata containing solutions have above average fitness. That is using a fixed representation, solutions of size 11 cannot be assembled from small conventional building blocks (low-order schemata of above average fitness).

9.5.3 Competition Between Programs of Size 12

Competition Between Program Shapes. Looking at the hyperspaces of size 12, i.e. schemata of order zero (see Figure 9.9), we see a similar picture to those of size 11. Hyperspaces containing solutions have a fitness above the average for programs of the same size (broken horizontal line), however there are other hyperspaces that are fitter (dashed line, order zero). That is, the search space is deceptive at the (size 12) hyperspace level.

Competition Between Schemata of the Same Order. Comparing schemata of the same order we see, apart from those of order 12 (i.e. programs), that there are always schemata outside the hyperspace with higher fitness. Also (unlike those of size 11), within the hyperspace the fittest schema (□) of each order for orders 5–10 does not contain a solution. There are many

[2] Prog3 is the only function that takes three arguments. If a program's shape is given and it contains three-way branches, then there must be a Prog3 at these points. I.e. the locations of Prog3 are fixed. Therefore, we do not study schemata or order 1 or 2.

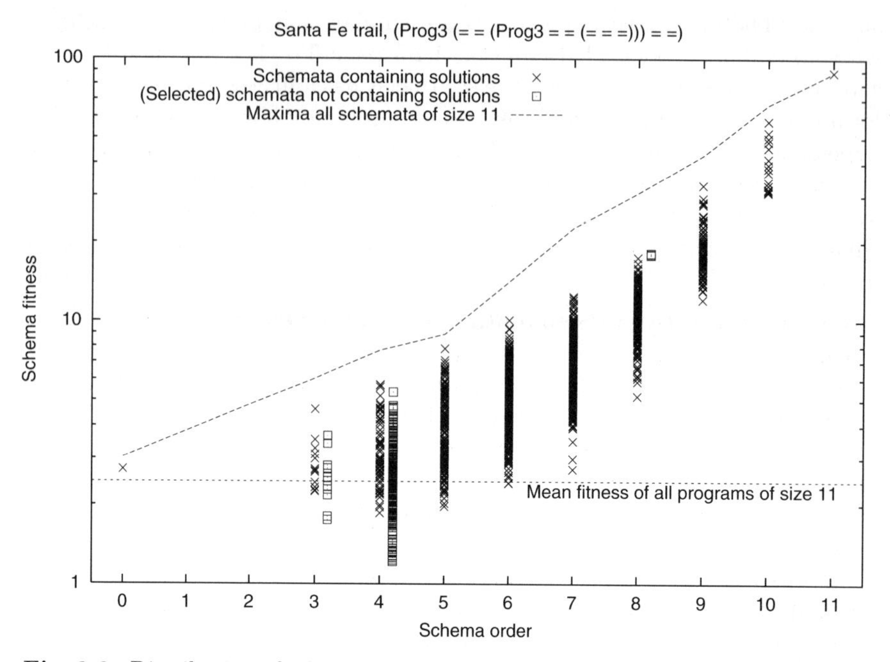

Fig. 9.8. Distribution of schemata fitness within a hyperspace of size 11 containing one of the solutions shown in Figure 9.13. □ indicates schemata within the hyperspace of order 3 or 4 which do not match any of the solutions, or in the case of order 8, have fitness higher than those of order 8 which do.

schemata that do not contain a solution and which are fitter than many of the same order that do (plotted with ×). Also ,there are small components of solutions with below average fitness. It is not until more than six components have been assembled that all schemata containing solutions have above average fitness. That is, again using a fixed size-12 representation, the Ant problem cannot be solved by assembling together low-order schemata of above average fitness.

9.5.4 Competition Between Programs of Size 13

Turning to programs of 13 nodes, the situation is complicated by the much larger number of solutions and their diverse nature. We have selected three hyperspaces which contain solutions of different types, see Figures 9.10, 9.11 and 9.12. (Only data for schemata of order 0, 3 and 4 are available).

Competition Between Programs Shapes. Looking at order zero, we see a similar picture to that for size 12: the three selected hyperspaces have fitness above the average for programs of the same size, however (apart from the hyperspace chosen because it has the highest fitness) there are other hy-

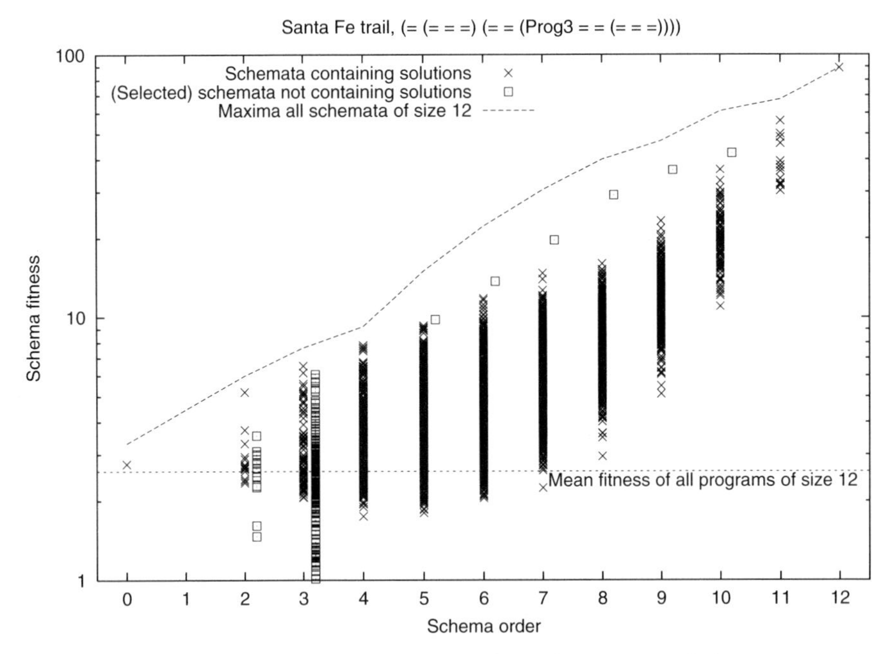

Fig. 9.9. Distribution of schema fitness within a hyperspace of size 12 containing one of the solutions shown in Figure 9.14. For order 3 or 4, □ indicates fitness of schemata within the Hyperspace which do not match any of the solutions. For orders > 4, □ shows only those schemata not matching the solutions which are of higher fitness than those of the same order which do.

perspaces that are fitter. That is, in most cases, the search space is deceptive in terms of the hyperspaces of size 13.

Competition Between Schemata of the Same Order. Comparing schemata of the same order we see there are always schemata outside the hyperspace with higher fitness (although the difference is small for the fittest hyperspace). As with schemata of sizes 11 and 12; in two hyperspaces the fittest schemata of order 0, 3 and 4 contain a solution (see Figures 9.10 and 9.12). However, in the fittest hyperspace (Figure 9.11) the fittest schemata of order 3 and 4 do not contain solutions. As with sizes 11 and 12, there are many schemata that do not contain a solution which are fitter than many of the same order that do. Also, there are small components of solutions with below average fitness. That is, again using a fixed size-13 representation, there are at least some solutions which cannot be assembled from low-order schemata of above average fitness.

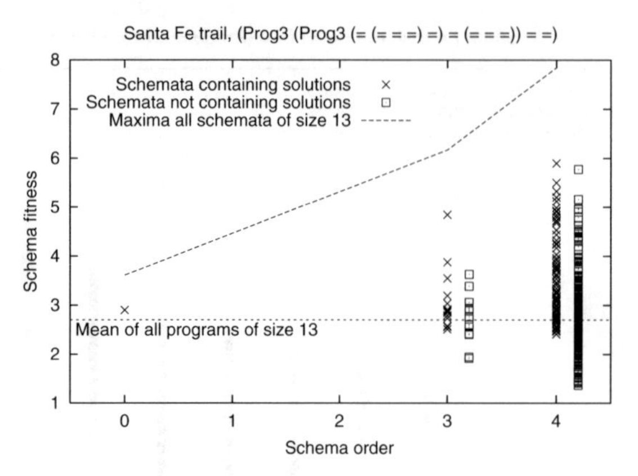

Fig. 9.10. Distribution of schema fitness within a hyperspace of size 13 containing a (two-move) solution, cf. Figure 9.16.

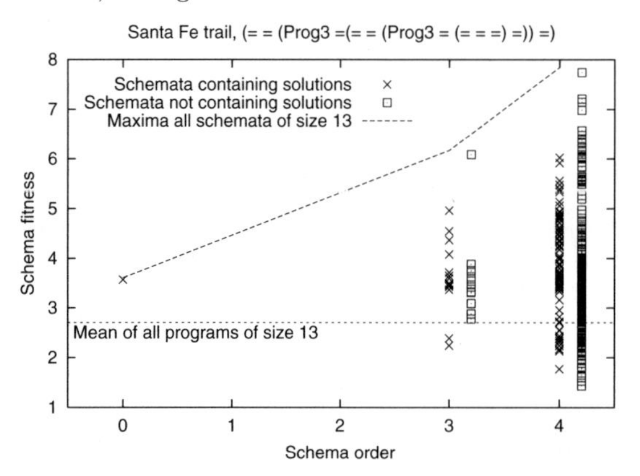

Fig. 9.11. Distribution of schema fitness within a highest fitness hyperspace of size 13 containing a solution.

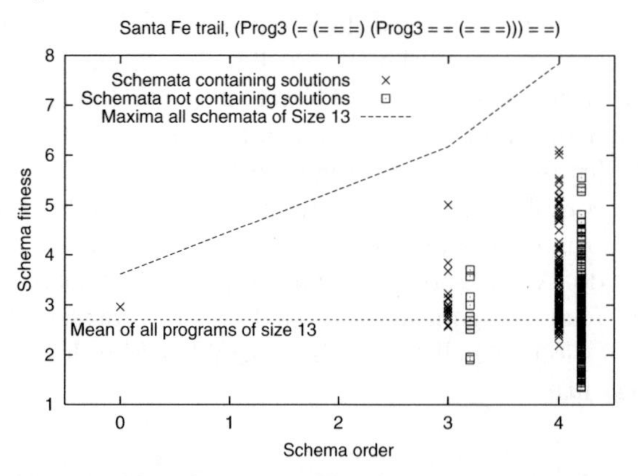

Fig. 9.12. Distribution of schema fitness within a hyperspace of size 13 containing a (intron) solution, cf. Figure 9.15.

9.6 The Solutions

We have analysed all the shorter solutions (see Figures 9.13–9.16). As we shall see, all the solutions of size 11 and 12 and most of those of size 13 are variations of each other.

Figure 9.13 shows the structure of all the solutions of size 11, of which there are twelve. Not only are they genetically distinct, but they cause the ant to exhinit different behaviour. That is, they are also phenotypically distinct. However, we can recognise certain symmetries. For example, they contain pairs of ant rotate operations, and it is no surprise that these can be either pairs of Left or pairs of Right terminals. Another symmetry is that the program consists of three parts which have to be performed in order, but the ant can start with any one of the three and still traverse the trail. Since the solution codes each of these as an argument of the root, the root's arguments can be rotated. Each rotation gives rise to a genetically different program, with slightly different behaviour. Each of these gives rise to a different tree shape and so the 12 solutions lie in three distinct hyperspaces.

The solutions of size 12 are the same as those of size 11. They are made one node longer by replacing a single Prog3 function with two Prog2 (see Figure 9.14). There are a total of four ways of doing this for each solution of size 11, giving rise to 48 solutions. While these are genetically distinct from each other and the solutions of size 11, they represent identical behaviour. There are 12 tree shapes (hyperspaces) each containing four solutions.

Extending this we can see that there must also be 48 solutions of size 13 created by replacing both Prog3 with Prog2 (there are four ways of arranging the Prog2). However, there are other ways to make use of the available space to represent the same solutions. This is done by adding introns. Each of the non-Prog3 nodes can be replaced by an IfFoodAhead, one of whose arguments is the previous node (and its arguments) and the other is either a terminal that is identical to the other argument or is never executed (see Figure 9.15). Most solutions of size 13 are of this type.

Thirteen nodes allow solutions of a different type which consecutively perform two moves before looking for food (see Figure 9.16). Again there is symmetry, in that the ant can be rotated either to the right or to the left, but whichever is done first the opposite must be done in the later part of the program. This gives rise to programs of the same shape with the same score. The program now consists of five parts that have to be executed in the correct order but, as with solutions of size 11, it does not matter which is first. Each of these five orderings gives rise to a different behaviour, but each traverses the trail. (However, they take slightly different amounts of energy to do so. Including energy as part of the fitness measure would give a means of breaking the symmetry of these solutions). Additionally there are three ways to arrange the arguments of the two Prog3 which are functionally identical. Each of these rearrangements yields solutions of different shapes.

Most of the other solutions of size 13 also perform two consecutive Move operations. These and the remaining solution of size 13 have less symmetry and are fewer in number.

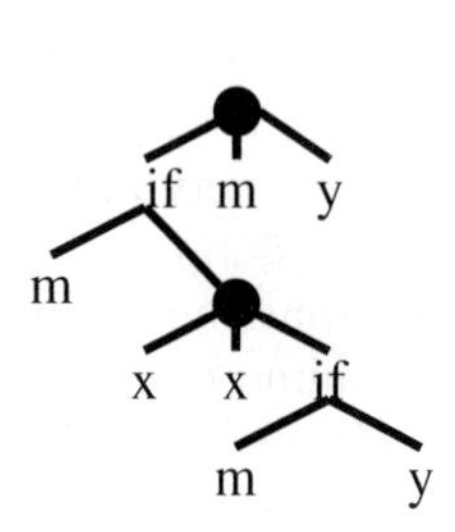

Fig. 9.13. Solutions of size 11. x and y can be either Left or Right, and the three arguments of the root can be rotated, giving 12 solutions. (Solid circles indicate Prog2 or Prog3, "m" Move and "if" IfFoodAhead).

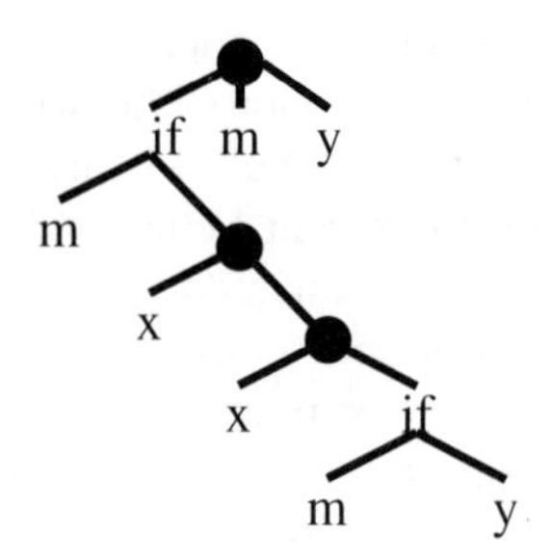

Fig. 9.14. Solutions of size 12. Like solutions of size 11 (x and y can be either Left or Right, and the three arguments of the root can be rotated) additionally one Prog3 is replaced by two Prog2, giving 48 solutions.

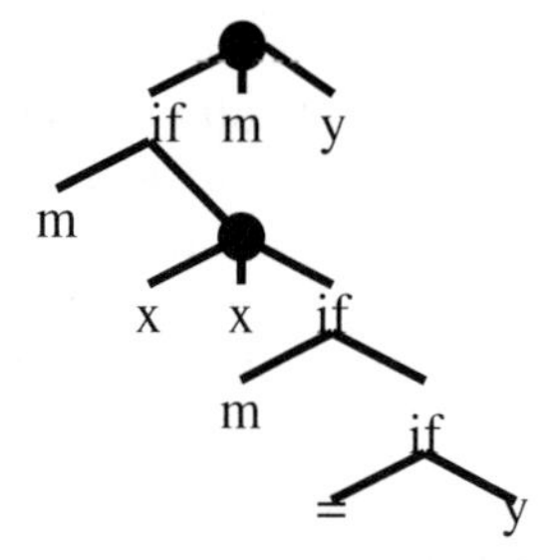

Fig. 9.15. Intron solution of size 13. Like solutions of size 11 (x and y can be either Left or Right, and the three arguments of the root can be rotated) and the = can be any terminal as it is never executed.

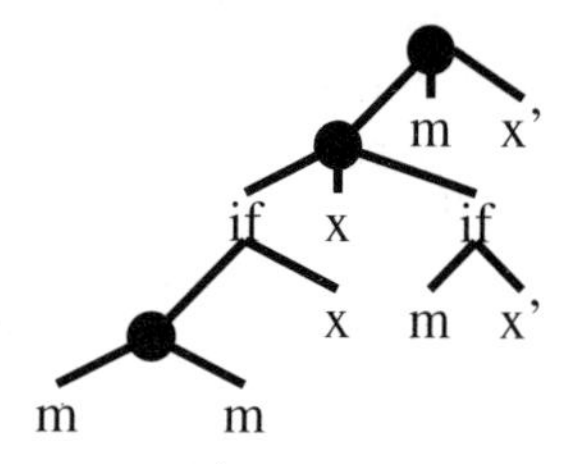

Fig. 9.16. Solutions of size 13 performing two Moves. x can be either Left or Right, but then x' must rotate in the opposite direction; again the arguments of the root can be rotated and there are equivalent ways to order the arguments of the two Prog3.

9.7 Discussion

As briefly introduced in Section 1.1.5, the No Free Lunch theorems [Wolpert and Macready, 1997] prove that averaged over all problems all search algorithms have the same performance. In particular, this means that averaged across all possible problems the performance of genetic programming is the same as random search. In adaptive search circles this has been frequently countered by arguing that we are not interested in all possible problems but in some ill-defined set of interesting ones (which, by implication, our favourite search technique is good at solving). This in turn implies there is a class of "badly behaved problems" (for our favourite technique) which we are not interested in solving, where our technique performs worse than random search. If the number of solutions is small, random search will not solve our problem, so it can be argued that, as our technique does, this implies our problem is "well behaved".

This chapter shows an example of a frequently studied interesting problem where GP performance is not dramatically better than random search. This suggests we in GP circles may be interested in "badly behaved problems" where GP performance lies close to random search. If such problems have unique solutions, then random search will not in practice find them. However, there are an exponentially large number of solutions to the Ant problem. If this is also true of other problems (in Chapter 7 we argue that it is), it may be the principal reason for the success of GP and other stochastic search techniques.

To explain the performance of adaptive search techniques, we need to consider the fitness landscape they see. The Ant problem has the features often suggested of real program spaces. The program space is large and, using the simplest neighbour relationship, forms a Karst landscape containing many false peaks and many plateaus riven with deep valleys. There are multiple distinct and conflicting solutions to the problem, some arising from symmetries in the primitive set and some from the problem itself. The landspace is riddled with neutral networks linking programs of the same fitness in a dense and suffocating labyrinth.

A limited schema analysis indicates that the problem is deceptive at all levels. Longer programs are on average slightly fitter but contain a slightly lower density of solutions. There are hyperspaces that do not contain solutions which are fitter than those of the same size which do. There are low- and middle-order schemata which are required to build solutions but which are below average fitness. Schemata typically have a high fitness variance. This means practical sized samples give noisy estimates of their fitness, leading GAs to choose between them randomly. However, the fitness of low-order schemata may be estimated more reliably (since GA populations can contain many instances of them). Where they are deceptive, this may lead a GA to discard them. (Extinction of complete primitives was seen in the list and stack problems [Langdon, 1998c, Chapters 6 and 8]).

We have not been able to find any building blocks (i.e. small components of a solution with above average fitness). We have only considered the simplest solutions using a fixed representation, but they cannot be assembled from building blocks. Indeed many constructs that a human programmer might use have below average fitness. However, it is possible that longer solutions might be constructed from fixed representation building blocks, or solutions might be constructed in a variable size representation (as used by GP) from building blocks. But as GP performance is similar to hill climbing, this suggests that either there are no building blocks for GP in this problem or that they give no benefit.

If real program spaces have the above characteristics (we expect them to have them but be still worse), then it is important to be able to demonstrate scalable techniques on such problem spaces. The Santa Fe trail provides a tractable problem for such demonstrations. From Table 9.3 it is obvious that current techniques are not doing well on it.

Current techniques do not exploit the symmetries of the problem. These symmetries lead to essentially the same solutions appearing to be the opposite of each other. For example, either a pair of Right or pair of Left terminals at a particular location may be important. If the search technique does not recognise them as the same thing it may spend a lot of effort trying to decide between them, when perhaps either would do (cf. the "competing conventions" problem in artificial neural networks). A possibly useful approach is to break this symmetry (e.g. by putting more of one primitive in the initial population) to bias the technique so that it chooses one option quickly. Alternatively, new genetic operators [Maxwell, 1996] might better exploit the semantics of the programs. The Ant problem is not atypical in the respect. Typically function sets contain some functions that have symmetries, such as commutativity and transitivity. There has been little analysis of their effects in GP. We might address the tangled network of programs with the same fitness, which consumes a lot of machine resources and promotes bloat by introducing a small bias. In the Ant problem we would expect a slight bias in favour of shorter programs to be beneficial as solutions are more frequent when programs are short.

The Ant problem appears to be difficult because of the large number of suboptimal peaks in the fitness landscape. These are created by the combination of the representation, the neighbour operator and the fitness function. While there may be improvements to the representation or better search techniques, we should also consider the fitness function, particularly how we reward partial solutions. In the next section we summarise experiments [Langdon, 1998a] that did this.

9.8 Reducing Deception

In Section 9.3 we have shown that the performance of several techniques is not much better than the best performance obtainable when using uniform random search. We suggested that this was because the program fitness landscape is difficult for hill climbers, and that the problem contains multiple levels of deception which also makes it difficult for genetic algorithms.

Analysis of high scoring non-optimal programs suggests many reach high rewards even though they exhibit poor trail following behaviour. Typically they achieve high scores by following the trail for a while and then losing it at a corner or gap. They then execute a blind search until they stumble into the trail at some later point and recommence following it (see Figure 9.17). The blind search may give them a better score than a competing program that successfully navigated the same bend or gap but later lost the trail (see Figure 9.18).

We redefine the problem. The same trail is used but we only place food on parts of the grid when the ant following along the trail nears them. This makes it difficult for an ant that moves away from the trail to find another part of it by blind search. This changes the fitness landscape. We anticipate that almost all optimal points within it will retain their scores, and that many previously high scoring non-optimal points will be given reduced fitness by the new training regime. (Of the 3916 solutions to the Santa Fe trail problem found by exhaustive search, uniform random and ramped half-and-half random search, all but at most two of them completed the trail when using the new fitness function). [Langdon and Poli, 1998c] considers other mechanisms to reduce deception. The changes to the fitness function are intended to make the landscape less deceptive and so easier for genetic algorithms. Removal of false peaks may also benefit hill climbing techniques.

Apart from the new training technique used, the problem is as in the rest of this chapter. The food pellets are numbered in the order we expect the ant to eat them. Only the first x are placed on the grid initially. As the ant moves and eats them new food pellets are added to the grid. When it eats food pellet n then we ensure all the uneaten food pellets up to pellet $n+x$ are on the grid. At least 50 independent runs where carried out in each experiment.

As expected, limiting the amount of uneaten food in front of the ant along the trail made the task of evolving suitable programs easier for GP. Figure 9.19 plots the "effort" versus how far along the trail the Ant can see. It is clear that *reducing* the fitness of programs that manage to eat a lot of food (in the original problem set-up), but do not follow the trial in the correct sequence, actually makes the problem *easier* for genetic programming.

9.9 Conclusions

There have often been claims that automatic programming is hampered by the nature of program spaces. These are undoubtedly large [Koza, 1992, page 2] and, it is often claimed, badly behaved with little performance relationship between similar programs [O'Reilly, 1995, page 8]. When we test these assumptions, we find that there is more to it. First, we can quantify how big search spaces actually are. Secondly, the assumption of ruggedness appears to hold only for small programs. Large programs are often interconnected by a dense network of paths connecting programs with the same

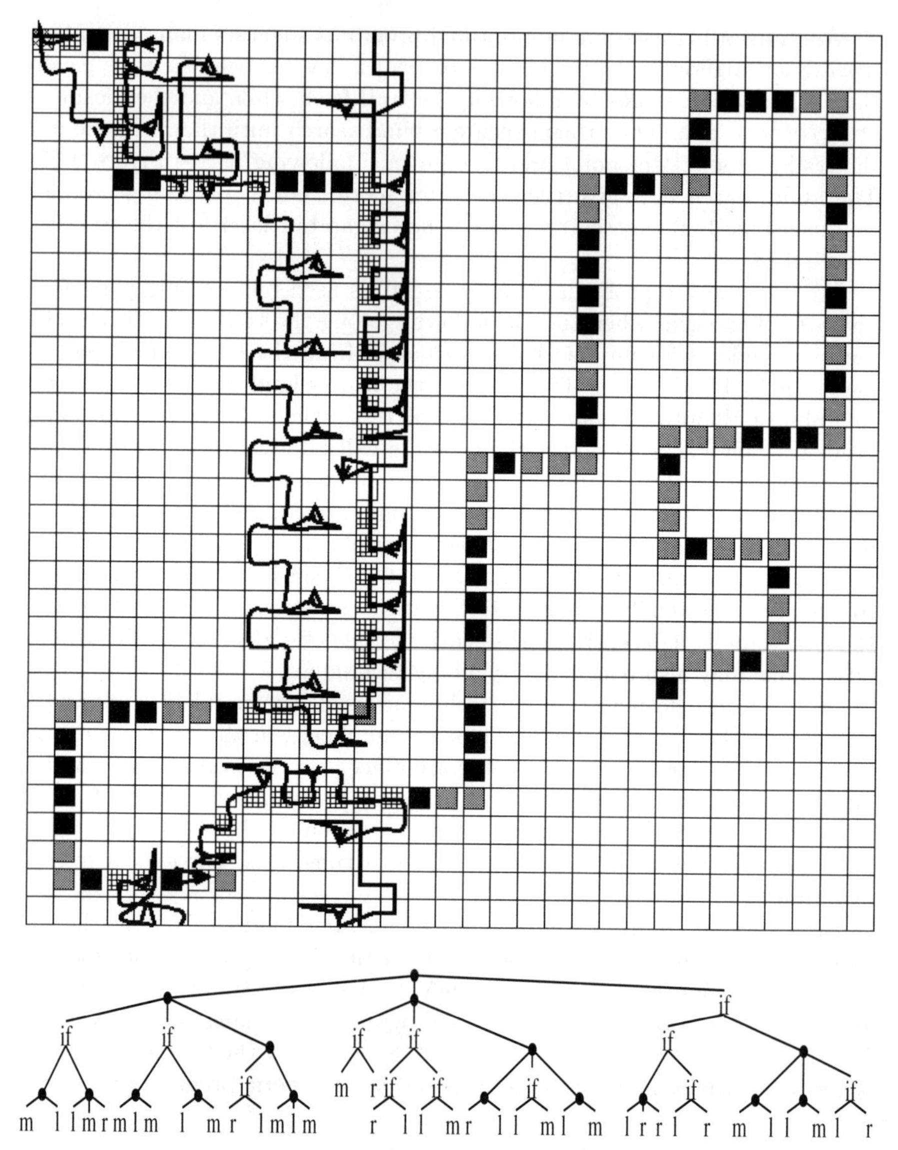

Fig. 9.17. Example program (bottom) from the initial generation which eats 40 pellets. The ant's path (top) is shown by the line with arrows, uneaten food pellets by black squares, eaten food by squares with horizontal crosses, and shaded squares show gaps in the trail.

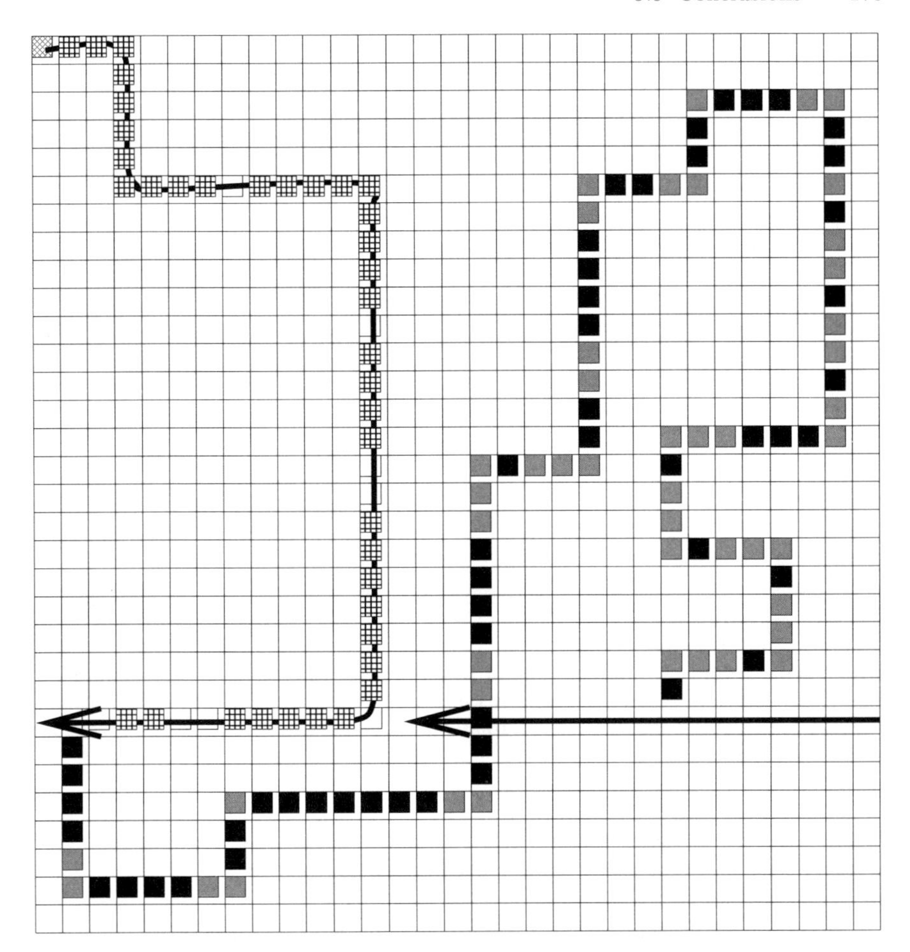

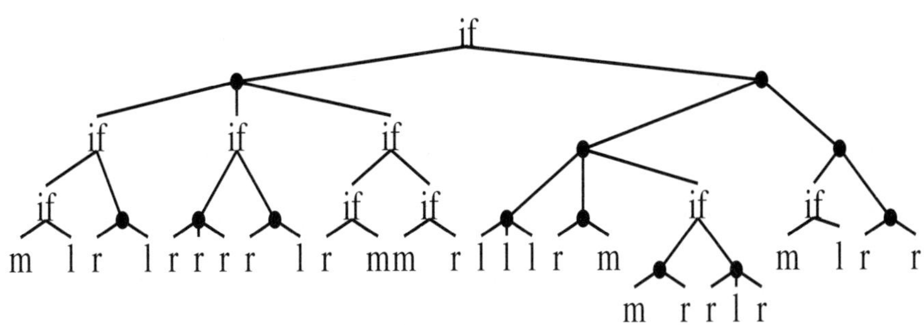

Fig. 9.18. Example program from the initial generation which eats 38 pellets

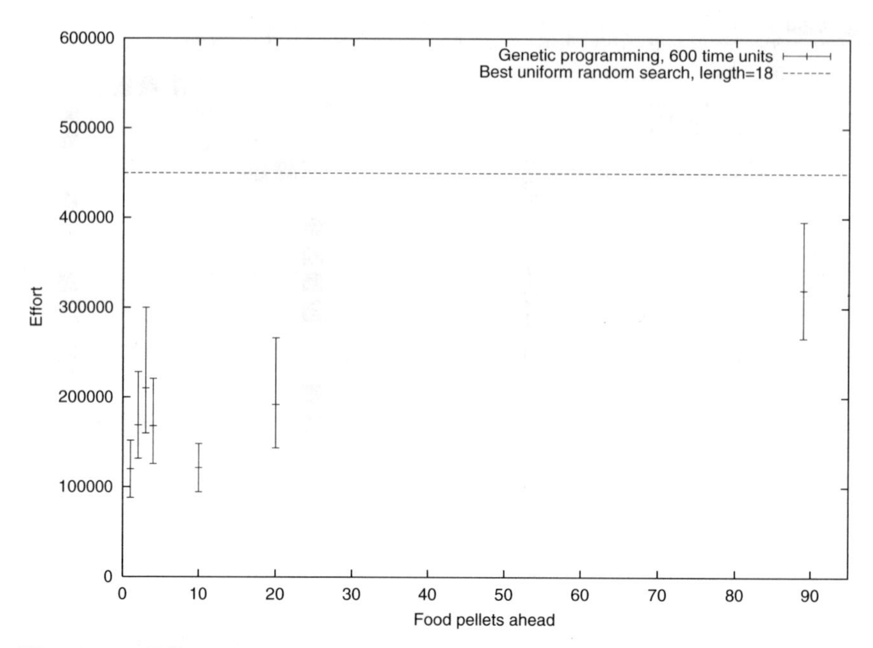

Fig. 9.19. Effort vs. number of food pellets ahead that the ant can see and eat. Error bars indicate estimates given by one standard deviation above and below the minimum measured figure. The original Santa Fe trail always allows all 89 food pellets to be seen.

fitness. (The so-called neutral networks). However, since these do not necessarily lead to programs with a better performance (i.e. high fitness), their utility is still an open question.

It is interesting that even a problem as simple as this Ant problem should have so many different solutions. We have highlighted that many of the solutions are simple transformations of each other. However, these symmetries in the search space do not match those of traditional GP genetic operators. This has similarities with the aliasing problem sometimes seen in artificial neural networks, where multiple solutions inhibit training as they have incompatible representations.

Using the program landscape and schema analysis, we have shown why the artificial ant following the Santa Fe trail problem is difficult for these search techniques. By making a modest change to the Santa Fe trail problem we have made it significantly easier for GP. The smallest solutions to the Ant problem are not symmetric. This suggests that full or bushy trees contain relatively few solutions. As described in Chapter 11, this suggests that limiting the size of programs, rather than their depth, might be beneficial to GP in *this* problem. We know [Langdon, 1998a] for the Ant problem, that the performance of GP is only marginally affected by a size limit, and is roughly constant for wide ranges in maximum size.

10. Example II: The Max Problem

In this second chapter on the application of our analysis techniques, we consider a second GP benchmark problem: the "Max Problem". Briefly, this problem is to find a program, subject to a size or depth bound, which produces the largest value [Gathercole, 1998]. This problem is unusual in several respects: we know in advance what the solution is, what value it returns and and exactly how many solutions there are. Also, the size or depth limit is fundamental to the problem. MAX is hard for GP because of the way deception interacts with size and depth bounds to make it difficult to evolve solutions.

[Gathercole and Ross, 1996] introduced the MAX problem to GP to highlight deficiencies in the standard GP crossover operator which result from the practical requirement to limit the size of evolved programs. They concentrated on the case where program trees are restricted to a maximum depth, but in [Gathercole, 1998] showed that similar effects occur when programs are restricted to a maximum number of nodes.

The MAX problem has known optimal solutions which are composed of regularly arranged subtrees or building blocks. Despite this, GP finds solving larger versions of the MAX problem difficult.

Here we extend [Gathercole and Ross, 1996] by considering bigger trees, different selection pressures, different initialisations of the population, measuring the frequency with which the depth limit affects individual crossovers, and measuring population variety and the number of steps required to solve the MAX problem. Qualitative models of crossover and population variety are presented and compared with measurements. We give an improved explanation for the premature convergence noted by [Gathercole and Ross, 1996], and this leads to the realisation that there are two separate reasons why GP finds the MAX problem hard. (1) the tendency for GP populations to converge in the first few generations to suboptimal solutions from which they can never escape. (2) convergence to suboptima from which escape can only be made by slow search similar to randomised hill climbing.

Section 10.1 describes the various MAX problems used in these experiments. Section 10.2 describes the GP used in our experiments, and Section 10.3 gives the results obtained and provides a detailed analysis and comparison with [Gathercole and Ross, 1996]. Section 10.4 shows the population typically retains a high level of diversity, and presents models of the variety in the initial population and its subsequent evolution. Section 10.5 considers the role of selection pressure and shows reduced performance with small tournament size. Section 10.6 presents experimental evidence showing the applicability of Price's Theorem (cf. Section 3.1) and shows it can be used to predict the behaviour of subsequent generations of the MAX problem.

10.1 The MAX Problem

In the MAX problem, "the task is to find the program which returns the largest value for a given terminal and function set and with a depth limit, D, where the root node counts as depth 0" [Gathercole and Ross, 1996, page 291]. In this chapter we use the function set $\{ +, \times \}$, and there is one terminal 0.5. For a tree to produce the largest value, the + nodes must be used with 0.5 to assemble subtrees with the value of 2.0. These can then be connected via either + or × to give 4.0. Finally, the rest of the tree needs to be composed only of × nodes to yield the maximum value of $4^{2^{D-3}}$. Every component of the evolved programs contributes to their fitness, and so no part can escape the effect of selection. That is, there can be no introns.

10.2 GP Parameters

Our GP system was set up to be the same as given in [Gathercole and Ross, 1996]. The details are given in Table 10.1, parameters not shown are as given in [Koza, 1994, page 655]. Fifty independent runs were conducted on each version of the problem (with six depths and two types of initialisation in the three bigger problems, and seven tournament sizes, this makes a total of 3150 runs).

In this chapter two sets of experiments are presented. In the first, the usual ramped half-and-half method [Koza, 1992, page 93] that creates random trees with depths between 2 and 6 (i.e. $D = 1, \ldots, 5$) was used. (However the initial trees obeyed the problem-specific height restriction). A second set of runs (with $D > 5$) were made with the maximum tree height in the initial population identical to the problem-specific limit. (Unless otherwise stated, the discussion refers to this second set of experiments). Gathercole and Ross created their initial populations "with no constraint on tree size (other than the overall size limit), i.e. not ramped" [Gathercole, 1997].

10.3 Results

Figure 10.1 shows the mean number of generations taken by GP to solve the MAX problem in the successful runs. The percentage of unsuccessful runs is given in Figure 10.2.

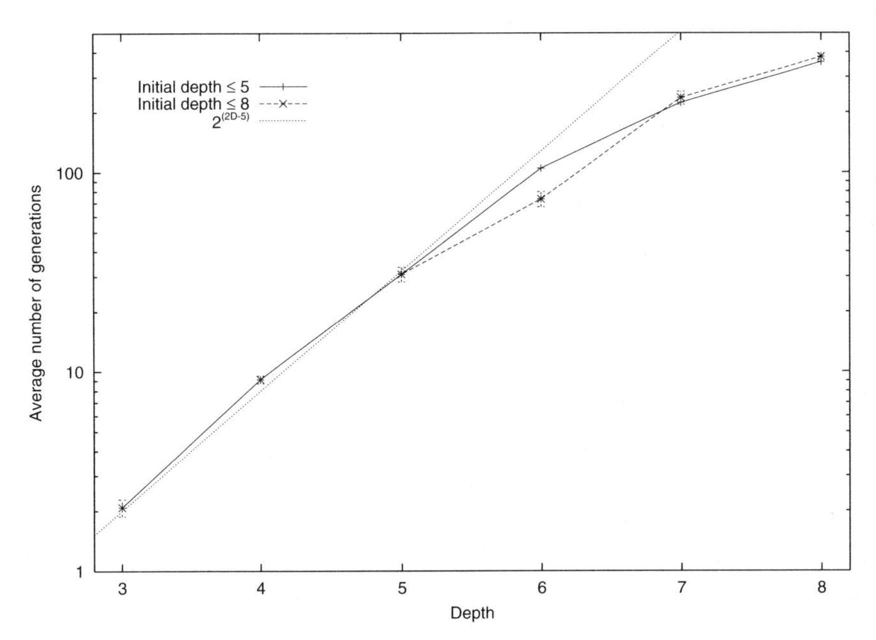

Fig. 10.1. Mean number of generations need by the successful runs. Error bars indicate standard error. The straight line is $2^{2D}/32$, which approximately fits the number of generations required, by successful runs, to evolve a suitable program, until the limit of 500 generations is approached.

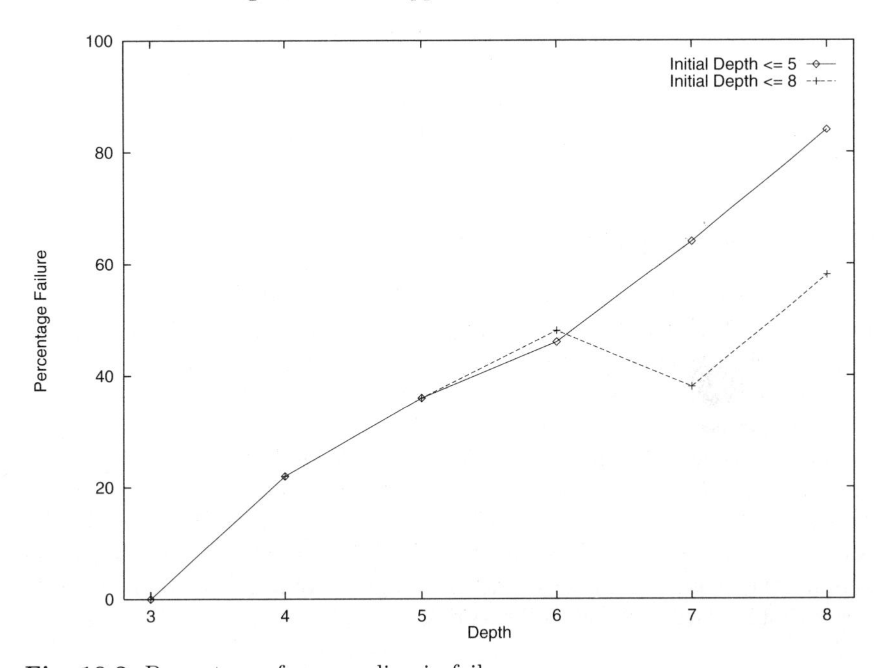

Fig. 10.2. Percentage of runs ending in failure

Table 10.1. GP Parameters for the MAX Problem

Objective	Find a program that returns the largest value
Primitives	$+, \times, 0.5$
Maximum depth	3, ..., 8 (NB root node is depth 0)
Initial depth	5 or maximum depth
Fitness	Value of tree
Selection	Tournament group size of 2–8, generational plus elitism.
Parameters	Population size = 200, G = 500, 99.5% crossover, no mutation. Crossover points selected uniformly between nodes.

10.3.1 Impact of Depth Restriction on Crossover

The size of programs within the population grows quickly until almost all individuals are full trees. Due to the depth restriction, crossover cannot make the trees any bigger and so crossover fragments are either the same size as the code they are replacing or smaller. In the case of full trees, this means that crossover can move code at the same level in the trees or higher, but not lower.

In the first run with $D = 8$, a third of crossovers are rejected because the subtree to be inserted would cause the offspring to violate the maximum depth. In these cases the roles of the two parents are reversed, and a shorter subtree is inserted instead. Of the remaining two-thirds of crossovers, half (i.e. $\frac{1}{3}$ of the total) result in replacing a subtree with one of the same height, so in total two-thirds of crossovers replace a subtree with a shorter one. Only 0.74% of crossovers resulted in a taller subtree replacing a shorter one. These occur throughout most of the run, with half occurring by generation 85 and the last in generation 441 (the mean generation is 105).

10.3.2 Trapping by Suboptimal Solutions

[Gathercole and Ross, 1996, page 295] suggested that the reason why GP finds the MAX problem hard is that the population quickly finds suboptimal solutions which contain + nodes near the root of the tree. To improve on such solutions, crossover must replace them with $\times$ nodes. They suggested that "once the trees have reached the depth limit, the only way the higher levels are affected is through the promotion of lower-level subtrees, which contain no $\times$ nodes, and the movement of subtrees within the same level". However the lack of $\times$ nodes in lower levels was not observed in our populations. For example, at the end of all but one of the 50 runs with $D = 8$, the population contains thousands of $\times$ nodes at levels 3 and 4, even in runs where levels 0, 1 or 2 contain large numbers of + nodes.

Crossover readily moves such nodes to higher levels. To explain why the suboptimal + nodes are able to remain in the population, we need to consider the fitness of the resulting offspring. Suppose both parents are near optimal (i.e. they consist mainly of +, × and 0.5 at the correct level of the tree) see Figure 10.3. The value of each subtree with a × as its root node is $4^{2^{D_\times - 3}}$ (where $D_\times$ is its depth), while if rooted with a + node it is $2 \times 4^{2^{D_+ - 4}}$, (Figure 10.4). If an otherwise optimal subtree with a + as its root node is replaced with an optimal subtree from a lower level (i.e. $D_+ > D_\times$), then the value of this part of the offspring is reduced. For example if an otherwise optimal subtree of depth 5 rooted with a + node (value 32) is replaced by an optimal subtree which is one level shorter (value 16). Then given the other subtrees at the same level are also near optimal (i.e. have values ≥ 1), the value (and hence fitness) of the offspring as a whole will be less than that of its parent by a factor of at least $\frac{1}{2}$. Therefore, the offspring is unlikely to have children in the next generation. In contrast, the $\frac{1}{3}$ of crossovers that exchange subtrees at the same level may produce offspring with the same fitness as their parents, and thus are more likely to have children themselves. Programs with a few + nodes near the root can readily reproduce copies of themselves but their children where a + is replaced with a × are unlikely, to survive and so the GP can remain trapped for long periods.

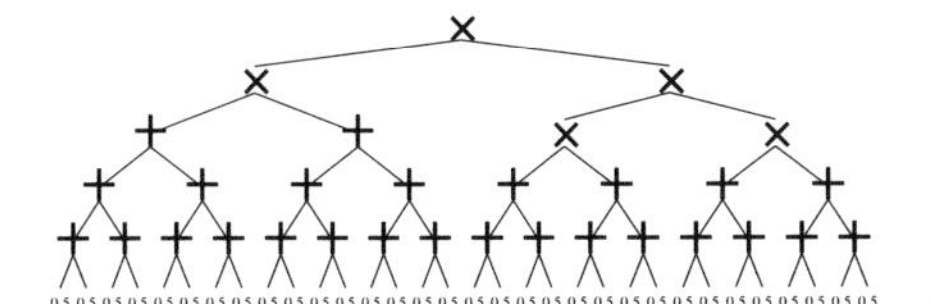

Fig. 10.3. An optimal subtree, $d = 5$. $4^{2^{d-3}} = 4^{2^2} = 4^4 = 256$. The two subtrees of the root node are both themselves one-level-shorter optimal subtrees (i.e. $d = 4$, and so have value $4^{2^{d-3}} = 4^{2^1} = 4^2 = 16$). Note the top two levels are all × notes. The next level can be a mixture of + and ×. The next two levels are all + nodes, and the bottom level are all leaves (0.5).

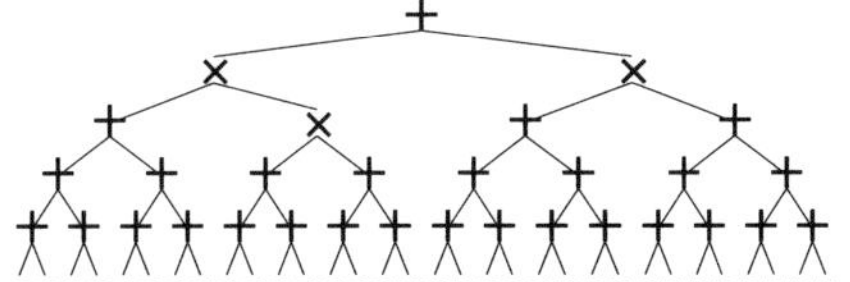

Fig. 10.4. As Figure 10.3 except root node is + rather than ×. So the two subtrees of the root are optimal subtrees ($d = 4$ value 16) but they are added together rather than multiplied. This radically reduces its value to ($d = 5$) $2 \times 4^{2^{d-4}} = 2 \times 4^{2^1} = 2 \times 4^2 = 32$.

10.3.3 Modelling the Rate of Improvement

[Gathercole and Ross, 1996, page 295] wrote: "if there are no × nodes in any tree at a particular high level, it [is] now impossible for crossover to introduce a × node to this level; the population has converged to being duplicates of a sub-optimal tree, and no further improvement is possible". However (taking the example with $D = 8$ again), in only seven cases does the population at the end of the run contain zero × nodes at any level from 0 to 3. That is, in 22 of the runs that failed, the population contained $\times$ nodes in all of the higher levels. (The impossibility of the seven runs succeeding was confirmed experimentally by running them again but this time to 5,000 generations. None succeeded, but in one case the population did improve by finding better suboptimal solutions by replacing + nodes with $\times$ nodes where these were available at the same level).

The reason why the 22 runs failed (and why the successful runs took so long), despite having $\times$ nodes available, is in part due to the low level of crossover activity near the root of the trees. That is, crossover is able to improve suboptimal trees but it takes a long time. We can estimate how long with the following approximate models.

Crossover at Critical Points. The model estimates how long it will take subtree crossover to replace a + node in one of the levels of the tree near the root with a $\times$ taken from the same level, if the + and $\times$ nodes are the crossover points. Assume that the population has converged to a suboptimal tree containing n_{+d} + and $n_{\times d}$ $\times$ nodes at level d ($d < D - 3$) (so $n_{+d}+n_{\times d} = 2^d$). Each individual will be a full (or nearly full) binary tree of height D, and so will contain $2^{D+1} - 1$ nodes. Therefore, the chance of selecting one of the + nodes at level d to be the crossover point is

$$\frac{n_{+d}}{2^{D+1} - 1}$$

and the chance of replacing it with a $\times$ node from the same level is

$$\frac{n_{\times d}}{2^{D+1} - 1} \tag{10.1}$$

First we note that the chance of improvement is much higher with large d. I.e. the nearer to the root the + node is, the harder it will be for crossover to shift. (In the case of the root node, this is impossible as there can be no $\times$ nodes also at the root). Secondly, improvement is easiest when the number of + nodes is equal to the number of $\times$ nodes, and the last + node at each level is the most difficult to remove. Also note that if the number of $\times$ nodes is small, it will be difficult to increase it.

Since replacing a + node at any of the higher levels will produce an improvement, by only considering crossovers which find improved solutions by moving subtrees at the same level as the + nodes we can form a lower bound on the overall chance of the offspring being an improvement:

$$\sum_{d=0}^{D-4} \frac{n_{+d} n_{\times d}}{(2^{D+1} - 1)^2} = \sum_{d=0}^{D-4} \frac{n_{+d}(2^d - n_{+d})}{(2^{D+1} - 1)^2} \tag{10.2}$$

As each crossover is independent, the number required to replace a + node with a $\times$ node has an exponential distribution. In an exponential distribution the mean and standard deviation are equal to each other and to 1/(probability of success in one step). We can therefore estimate the mean and standard deviation using Equation (10.2) as:

$$\text{crossovers to improvement} \leq \frac{(2^{D+1}-1)^2}{\sum_{d=0}^{D-4} n_{+d}(2^d - n_{+d})}$$

Thus the expected number of generations until the next improved solution is found in a population which has converged towards a good but suboptimal solution is

$$\text{generations to improvement} \leq \frac{(2^{D+1}-1)^2}{Mp_{xo}\sum_{d=0}^{D-4} n_{+d}(2^d - n_{+d})} \tag{10.3}$$

where M is the population size and p_{xo} is the crossover probability (assuming no mutation).

Crossover at Critical Points and Nearer Root. We can refine the model slightly. Crossover closer to the root can also find improved solutions by replacing trees containing + nodes with others where they are replaced by × nodes, including where the inserted subtree contains + nodes further from the root. In general, where there are + nodes at different levels, deciding how many crossover points will yield an improved solution becomes complex. However, if there are few incorrectly positioned + nodes, we need only consider crossovers where the crossover points in both parents are at the same level. For the special case of a single + node at level d, then crossover at any node connecting it to the root with a different node at the same level will yield the optimal solution. There are

$$\sum_{i=1}^{d} 2^i - 1 = 2^{d+1} - d - 2 \tag{10.4}$$

such crossover point pairs. Note in this case that the exact figure is about twice the approximation of $n_{\times d}$ used in Equations (10.1) and (10.2).

The predictions of Equation (10.4) where tested by measuring the number of generations required to replace the last misplaced + node with $D = 8$ (this required extending 14 runs up to 5,000 generations). Figure 10.5 shows good agreement between prediction and measurement. At every depth the mean lies within 1.4 standard errors of the predicted value.

To test further the model we looked at multiple + nodes. Since this is more complex, we looked at only one run. We selected an unsuccessful run for $D = 8$. In this run the population had converged towards a tree with one + node at level 3 and four at level 4. It was run beyond generation 500, and the generations where improved solutions were found were noted. The results are shown in Table 10.2. The last-but-one column of Table 10.2 gives the exact count of crossovers that would generate an improved solution. The rightmost column gives the corresponding expected number of generations to find an improved solution. There is reasonable agreement with the actual number of generations, recorded in column 3. Whereas the simpler model given by Equation (10.3), columns 4 and 5, consistently overestimates the time required to find the next solution.

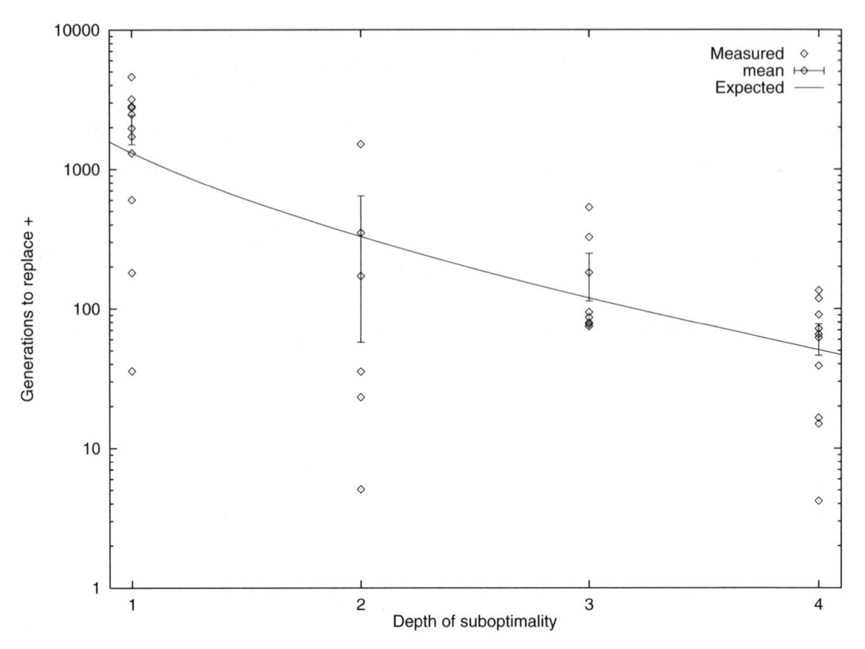

Fig. 10.5. Number of generations to displace last misplaced + ($D = 8$). Error bars indicate standard error.

Table 10.2. Generations (column 2) and location (column 1) at which improved solutions were found in an extended run ($D = 8$). Columns 4 and 5 give the estimated number of crossover pairs that give improved solutions, and the corresponding expected number of generations between improved solutions, for the simple $\sum n_+(2^d - n_+)$ model, cf. Equation (10.3). The last two columns refer to the detailed counting model. The simple model consistently over estimates ΔGen.

Measurements			Simple model		Detailed model	
Depth	Gen	ΔGen	Crossover pairs	Predicted ΔGen	Crossover pairs	Predicted ΔGen
4	523	*23*	55	*24*	73	*18*
4	542	*19*	46	*29*	65	*20*
4	569	*27*	35	*37*	51	*35*
3	584	*15*	21	*62*	34	*39*
4	611	*27*	15	*87*	26	*50*

10.3.4 Number of Steps to Climb the Hill

The number of improved solutions found before the optimal solution is found is plotted in Figure 10.6. In successful runs, it takes about as many improvement steps as there are functions in the optimal solution (i.e. $2^D - 1$). This supports the notion that, in this problem, GP finds solutions one step at a time and is not benefiting from the implicit parallelism expected when complete solutions can be assembled from building blocks. From Equation (10.3) we see that each step takes $O(2^{2D})$ generations, so we expect the total number of generations required, in successful runs, to grow as $O(2^{2D})$, i.e. parallel to the straight line in Figure 10.1. Figure 10.1 shows good agreement until the limit of 500 generations per run acts to cut short runs which would otherwise have eventually found a solution.

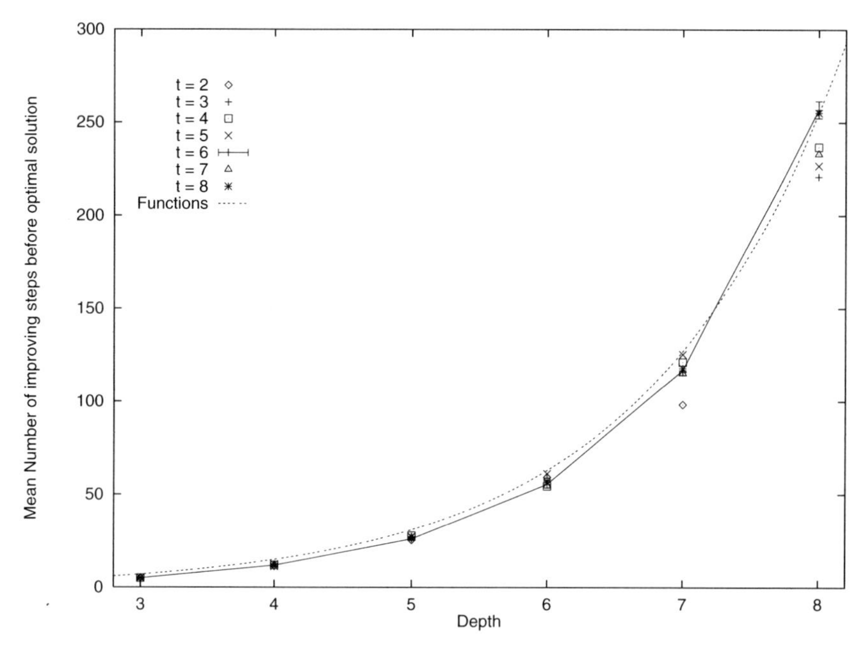

Fig. 10.6. Mean number of improved solutions found before optimal solution in successful runs. Error bars show standard error for a tournament size of six.

10.4 Variety

10.4.1 Variety in the Initial Population

The rightmost column of Table 10.3 contains the average number of different programs in the initial populations. It highlights another potential problem with GP when it is applied to the MAX problem: in contrast with normal GP practice, the initial populations contain a large number of duplicates (see also Figures 10.7 and 10.8). This is inherent in the ramped half-and-half method when used in a problem with such small function and terminal sets. Recall that with the ramped half-and-half method, random trees are created with a maximum depth evenly distributed between 1 (counting the root node as zero) and the maximum depth allowed in the initial population. Half are created as full trees of this maximum depth and half are created using a grow method which creates trees of different shapes. ([Iba, 1996, Bohm and Geyer-Schulz, 1996] and [Langdon, 2000] discuss alternative means of generating random trees for use as the initial population in GP runs).

With $D = 8$ about $\frac{1}{8}$ (i.e. $\frac{1}{D}$) of trees will be created with a maximum depth of 1 and therefore will contain a single function. That is, about 25 trees will consist of a single + node and two 0.5 terminals, or a $\times$ node and two 0.5 terminals, and therefore they are very likely to be identical to at least one other member of the population. $\frac{1}{D}$ of trees are created with a maximum depth of 2, and so will contain no more than two or three functions. As there are 16 possible trees with two or three functions, it is likely that most trees with a depth of two will not be unique. With larger trees the number of possible trees grows rapidly and so most will be unique. One-quarter of trees generated by the grow method will not grow either branch from the root, and so a further $\approx \frac{1}{2}\frac{D-2}{D}\frac{1}{4}$ (i.e. 18.75) trees will be very small and so are unlikely to be unique.

In summary, we expect at least $\frac{1}{D} + \frac{1}{D} + \frac{1}{8}\frac{D-2}{D} = \frac{1}{8} + \frac{7}{4D}$ duplicates in the initial population (unless the initial population is rather less than 200). Table 10.3 shows that this is close to the mean number of duplicate individuals found in 50 initial populations. In problems with larger function or terminal sets, the chance of randomly generating duplicates will be much smaller, but a high proportion of small programs can still be expected.

Table 10.3. Variety in initial MAX problem populations created by ramped half-and-half. D (column 1), problem size; column 2, predicted fraction of duplicates; column 3, predicted number in population of 200; column 4, predicted variety (200 – column 3); column 5, mean variety in 50 initial populations.

D	$\frac{1}{8} + \frac{7}{4D}$	$\times$ population size	Predicted	Measured
3	0.7083	141.67	58.33	66.3
4	0.5625	112.5	87.5	94.1
5	0.475	95	105	111.2
6	0.4167	83.33	116.67	120.6
7	0.375	75	125	128.9
8	0.3438	68.75	131.25	136.4
∞	0.125	25	175	

10.4.2 Evolution of Variety

While the populations converge to the extent that most of the trees in the population are similar to each other, the population does not converge to the extent that all of the population are identical. Where the maximum depth is large, on average after 250 generations, the number of different programs in the populations is about two-thirds of the total population size (variety

≈ 133, cf. Figure 10.7). That is, one-third of the population is composed of programs that are identical to at least one other in the population. Figure 10.8 shows that in the case of smaller trees, population variety is lower still.

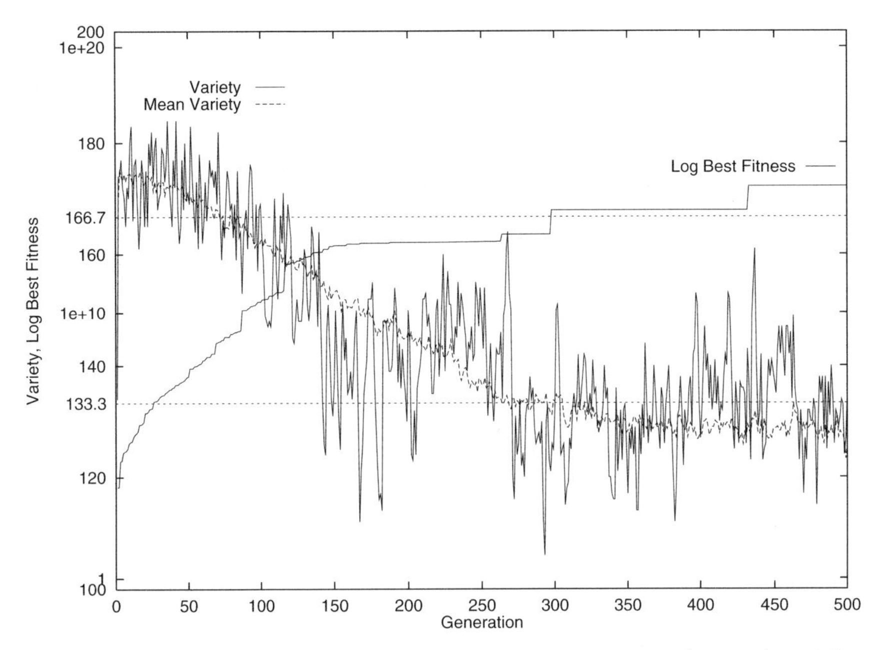

Fig. 10.7. The number of different trees in the population (variety) and best fitness in the first run and mean of 50 runs ($D = 8$).

10.4.3 Modelling Variety

Recall from Section 10.3.1 that about two-thirds of children are of a different length than their first parent (i.e. the one they inherit their root from). With a relatively small population it seems reasonable to assume that children which are different from their parent are also likely to be different from every other child in the population. That is two-thirds of children are likely to be unique. On the other hand, due to the convergence of the population, children produced by the one-third of crossovers that swap subtrees at the same level are likely to be the same as one parent. With a high selection pressure, such as when using tournament selection, these children are likely to be the same as another child of the same parent or of an identical individual in the previous generation. Therefore, we expect the population to be composed of one-third copies and two-thirds unique individuals. This is approximately the case for $D = 8$.

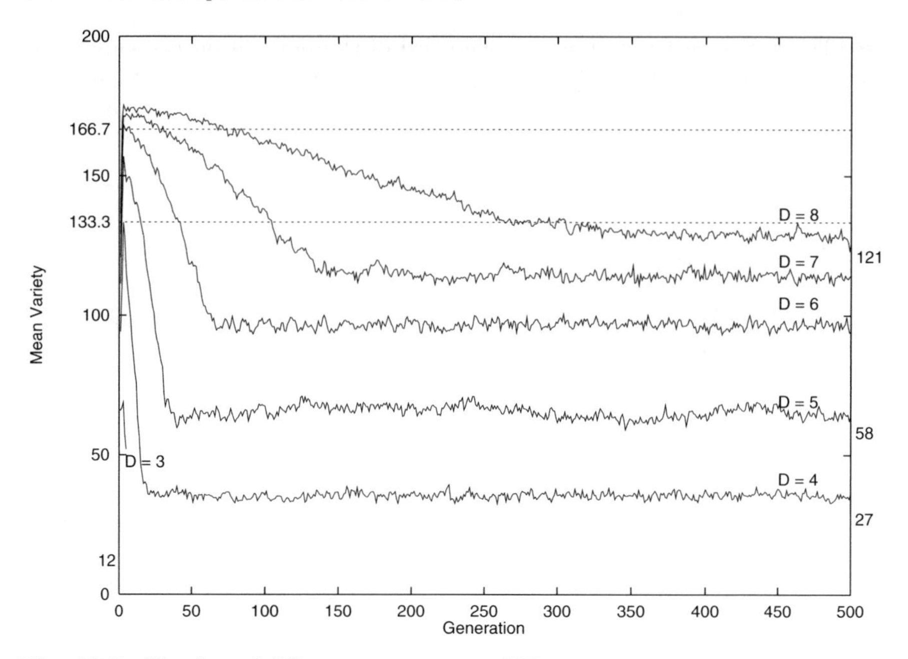

Fig. 10.8. Number of different trees, mean of 50 runs

The lower variety measured for $D < 8$ can be explained if one considers the breeding population to be composed mainly of copies of a full binary tree with only one node type at each level. The number of different trees that can be produced by crossover between such trees is limited (see Table 10.4). In the cases of $D = 4$ and 5, variety is on average close to the figure given in Table 10.4 while for $D = 6$ and 7 the two-thirds limit is also important.

Table 10.4. Number of different trees that can be produced by crossing over two identical full binary trees, where each level contains only one type of node

D	3	4	5	6	7	8
Number	12	27	58	121	248	503

10.5 Selection Pressure

In the MAX problem, with high selection pressure we see a rapid convergence of the population towards solutions that are beneficial in the first few generations. However, early solutions may not readily evolve to acceptable solutions. We suspect that this is common to many GP problems. In the case of the MAX problem, in the early generations individuals may use + functions to form comparatively large values (and hence have high fitness). Later generations can then yield still larger values by joining subtrees composed of + nodes with × functions. However, as [Gathercole and Ross, 1996] pointed out, the depth restriction and the mechanics of the crossover operator make the insertion of × nodes near the root of large trees difficult. This section was motivated by the suggestion that the selection pressure was too high (i.e. the tournament size was too big), and that this was responsible for driving the initial generations towards convergence too quickly.

The chance of an individual i being selected by a tournament is given by its rank r_i within the population of size M and the tournament size T according to the following formula [Blickle and Thiele, 1995, Langdon, 1998c] (cf. Section 3.1.3, page 31):

$$\frac{(r_i/M)^T - (r_{i-1}/M)^T}{r_i - r_{i-1}}$$

The expected number of children produced by individuals of rank R or less is

$$M \sum_{i=1}^{R} \frac{(r_i/M)^T - (r_{i-1}/M)^T}{r_i - r_{i-1}} \approx M(R/M)^T$$

That is, the worst $\approx \sqrt[T]{1/2}$ of the current generation produce half the children of the next generation, as do the best $1 - \sqrt[T]{1/2}$. With a tournament size of 6 this means on average that the best 11% of the population (i.e. 22 individuals) produce half the next generation (i.e. 100 children). Given that one-third of these are likely to be identical to their parents, the elite members of the population can readily pass identical copies of themselves to the next and succeeding generations.

Figures 10.9 and 10.10 give the average number of generations in successful runs and the proportion of runs that failed, respectively, for different tournament sizes. Each point represents the mean of 50 independent runs. For clarity, only the data for the runs where the maximum height in the initial population is the same as that in later generations are displayed. We see that performance is essentially independent of selection pressure, except for tournament sizes of 2 and 3.

With lower selection pressure ($T = 2$ or 3), GP performs worse on the MAX problem. This may be because the fitter members of the population have on average only two or three (i.e. T) children. Since only one-third of these will be identical, it becomes impossible for them to keep passing on identical copies of themselves to later generations. (The GP remains elitist and so passes on one copy of the best individual to the next generation without crossover). Therfore, lower selection pressure increases the diversity of the population. This permits greater exploration but at the expense of making the GP perform worse as a hill climber. The anticipated benefit of reduced premature convergence does not happen (cf. Figure 10.11). Perhaps selection pressure below that given by a tournament size of 2 is needed?

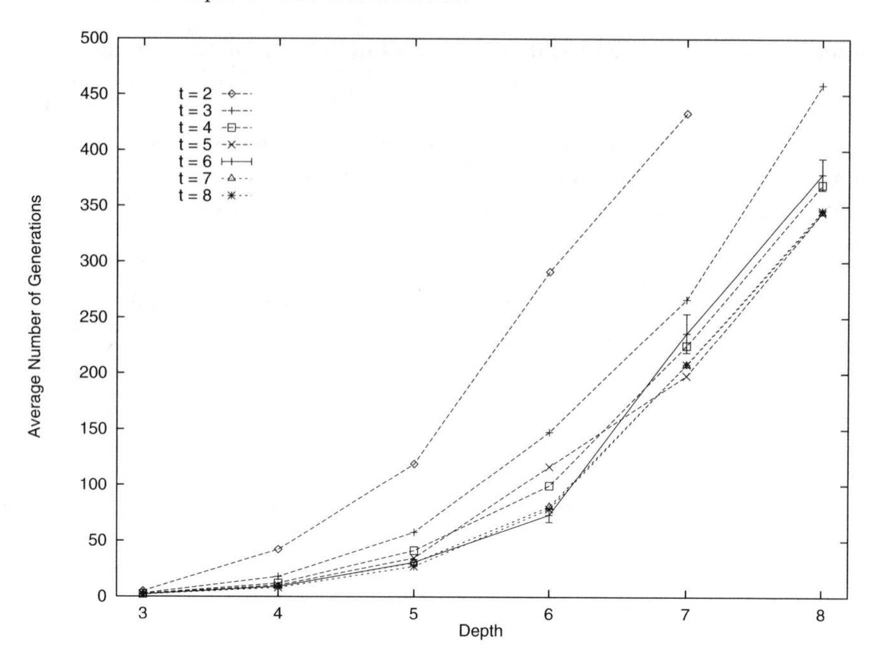

Fig. 10.9. Mean number of generations need by the successful runs with tournament sizes from 2 to 8. Error bars indicate standard error. With a tournament size of 2 (and to a lesser extent 3), GP performance decreases on the MAX problem.

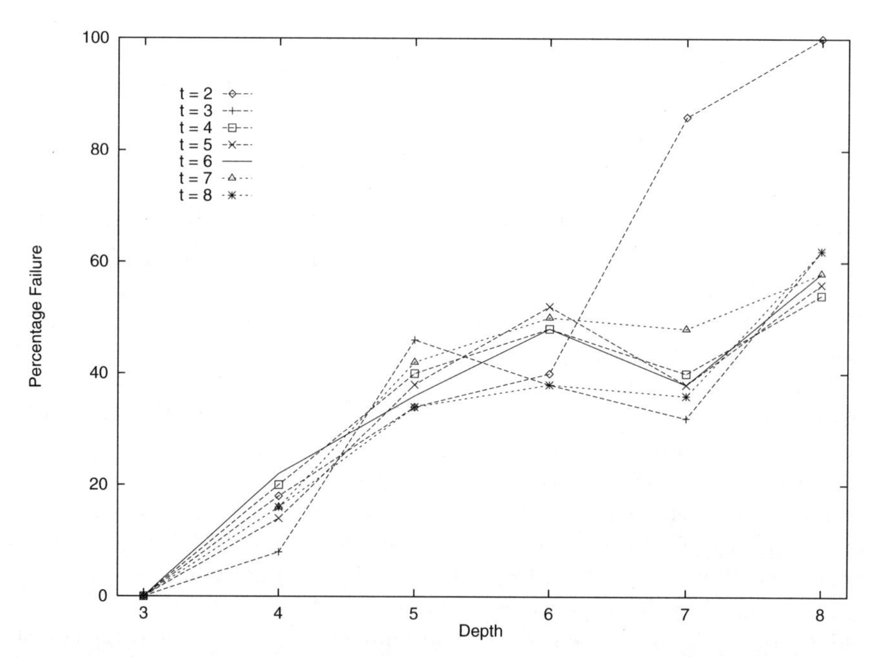

Fig. 10.10. Percentage of runs ending in failure with tournament sizes from 2 to 8.

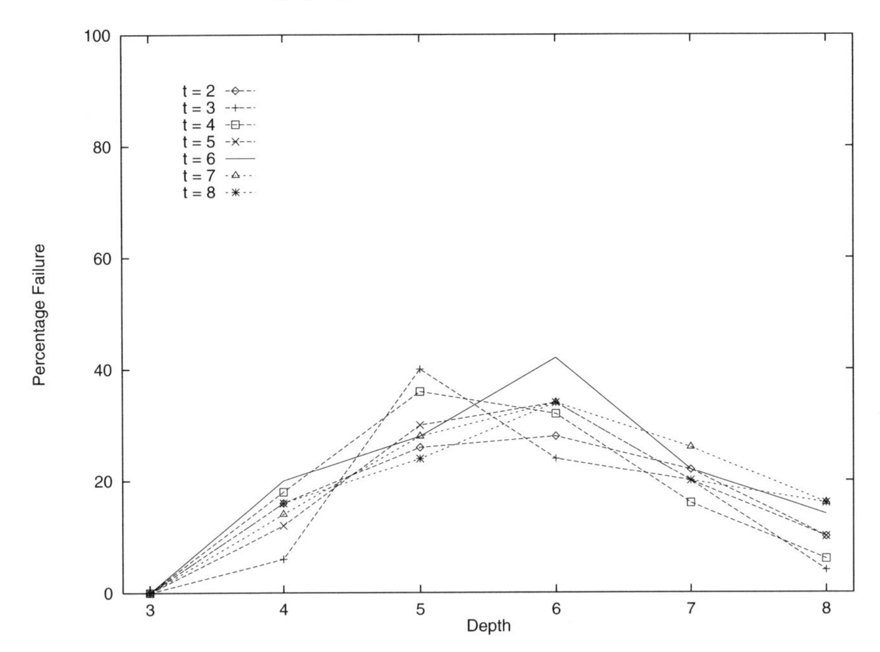

Fig. 10.11. Percentage of runs ending in failure with no × nodes in a higher level. Note there is no appreciable difference between tournament sizes from 2 to 8.

10.6 Applying Price's Covariance and Selection Theorem

Recall (cf. Section 3.1) that Price's covariance and selection theorem [Price, 1970] from population genetics relates the change in frequency of a gene in a population from one generation to the next, to the covariance of the gene's frequency in the original population with the number of offspring produced by individuals in that population.

Since in the MAX problem we use tournament selection, we can adapt Approximation (3.5) (page 31). As the size of the population is fixed, and by considering only the relationship between the first parent and its offspring, we can set $\overline{z} = 1$. So, as discussed in Section 3.1.4, we expect the change in frequency of a gene to be given by covariance$(q, T(r_i/M)^{T-1})$, as long as crossover is random.

Figure 10.12 shows good agreement between theory and measurement for + nodes near the root, but at the maximum allowed depth for + nodes (cf. Figure 10.13) the crossover depth restriction acts to depress the change in frequency. After three generations the number of + nodes no longer increases dramatically and, instead clusters near zero even though the covariance is positive. Similar effects are seen with × nodes and the opposite with 0.5 nodes, where the change in frequency tends to be larger than predicted later in the run when most trees have reached near the maximum size.

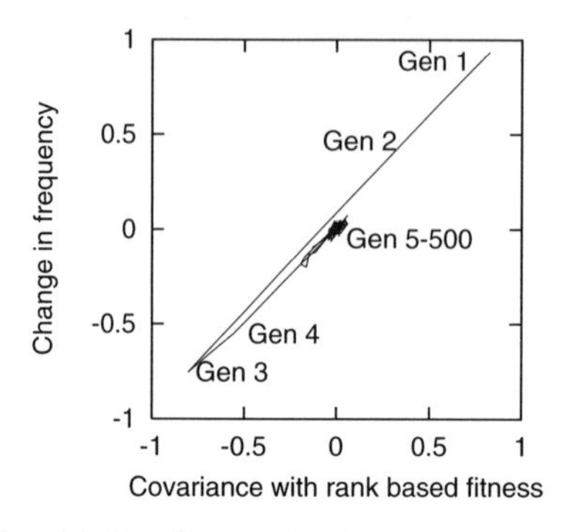

Fig. 10.12. Change in frequency vs. covariance for + at depth 2 in the first $D = 8$ run

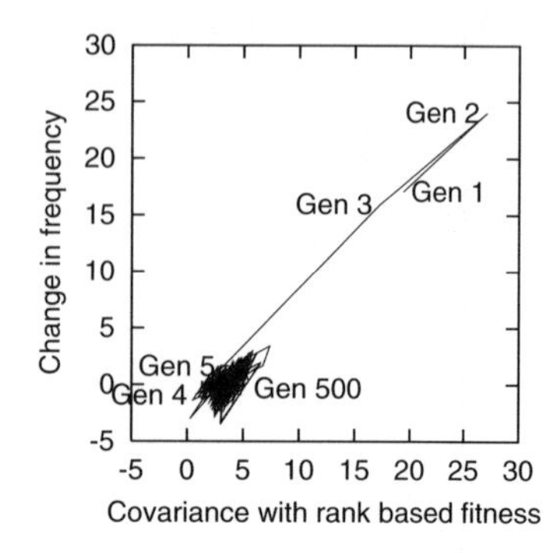

Fig. 10.13. Change in frequency vs. covariance for + at the maximum depth in the first $D = 8$ run.

Figure 10.14 considers three runs, the first one (which failed but does find a solution if run for long enough), a successful run and one that can never find the optimal solution. The three are principally separated by the number of $\times$ nodes at the second level of the tree (i.e. $d = 1$). In the successful run, this rises rapidly so that by generation four there are on average nearly 2.0 such nodes per individual in the population. This remains true throughout the rest of the run.

In the first run, the number of $\times$ nodes at level 1 rises in generations 1 and 2, as it does in the successful run but then its covariance with fitness drops to near zero and then its frequency converges towards 1.0 where it remains until the end of the run. The unsuccessful run starts like the other two runs, but then in generation 3 the covariance becomes negative and frequency falls to near zero in the next generation. The population eventually converges to zero; i.e. there are no $\times$ nodes at level 1 anywhere in the population and (as explained above) it cannot escape from this trap. That is, the eventual outcome of these three runs 500 generations later can be predicted from the covariance of gene frequency with (rank-based) fitness in generations 2 and 3.

Figure 10.15 considers the fate of $\times$ nodes in successful and unsuccessful runs with $D = 5$. We see from Figure 10.15 that the two sets of runs start from approximately the same point as their initial population, but diverge radically after the first generation created by crossover. The mean covariance in successful runs is positive (and remains positive) with the frequency continuing to rise. In contrast, in unsuccessful runs the covariance remains low and the frequency of $\times$ nodes fails to rise, eventually being totally displaced by + nodes.

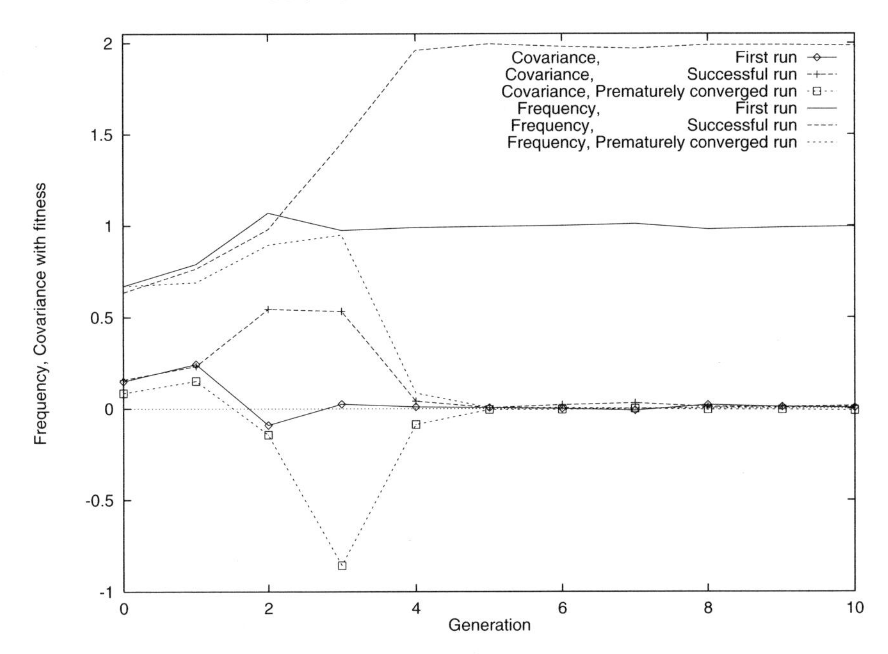

Fig. 10.14. Covariance with fitness and frequency for × in the second level of the tree in the first run, a successful run and a prematurely converged run ($D = 8$).

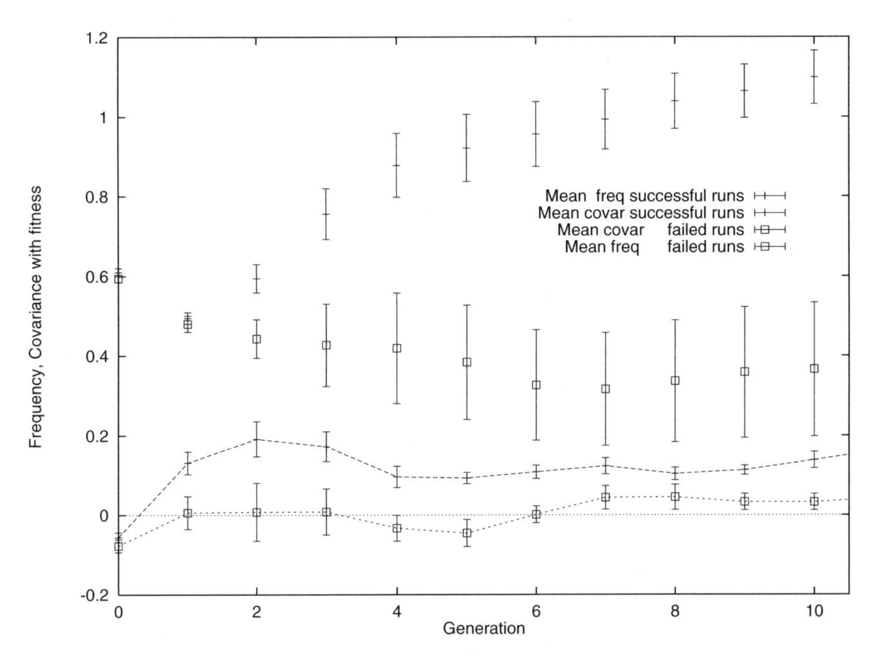

Fig. 10.15. Covariance with fitness and frequency for × in the second level of the tree. Means of successful and unsuccessful runs. Error bars indicate standard error ($D = 5$).

10.7 Conclusions

The analysis shows on the larger MAX problems that GP has two serious problems:

1. GP populations have a significant risk of losing vital components of the solutions at the very beginning of the run, and these components cannot be recovered later in the run. We have used Price's Theorem to analyse why this happens and which runs will be affected. (Similar effects are reported in [Langdon, 1998c]).
2. Where solution is possible, the later stages of GP runs are effectively performing randomised hill climbing, and so solution time grows exponentially with depth of the solution.

We have extended the analysis of the difficulties that crossover experiences presented in [Gathercole and Ross, 1996] to include a quantitative model of the later evolution of MAX problem populations. Comparison with experimental results indicates that the model gives a reasonable approximation of the rate of improvement.

While [Gathercole and Ross, 1996] suggest that MAX problem populations converge, measurements indicate that after many generations up to two-thirds of the population are unique, depending upon the maximum depth and selection pressure. A model of this based on the interaction between crossover and the depth restriction has been presented. We shall return to convergence in GP in Chapter 11. The interaction of depth (and size) limits has been discounted in GP; the MAX problem clearly shows they can have an important impact (we return to this in Section 11.4). Additionally, a model of the number of duplicate individuals in the initial populations has been presented which highlights a potential problem with the standard technique for generating the initial random population.

In Section 10.6 we used the MAX problem to demonstrate the applicability of Price's covariance and selection theorem of gene frequencies to GP populations, but noted that GP's depth restriction influences crossover and the consequent implications on the changes in gene frequencies. Gene covariance was analysed to help explain why GP populations get locked into suboptimal solutions in the first few generations.

11. GP Convergence and Bloat

While genetic programming with one-point crossover behaves like a genetic algorithm (see Sections 4.4.6–4.4.8), the question of convergence in genetic programming, with standard subtree crossover has a mixed history. Using experimental evidence, we will suggest that bloat in GP is a manifestation of convergence.

After Section 11.1 which discusses what it means for evolutionary algorithms to converge, in Section 11.2 we give a brief description of bloat and its history in GP including an example and a summary of the theories proposed to explain bloat. Most of the chapter (Section 11.3) concerns our a semi-quantitative predictions regarding bloat. Section 11.3 includes discussions of the evolution of size and shape, and presents two experiments to test the theory in extreme conditions not previously encountered in GP. This is followed by a discussion which shows that commonly used size and depth limits have an impact after surprisingly few generations. This and other GP bloat issues are discussed in Section 11.5. Prior to the chapter's conclusions (Section 11.7), Section 11.6 summarises techniques that can be used to combat bloat.

As already pointing out, on page 59, the trees produced by one point crossover cannot be deeper (or shallower) than their parents. Therefore one-point crossover by itself cannot support unconstrained search and so cannot give rise to bloat. Instead (in the absence of other size-changing genetic operations), the maximum (and minimum) size of programs in the GP population are defined by the maximum (and minimum) depths of programs in the initial population (and the maximum, and minimum, number of arguments any function can take).

11.1 Convergence

The term "convergence" is open to wide interpretation or misuse. In evolutionary algorithms we use it to mean that the population contains substantially similar individuals. However, it is often used loosely to mean that the algorithm is not progressing (sufficiently fast) or even that it has terminated, e.g. due to finding an acceptable answer (cf. premature convergence).

GP convergence is more difficult to quantify than traditional GAs. First the traditional measures of convergence, such as Hamming distance, cannot be directly applied to variable length GP. "Edit" (or Levenshtein) distance between trees offers an analogue of Hamming distance between bit strings, but it is less than ideal [Keijzer, 1996, pages 269–270]; [O'Reilly, 1997]. Another possibility is to measure and use the entropy of a population of trees. For example, Rosca defines entropy in terms of program scores, i.e. phenotypes [Rosca, 1995, Rosca and Ballard, 1996, Rosca, 1997c]. Secondly traditional measures of genotypic diversity, such as variety [Koza, 1992], say that nearly every tree in the population is unique, but overlook the fact that they are strongly related in three ways: (1) they have common ancestors, (2) the trees are similar (though seldom identical), (3) many of them have similar phenotypic behaviour. However, [Keijzer, 1996] shows that the number of distinct subtrees in the population can be counted and used to give an indication of convergence of GP populations.

Genetic programming populations do sometimes converge like fixed length GAs when they find a small tree with fitness above that of its neighbours. With crossover (and some types of mutation) there are a small number of trees that can be created from this point in the search space. If (almost) all of them have a lower fitness, the GP population may converge in the genotype space to this one point. Hence it will also converge in the phenotype space. In this case the GP does not bloat. (Section 11.3.3 contains an example of this). However convergence in traditional GP and traditional GA are normally different.

Figures 11.1 and 11.2 highlight the similarities and differences between convergence seen in fixed representation genetic algorithms and genetic programming with genetic operators (crossover or mutation) which allow the representation to evolve. (These represent typical situations, but things are not always this simple).

In GAs there is a one-to-one mapping between genotype and phenotype, and so in Figure 11.1 the phenotype and genotype rectangles and ellipses (at each generation) are the same size and shape. To stress the more complex mapping in GP, in Figure 11.2 the genotype box has a different size and shape to that of the phenotype.

Figures 11.1 and 11.2 have three elements:

1. In evolutionary algorithms, such as genetic algorithms and genetic programming, selection acts on phenotypes.
   Individuals in the population with the better (fitter) phenotypes are selected to have more children than the less fit. By itself, fitness selection reduces the spread of phenotypes in the population. The spread in the population is unequally reduced, so that individuals are more tightly clustered about the better phenotypes discovered so far. In Figures 11.1 and 11.2 this is shown by the reduction in size and movement of the ellipses between the top left and second left frames.

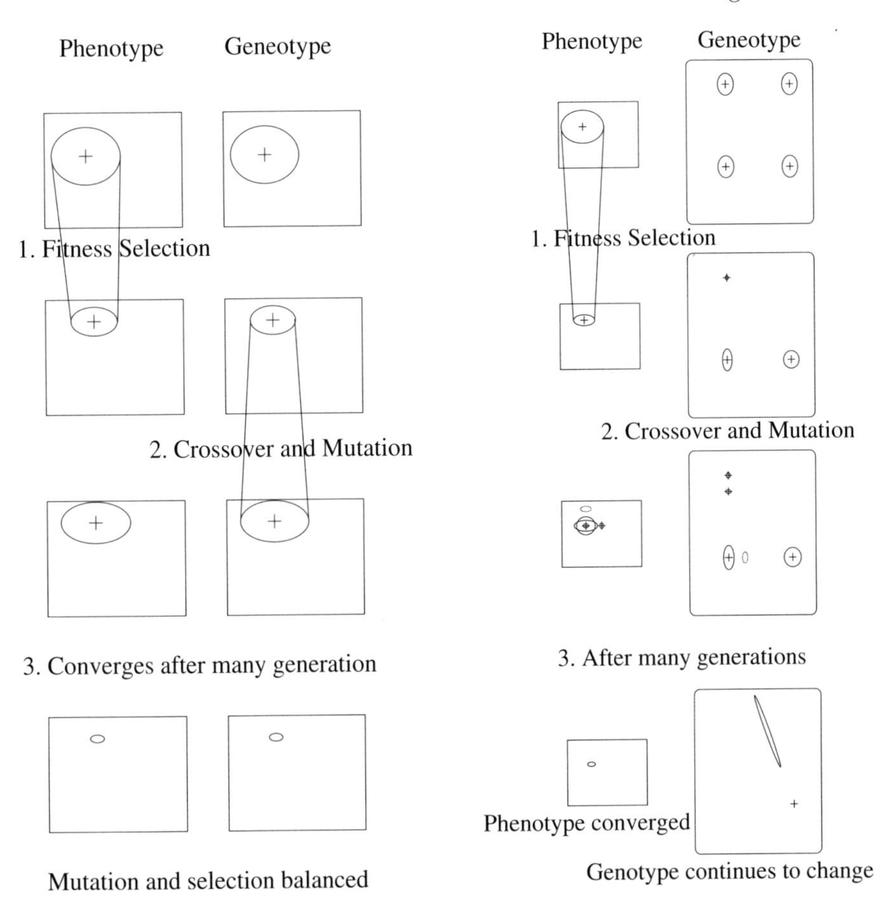

The search space and location of the population within it. at various times. (going down the page). 1. and 2. early generations, 3. after many generations. The search space is shown as a rectangle, while the location of the population is given by ellipses. The centre of each ellipse is often highlighted by a small cross. At each time step the population is viewed both as a population of phenotypes and as a population of genotypes.

Fig. 11.1. Convergence in Genetic Algorithms.

Fig. 11.2. Convergence in Genetic Programming.

In the case of fixed representation GAs, there is typically a one-to-one mapping between phenotype and genotype. Thus selection produces a corresponding reduction in GA genotypes. This is shown in Figure 11.1 by an identical change to the genotype (right-hand side) ellipse.
Early in the evolutionary process, the complex one-to-n phenotype-to-genotype mapping used in GP means that diverse parts of the (genotype) search space, which map to similar phenotypes, may be occupied. Figure 11.2 1. shows the effect of selection on early GP generations. As with GAs, selection produces a tighter distribution of genotypes. In finite populations, we may expect that, due to the stochastic effects associated with strong fitness selection, the number of distinct genotypes with the same phenotype will fall (cf. genetic drift). So, in addition to the tightening of the distribution about the best phenotypes, the GP genotypes tend to rapidly concentrate upon a few clusters which map to the better phenotypes. This is depicted in Figure 11.2 (right-hand side), by removal of some ellipses and uneven movement and reduction of others.

2. Crossover and mutation create new genotypes which are different from their parents; i.e. they spread the genotypes.
   In genetic algorithms the fixed mapping means this leads to a corresponding spread of the phenotypes. This is shown in the right-hand side of Figure 11.1 by an increase in the ellipse size, about the same centre (and identical change to the phenotype ellipse in Figure 11.1, left-hand side).
   In genetic programming, particularly in early generations, the same thing happens to some extent. In fact, initially subtree crossover may yield more diverse genotypes which in turn correspond to diverse phenotypes. In the right-hand side of Figure 11.2 we attempt to show the creation of new genotypes by the introduction of new ellipses lying between the originals. These are shown as being smaller, because subtree crossover tends to explore locally and so produces few very different genotypes. In contrast, the original ellipse are slightly changed in size and shape.
   The left hand side of Figure 11.2 2. depicts the corresponding change to the phenotype. First note the genotype-phenotype mapping is fixed so the slightly changed original ellipses still map back onto almost the same phenotype as before (shown as overlapping ellipses of various sizes). While radically new genotypes are shown by new (small) phenotype ellipses.
3. After many generations in bit string genetic algorithms, selection and drift cause both phenotypes and genotypes to become concentrated. As the genotypes are similar, crossover causes little disruption and a dynamic equilibrium is established between mutation and selection. At equilibrium, the population spread caused by mutations is balanced by selection. The concentrated phenotype and genotype are depicted in Figure 11.1 3. by identical small ellipses.

Typically in genetic programming something similar happens to the phenotype, but the genotype behaviour is very different. Over a number of generations the GP genotypes concentrate upon just one cluster that maps to the best phenotype. There are two reasons for this. First, the random nature of fitness selection in a finite population, cf. genetic drift. Secondly, some genotypes find it easier to resist the effects of crossover and mutation and "breed true"; i.e. their offspring have the same phenotype. Where this phenotype has the highest fitness, these genotypes quickly dominate the others. Therefore the population convergences to contain just the descendents of one phenotype-genotype mapping. So while the GP representation allows many many mappings between genotypes and phenotypes, the GP population rapidly converges to contain only a tiny fraction of these, and these are closely related.
Figure 11.2 3. shows the phenotypes tightly clustered about the best solution found so far. (As with GAs, a dynamic equilibrium is established between the disruptive effects of crossover and mutation and the concentrating effect of selection. However, over time the genetic operators tend to get less disruptive and so the phenotype cluster tends to contract). However, the genotype cluster does not stabilise but continues to evolve from a single point. The population's ancestor, i.e. the individual program where most of its genetic material came from, is shown with a + in Figure 11.2, bottom right. Since each fit child's genotype tends to be bigger than its parents, there is a progressive increase in size, which we know as bloat.

Given that genetic programming populations converge in this way we are tempted to suggest that bloat is intimately connected with GP convergence. Indeed, perhaps bloat is one of the mechanisms that allows convergence, for without bloat, the population consists of trees that are too small for standard crossover to be non-destructive, so that instead it continues to spread the population.

11.2 Bloat

The rapid growth of programs produced by genetic programming is a well-documented phenomenon [Koza, 1992, Blickle and Thiele, 1994, Nordin and Banzhaf, 1995, McPhee and Miller, 1995, Soule *et al.*, 1996, Greeff and Aldrich, 1997, Soule and Foster, 1998a, Langdon *et al.*, 1999]. Indeed, in the absence of countermeasures, bloat almost always happens. More recent empirical work includes [Rosca, 1997b, Smith and Harries, 1998, Iba, 1999, Kennedy and Giraud-Carrier, 1999, Luke, 2000a, Luke, 2000b], while [Langdon and Banzhaf, 2000, Poli, 2001b, McPhee and Poli, 2001] are more analytic. This growth, often referred to as "code bloat", need not be correlated with increases in the fitness of the evolving programs, and consists primarily of code that does not change the semantics of the evolving program. The rate of growth appears to vary depending upon the particular genetic programming paradigm being used, but exponential rates of growth have been claimed [Nordin and Banzhaf, 1995].

Code bloat occurs in both tree based and linear genomes [Nordin, 1997, Nordin and Banzhaf, 1995, Nordin *et al.*, 1997] and with automatically defined functions [Langdon, 1995]. Recent research suggests that code bloat will occur in most fitness based search techniques that allow variable length solutions [Langdon, 1998b, Langdon and Poli, 1997b].

Clearly, a large rate of growth precludes the extended use of GP or any other search technique that suffers from code bloat. Even linear growth seriously hampers an extended search. This alone is reason to be concerned about code growth. However, the rapid increase in solution size can also decrease the likelihood of finding improved solutions. Since no clear benefits offset these detrimental effects, practical solutions to the code bloat phenomenon are necessary to make GP and related search techniques feasible for real-world applications.

Many techniques exist for limiting code bloat (see Section 11.6 for a summary). They tend to be highly effective at preventing programs growing bigger or deeper, but their effect on GP's ability to successfully evolve useful programs is less clear. Without definitive knowledge regarding the causes of code bloat, any solution is likely to have serious shortcomings or undesirable side effects (such as we saw in Chapter 10). A robust solution to code bloat should follow from, not precede, knowledge of what actually causes the phenomenon in the first place.

11.2.1 Examples of Bloat

With ramped half-and-half initialisation, there is often a size decrease in the first few generations. However, typically after about ten generations, bloat starts. The increase in average size is accompanied by corresponding increases in the size of the largest program and the standard deviation. (Figures 11.3 and 11.4 refer to GP runs on the sextic polynomial regression problem [Koza, 1992]). Note that the average curve conceals wide variation between runs. The standard deviation indicates there is also wide variation within each population.

11.2.2 Convergence of Phenotype

Figure 11.5[1] shows the evolution of the behaviour of the best of generation individual in the first run. This shows the typical behaviour that the best of the initial random population is a constant. After a few generations, GP

[1] `http://www.cs.ucl.ac.uk/staff/W.Langdon/seminars/aigp3/` contains an animation of Figure 11.5.

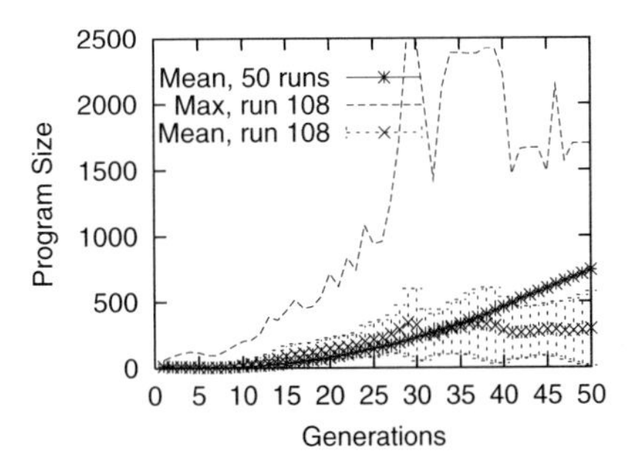

Fig. 11.3. Evolution of program size in GP sextic polynomial symbolic regression runs. On average, size increases steadily but the increase is more erratic in individual runs.

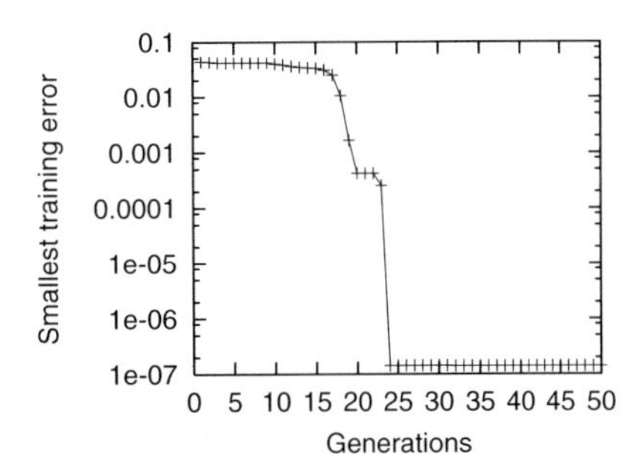

Fig. 11.4. Evolution of the best program in the population's average error (run 108). Note that the rate of finding improvements in performance falls with time.

typically finds more complex behaviours which better match the fitness test cases. Later, more complex behaviours often "misbehave" between points where the fitness test cases test the program's behaviour. In fact, the behaviour of the best of generation individual (including its misbehaviour) is remarkably stable. Note that, this is the behaviour of single individuals and not an average over the whole population. We might expect more stability from an average. This stability stresses that GP is an evolutionary process, making progressive improvements on what it has already been learnt.

11.2.3 Theories of Bloat

There are a number of different theories of why bloat happens. Since these have been extensively discussed in the literature (for example [Langdon *et al.*, 1999]), we shall simply summarise the main ones here.

The increase in program size without a corresponding change in performance suggests the extra code has no purpose. The extra code may have either no effect when it is executed or may never be executed. This code is called variously introns [Angeline, 1994], junk code, fluff, ineffective or inviable code ([Langdon *et al.*, 1999] explains the significance and importance of the last two).

In standard GP bloat arises from the interaction of genetic operators and fitness selection; i.e. fitness pressure is required for bloat (cf. Section 11.2.4). GP crossover by itself does not change the average program size. Bloat can arise with other types of genetic operator such as mutation, and in other types of stochastic search, such as simulated annealing and hill climbing [Langdon, 1998b]:

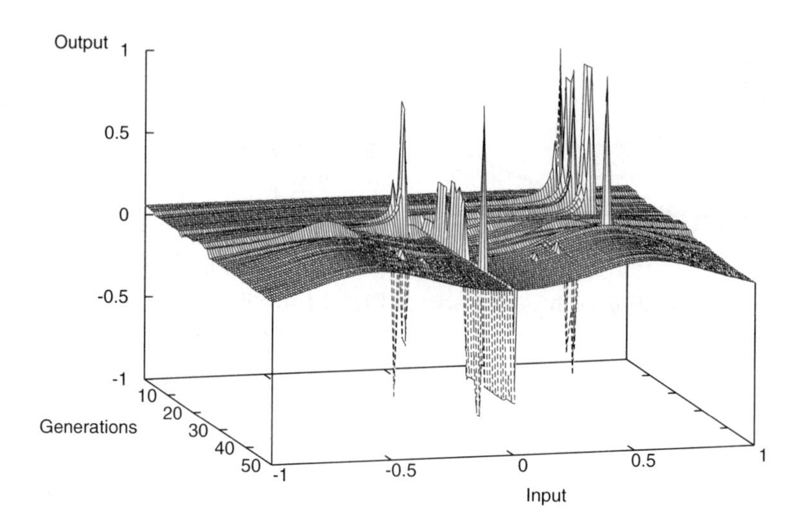

Fig. 11.5. Evolution of phenotype. Value returned by the "best" program in the population. (Showing the first of 50 GP runs of the sextic polynomial problem).

1. The oldest theory (perhaps first suggested by Singleton [Tackett, 1994]) is that bloat is evolved by GP, as junk code increases, to protect useful code from the effects of crossover. Fit individuals with more junk code are less likely to be disrupted by crossover, so their children are more likely to be as fit as they are. Thus more of their children survive to themselves have children. Since the protective advantage of being larger is both inherited and increases with size there is a continuous tendency to increase in size [McPhee and Miller, 1995, Nordin and Banzhaf, 1995, Blickle and Thiele, 1994, Banzhaf *et al.*, 1998a].
2. Terry Soule [Soule and Foster, 1998b, Soule, 1998, Soule and Foster, 1998a] has suggested that there are multiple causes of bloat, one of which is "removal bias" in crossover (and other genetic operators [Langdon *et al.*, 1999]). Briefly, this suggests junk code lies towards the tips of program trees. Thus a crossover that removes a small subtree is likely to have less effect than one that removes a large subtree. However, there is no corresponding fitness bias on the size of the subtree added. Therefore, children produced by removing a small amount of their first parents are likely to have better fitness. Since the inserted subtree is randomly chosen, it will be on average bigger than that removed, and hence the child program will also be bigger than its parents, on average. Correlating the size of the removed and inserted subtrees, as is done in size fair crossover [Langdon, 2000], can be effective against "removal bias" as a cause of bloat.

3. Finally, bloat has been described as a random spread along neutral network; i.e. a network of programs of the same fitness connected by genetic operations [Langdon and Poli, 1997b, Langdon, 1998b, Langdon and Poli, 1998a, Langdon and Poli, 1998b, Langdon *et al.*, 1999]. As Chapter 7 has shown, above a certain size threshold there are exponentially more longer programs of the same fitness as the current best program than there are of the same length (or shorter). If the genetic operators were to sample evenly around the current best program, the children they produce of the same fitness, would tend to be longer. Of course, the sampling in an infinite space cannot be uniform, and we are beginning to obtain more precise information about the size biases in our genetic operators [McPhee and Poli, 2001]. Common genetic operators do tend to sample the nearby search space.

11.2.4 Fitness Variation is Needed for Bloat

[Tackett, 1994] pointed to the importance of selection pressure on bloat. Price's covariance and selection theorem allows us to be more quantitative. In GP program size is an inherited property, so we can use Price's theorem (see Section 3.1) to predict the change in average program size from one generation to the next. (Since a program's size is the sum of its genes, i.e. a linear combination, and Price's theorem applies to each gene in the population, it also applies to their sum. That is, Price's theorem applies to the total number of genes in the population and hence to the average program size). Using Equation 3.1 (page 28), the expected change in mean program size between the current and next generation, Δsize, is given by the covariance of size variation with fitness (number of children, z) variation in the current population:

$$\Delta\text{size} = \frac{\text{cov}(z, \text{size})}{\bar{z}}$$

Equation (3.1) holds if genetic operations are random with respect to the gene. This is generally true in GP [Langdon, 1998c] but will fail when the population runs into size or depth limits, since then crossover etc. must take into account program size and so is no longer completely random [Langdon and Poli, 1997a].

As mentioned in Section 10.5, when using tournament selection in large populations, Equation (3.5) (page 31) can be used to give:

$$\Delta\text{size} = \frac{T}{\bar{z}}\text{cov}((r/M)^{T-1}, \text{size})$$

where T is the number of individuals in the tournament and r is the program's position or ranking in the population (of size M) [Langdon and Poli, 1998a]. Figure 11.6 shows that the covariance between programs' rank to the power $(T-1)$ and size does indeed predict the change in mean program size from one generation to the next. (There is a wide variation between runs. While Figure 11.6 appears to show the measured change in size diverges from the prediction in later generations, in fact it remains within about one standard error).

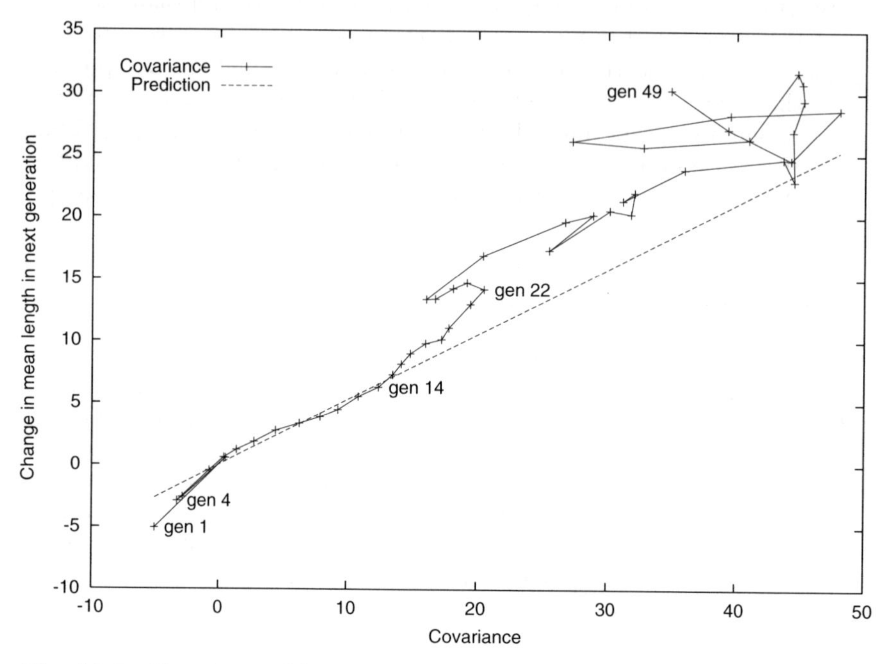

Fig. 11.6. Covariance of fitness and program size gives change in mean size from one generation to the next. Positive increase (e.g. bloat) requires positive covariance, i.e. fitness variation in current generation. GP Sextic polynomial, mean of 50 runs.

11.3 Subquadratic Bloat

Most of the remainder of this chapter discusses a theoretical argument and describes experimental testing of it: that program size increases on average at less than $O(\text{time}^2)$. Although rapid, this is clearly very much less than the exponential rate often assumed. While the mathematical analysis assumes binary trees (i.e. all functions take exactly two arguments), in practice we would expect the power-law predictions to approximately hold where the population contains functions with a mixture of numbers of arguments.

First Section 11.3.1 says why we predict that on average subtree crossover would tend to cause programs to grow with size $< O(t^2)$, but will approach a square power law, $O(t^2)$, as the programs get bigger. Up to generation 50,

[Langdon, 2000, Table 5] reports good agreement on average. Section 11.3.2 describes two experiments which look for the proposed quadratic limit. Their results are given in Section 11.3.3. Section 11.3.4 shows measurements of GP specific convergence, which explains why, in one case, the quadratic limit is not reached. At up to a million elements, these may be amongst the largest programs deliberately evolved so far.

11.3.1 Evolution of Program Shapes

In addition to changing size, programs with a tree-structured genome can also change shape, becoming bushier or sparser as they evolve. The study of this aspect of GP evolution was initiated by [Soule and Foster, 1997]. In this section we consider the size and depth of the trees. While the density of trees affects the size of changes made by subtree crossover and many mutation operators [Rosca, 1997a, Soule and Foster, 1997], experiments show that bloating populations evolve towards shapes that are of intermediate density. As Figure 11.7 shows to a first approximation, this can be explained as simple random drift towards the most popular program shapes [Langdon *et al.*, 1999]. Figure 11.8 shows that the whole population evolves and spreads out like a random cloud, rather than following the neat line indicating the average.

Although programs clearly evolve away from their initial shape towards the ridge line, they do not appear to converge to it. We can advance two competing hypothesis for this:

1. If we plot the number of programs of each size and depth on top of Figure 11.7 and then calculate its gradient, we discover between the 5% line and the peak line, the gradient is almost vertical, with only a small component towards the peak line. That is, if the population were always to move in the local direction where the largest number of programs are, it would remain to one side of the peak line.
2. Standard crossover has a bias towards choosing crossover points that are not leaves. If this bias is removed and crossover points are chosen uniformly, on average populations evolve closer to the ridge [Langdon *et al.*, 1999, Figure 8.9].

While we have not yet completed a mathematical analysis of the rate of tree growth with crossover between random trees, such an analysis may be tractable. Figure 11.9 gives a strong indication that the average depth of binary trees in a population grows linearly at about one level per generation. Clearly if we know tree depth and the relationship between tree depth and tree size, we can infer tree size. The relationships between size and depth for random binary trees were described in Section 7.6. Using these with the linear depth assumption and evolution towards the ridge line, we can predict growth in size of $O(\text{generations}^{1.6})$ for programs of a reasonable size rising to a limit $O(\text{generations}^2)$ for programs of more than 32,000 nodes. Note that

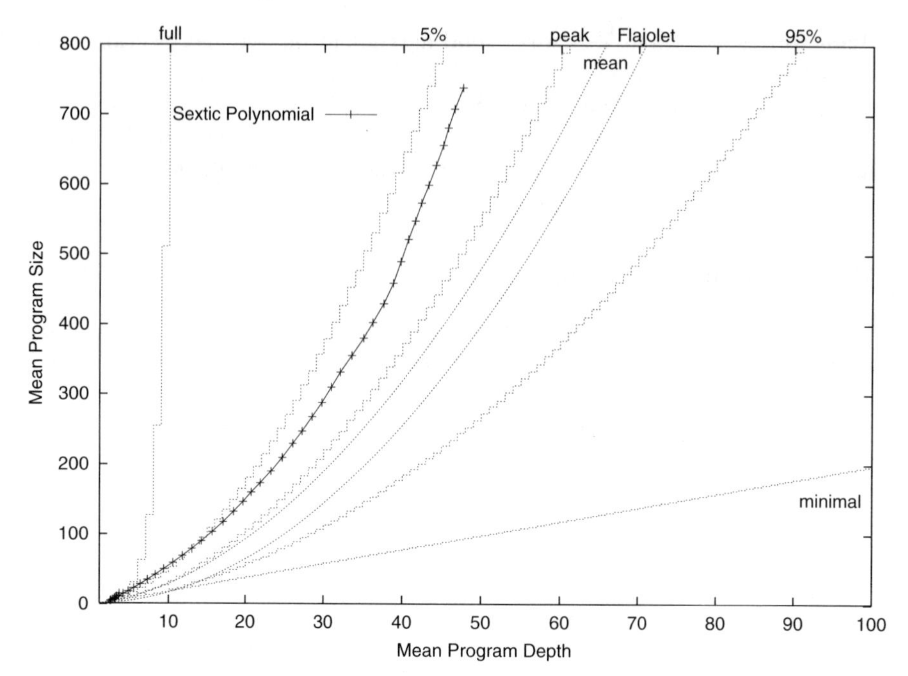

Fig. 11.7. Evolution of program tree shapes for the sextic polynomial problem. Each point represents the mean size and mean depth of trees in a single generation (averaged over 50 runs). For comparison, full and minimal trees are plotted with dotted lines, as are the most popular shapes (peak) and the boundary of the region containing 5% and 95% of trees of a given size. The Flajolet line refers to the parabolic approximation to the mean depth $= 2\sqrt{\pi \lceil \text{size}/2 \rceil}$ [Flajolet and Oldyzko, 1982]. The initial population is produced by ramped half-and-half, and so contains predominantly "bushy" trees. However, with time the population evolves towards programs with a random shape (i.e. near the peak).

this indicates quadratic or subquadratic growth rather than an exponential growth. Also, the actual program sizes will depend upon their depth when the linear growth begins, and so will be problem dependent.

This analysis and these data only looked at tree-based genomes. It is clear that shape considerations will not apply to linear genomes. However, it is possible that the linear distribution of viable and inviable nodes are subject to some similar considerations. For example, a very even distribution of viable nodes in a linear genome may make it more likely that at least a few viable nodes will be affected by most operations. In which case an even distribution of viable nodes is unlikely to be favoured evolutionarily. More complex genomes, such as graph structures, do have shapes and it seems likely that they are also subject to the evolutionary pressures discussed here.

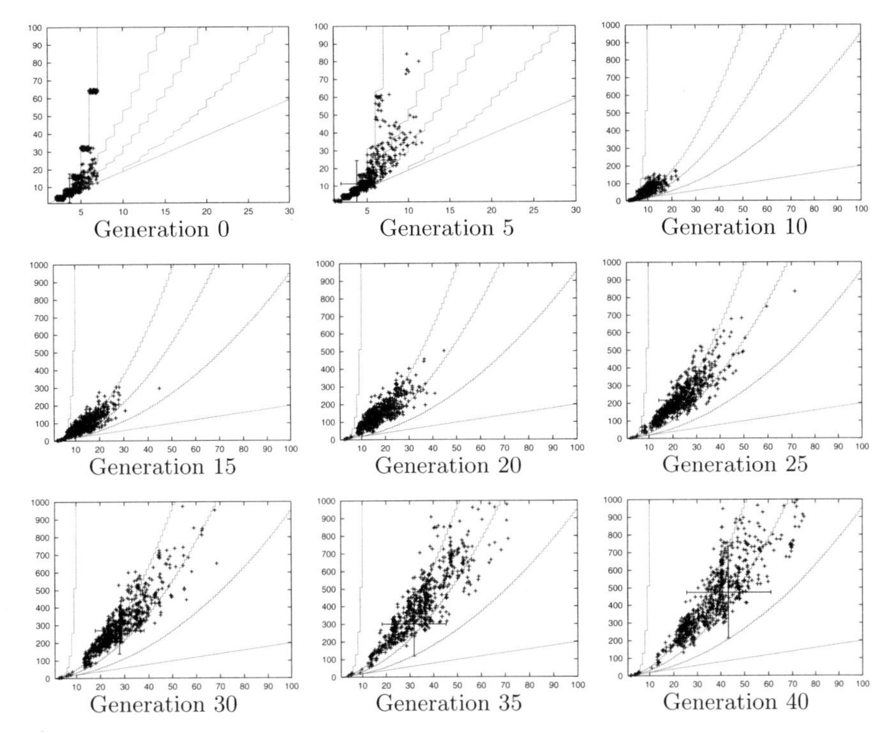

Fig. 11.8. Depth (horizontal) and size (vertical) of individual programs in a quintic GP population every five generations. (Note rescaling of x- and y-axes). The crosswires indicate the location of the population mean and the standard deviation of depth and size.

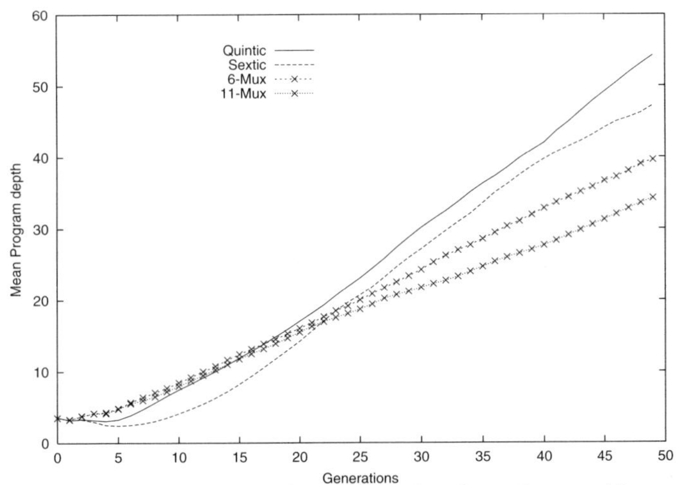

Fig. 11.9. Near-linear growth in depth with time in various problems. Means of 50 runs shown on each problem.

11.3.2 Experiments

As we intend to run artificial evolution for hundreds of generations on rapidly bloating populations, here we restrict ourselves to two problems: one Boolean (6-multiplexer [Koza, 1992, page 187]) and one continuous (symbolic regression of the quartic polynomial [Koza, 1992]). Note that for speed, the quartic problem is simplified and only uses ten test cases. As the number of programs of each size and shape is known for binary trees, we use only binary functions. Accordingly, we introduce a new variation on the Boolean 6-multiplexer problem by replacing Koza's function set with his usual four binary Boolean operators. Using a 64-bit C++ compiler, we can evaluate all the 6-multiplexer fitness cases in parallel [Poli and Langdon, 1999].

Apart from the binary function set, quartic test-set size, the absence of size or depth restrictions, and the use of tournament selection, our GP runs are essentially the same as those in [Koza, 1992]. The parameters are summarised in Tables 11.1 and 11.2.

Table 11.1. GP parameters for the quartic symbolic regression problem

Objective	Find a program that produces the given value of the quartic polynomial $x^2(x+1)(x-1) = x^4 - x^2$ as its output when given the value of the one independent variable, x, as input
Terminal set	x and 250 floating point constants chosen at random from 2001 numbers between -1.000 and $+1.000$
Functions set	$+ - \times$ % (protected division)
Fitness cases	10 random values of x from the range $-1, \ldots, 1$
Fitness	The mean, over the 10 fitness cases, of the absolute value of the difference between the value returned by the program and $x^4 - x^2$
Hits	The number of fitness cases (between 0 and 10) for which the error is less than 0.01
Selection	Tournament group size of 7, non-elitist, generational
Wrapper	None
Pop Size	50
Max program	10^6 program nodes
Initial pop	Created using "ramped half-and-half" with depths between 8 and 5 (no uniqueness requirement)
Parameters	90% one child crossover, no mutation. 90% of crossover points selected at functions; remaining 10% selected uniformly between all nodes.
Termination	Maximum number of generations 600 or maximum size limit exceeded

Table 11.2. GP Parameters for multiplexor problem (as Table 11.1 unless stated)

Objective	Find a Boolean function whose output is the same as the Boolean 6-multiplexor function
Terminal set	D0 D1 D2 D3 A0 A1
Functions set	AND OR NAND NOR
Fitness cases	All the 2^6 combinations of the six Boolean arguments
Fitness	number of correct answers
Pop size:	500
Max program	10^6 program nodes
Initial pop	Ramped half-and-half maximum depth between 2 and 6

11.3.3 Results

Quartic Symbolic Regression. In nine of ten independent runs, the population bloats. (In run 102 at generation 7, GP finds a high-scoring program of one function and two terminals, which it is unable to escape from and the population converges towards it. After generation 35, approximately 90% of the population are copies of this local optima. Similar trapping is also reported in [Langdon, 1998c]). All populations complete at least 400 generations. However three runs stop before 600 generations when they reach the size limit (one million).

As expected, in all runs most new generations do not find programs with a better fitness than found before. That is, changes in size and shape are due to bloat. Figure 11.10 shows that, while there is variation between the remaining runs, on average each population evolves to lie close to the ridge and moves along it, as predicted.

Figure 11.11 shows that the average population depth varies widely between runs, and in several runs the mean depth does not increase monotonically at a constant rate. However the mean of all ten runs is better behaved and increases at about 2.4 levels per generation.

Figure 11.12 shows the coefficient obtained by fitting a power law to the evolution of mean size of programs from generation 12 to later generations. Again, there is wide variation between runs, but on average the exponents start near 1.0 (generations 12–50) and steadily rise to 1.9 (between generations 12 and 400). That is, towards the predicted quadratic limiting relationship between size and time.

Binary 6-Multiplexor. In all runs, most new generations do not find better programs, and changes in size and shape are due to bloat. Figure 11.13 shows that, while there is variation between runs, on average each population evolves to lie close to the ridge and moves along it, as predicted.

Figure 11.14 shows that the average population depth varies widely between runs, as with the continuous problem, and in several runs the mean depth does not increase uniformly at a constant rate. However, the mean of all ten runs is better behaved and increases at about 0.6 levels per generation.

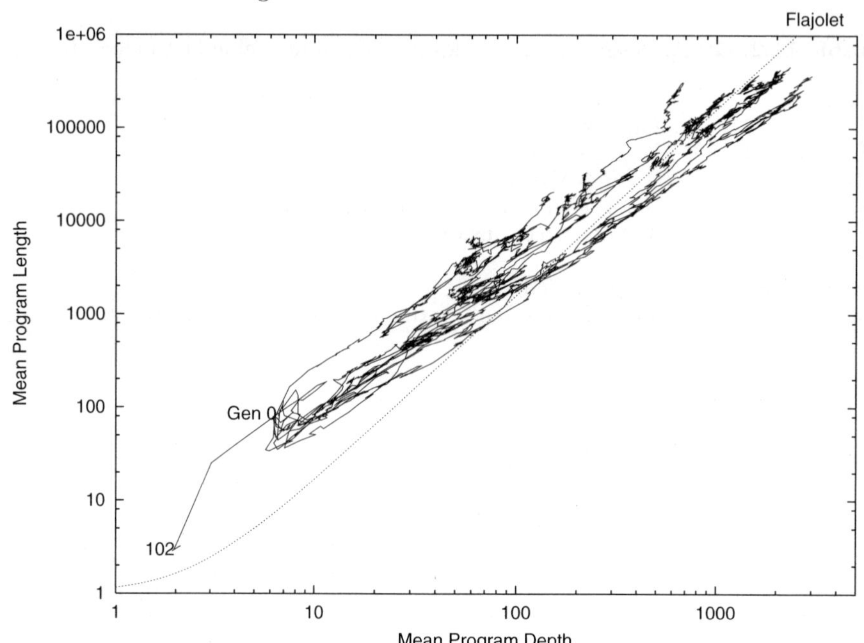

Fig. 11.10. Evolution of tree shape in ten runs of the quartic symbolic regression problem. Note log scales.

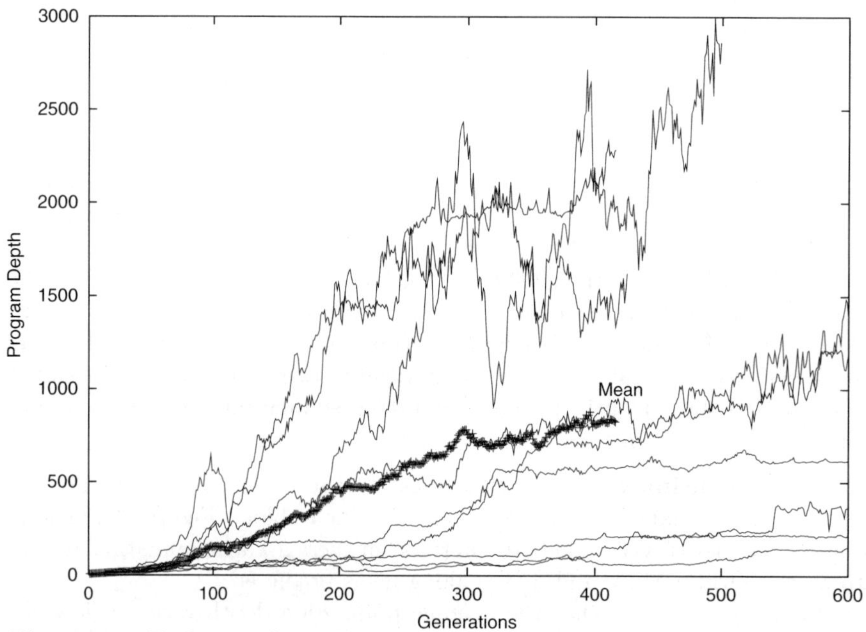

Fig. 11.11. Evolution of tree depth in ten runs of the quartic symbolic regression problem.

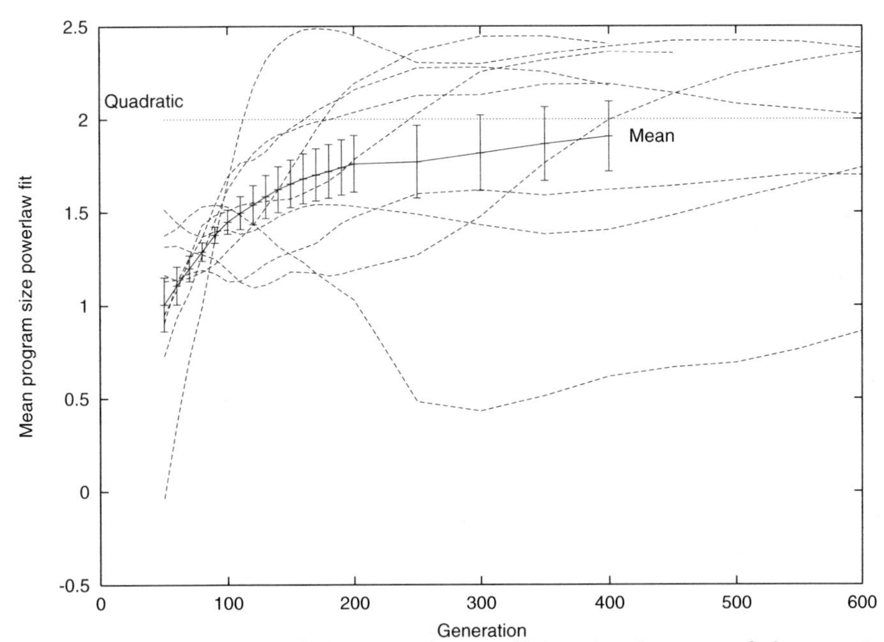

Fig. 11.12. Evolution of the power-law coefficient in nine runs of the quartic symbolic regression problem (excludes run 102). Error bars show standard error.

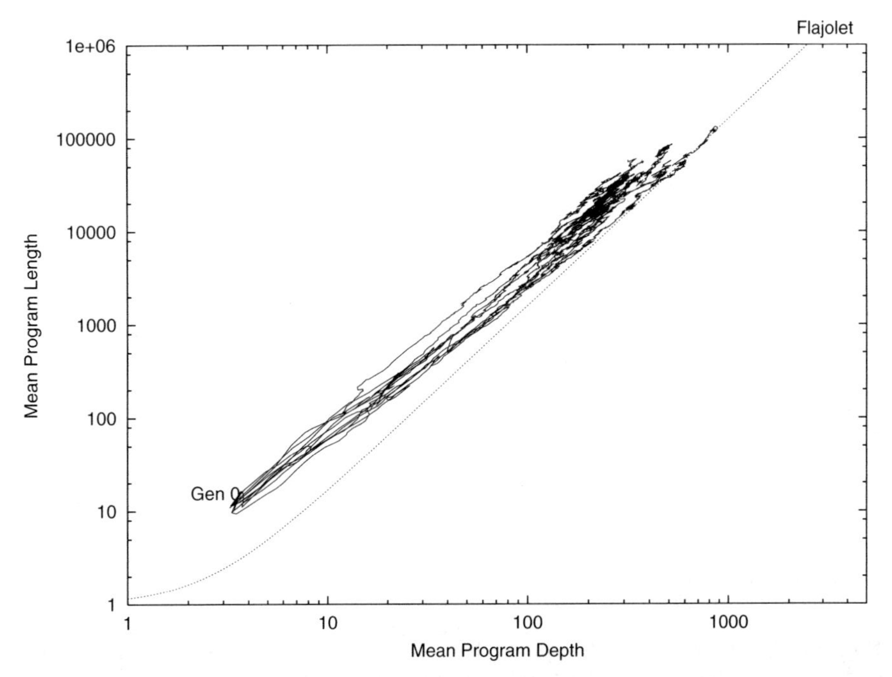

Fig. 11.13. Evolution of tree shape in ten runs of the binary 6-multiplexor problem. Note log scales.

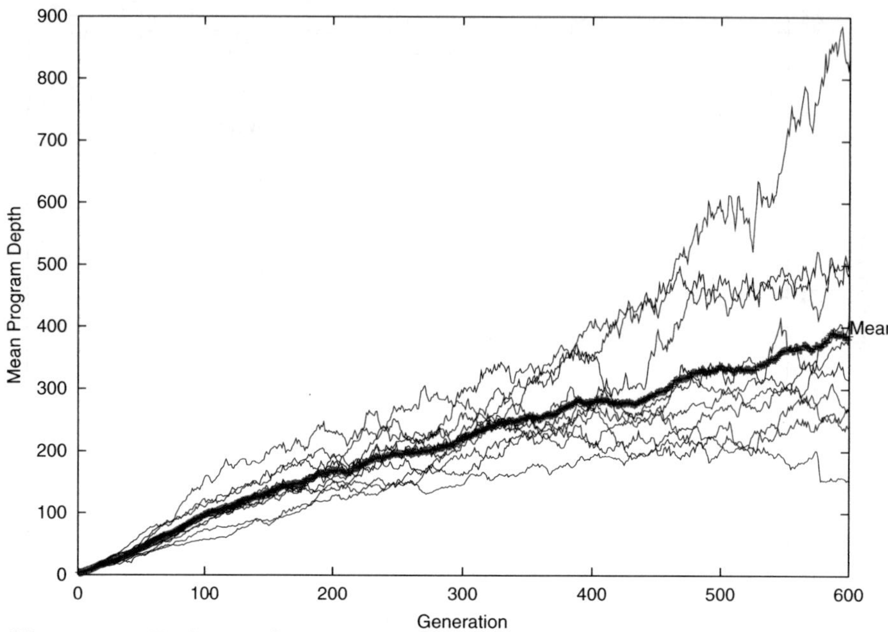

Fig. 11.14. Evolution of tree depth in ten runs of the binary 6-multiplexor problem.

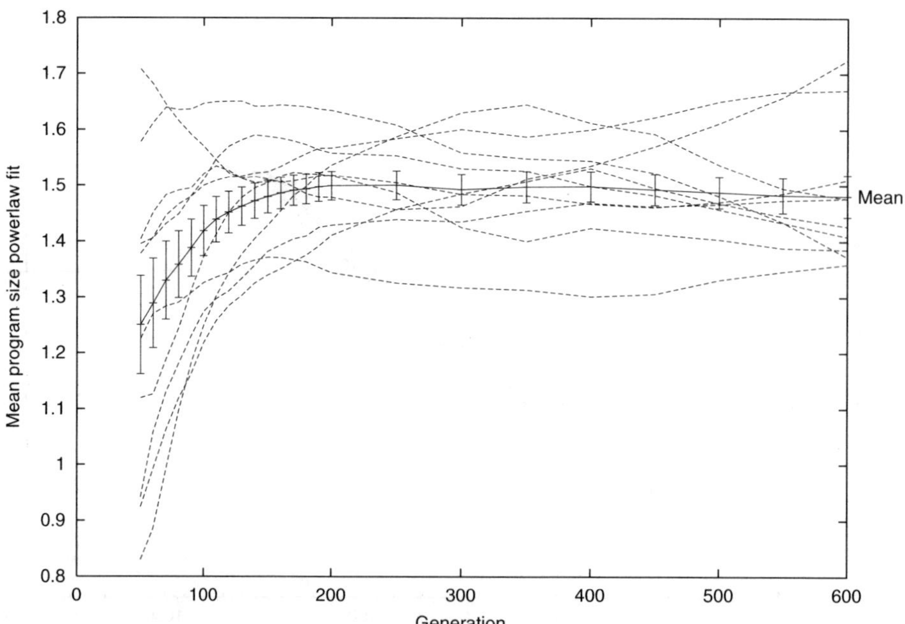

Fig. 11.15. Evolution of the power-law coefficient in ten runs of the binary 6-multiplexor problem. Error bars indicate standard error.

Figure 11.15 shows the coefficient obtained by fitting a power law to the mean size of programs. Again, there is wide variation between runs, but on average the exponents start at 1.25 (generations 12–50) and rise. E.g. between generations 12 and 100 the average has reached 1.4. By the end of the runs (generations 12–600) it reaches 1.5. This is approximately the same as for the 6-multiplexor with the traditional, multi-arity, function set [Langdon, 2000].

11.3.4 Convergence

We consider but reject two possible explanations for the failure of the 6-multiplexor runs to reach a quadratic exponent:

1. The power law coefficients for the two problems are statistically different. That is, the difference (1.5 vs. 2.0) is unlikely to be due to random variations.
2. The multiplexor programs are shorter, so the Flajolet limit does not apply. However, [Flajolet and Oldyzko, 1982, Table II] shows that the parabolic estimate is within 10% of the actual mean for programs of more than 1000 nodes. By generation 90, on average the bulk of the populations exceed 1000. That is, for at least 500 generations the bulk of the 6-multiplexor populations are reasonably close to the Flajolet limit.

Having rejected these, our proposed explanation is, in discrete problems, that crossover may cease to be disruptive when the programs become very large. (Similar inability to affect big trees is reported in [Langdon and Nordin, 2000, Section 2]). In fact, there are whole generations when every program in the population has identical fitness. Therefore, the selection pressure driving bloat falls as the populations grow in length, and we suggest that this is why the quadratic limit is not reached. Figure 11.16 shows that the fraction of parent programs selected entirely at random rises towards 100% in all ten runs.

11.4 Depth and Size Limits

In this last experimental section we present an experiment on the quintic symbolic regression problem (parameters as in [Langdon, 2000]). This shows that we can approximately predict when depth and size limits commonly used in GP will have an impact on the evolving population.

Using the average depth of the initial population (3.64) and rate of increase in depth (1.2 per generation) from [Langdon, 2000, Table 4] we can estimate how long it will take for bloat to take the population on average to the depth limit (17). Number of generations $= (17 - 3.64)/1.2 = 12$.

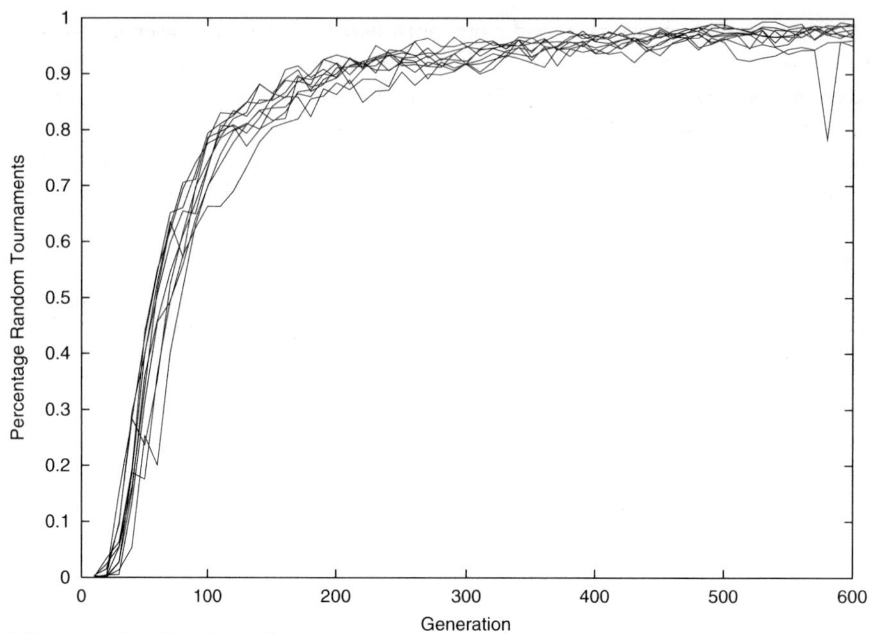

Fig. 11.16. Fraction of selection tournaments where all seven potential parents have the same fitness. For clarity we plot the mean of ten generations at a time. Ten runs of the binary 6-multiplexor problem.

Figure 11.17 plots the mean sizes and depths for: no limits, a conventional depth limit (17) and a size limit of 200. They lie almost on top of each other until generation 12, when they start to diverge. Coincidentally, at generation 12 the size-limited population also diverges from the unlimited population. Given the variability between runs, the agreement between the prediction and measurements is surprisingly good.

11.5 Discussion

The various theories of bloat (cf. Section 11.2.3) have up till now been qualitative descriptions; GP is now advanced enough to start making at least semi-quantitative predictions (cf. Section 11.3; [McPhee and Poli, 2001]). In the process of testing these we make new discoveries.

In Section 11.4 we used our theory to predict when standard GP populations run into common size and depth limits, and found the predictions to be surprisingly good. GP populations run into common size or depth limits very quickly. In fact so fast, that we anticipate that size or depth limits have an impact in many GP runs. This is in contrast to statements that they can be neglected. (E.g. [Koza, 1992, page 105] uses a depth limit of 17). Depth and size limits have contrasting effects. Depth limits tend to encourage the

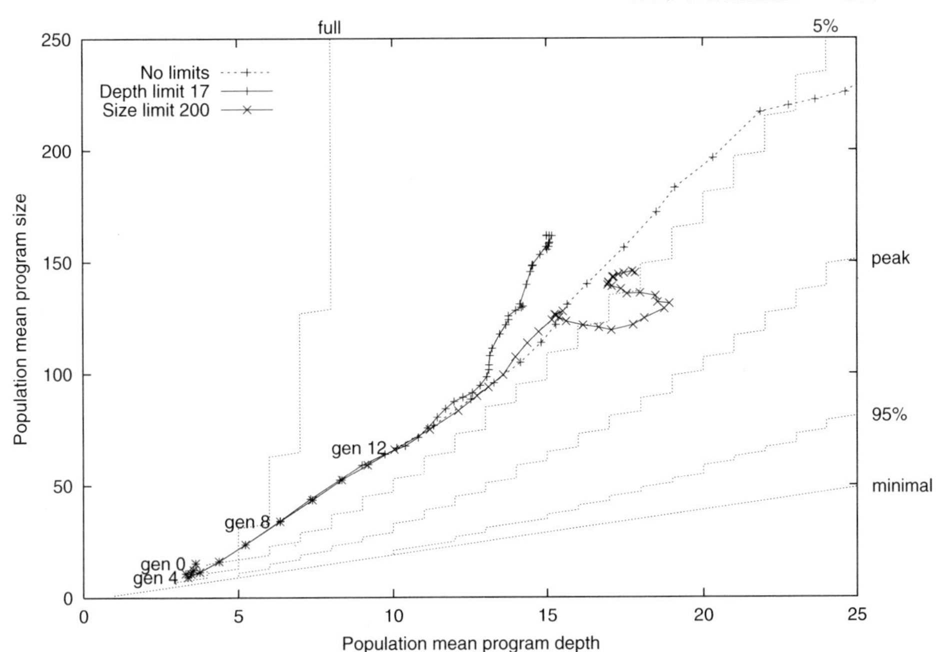

Fig. 11.17. Evolution of population size and depth in a run of the quintic regression problem. (Tick marks at every generation).

population to evolve bushier trees. This might be beneficial in some problems. For example, we can expect solutions to problems (such as the parity problems) which require all inputs to be processed to a similar extent may be far commoner in bushy trees than they are in random trees. Thus a depth limit might help GP. However, Chapter 10 and [Gathercole, 1998] show when the whole population presses against size or depth limits, they constrain subtree crossover possibly resulting in premature convergence.

There are problems (e.g. data mining) where some inputs are more important than others. Indeed, some might be omitted from the solution altogether. (In data mining this is called feature selection). In these cases a size limit might be beneficial as it may encourage the evolution as asymmetric trees.

The ridge in the depth versus size graphs divides the search space in half (Figure 11.7). Populations initially created one side of the ridge tend to evolve both towards bigger trees and towards the ridge, but they tend not to cross it (cf. Figures 11.10, 11.13 and 11.8; [Soule, 1998, Langdon *et al.*, 1999]). So GP does not explore the search space on the far side of the ridge. That is, it ignores half the search space.

Traditionally, populations are created with ramped half-and-half [Koza, 1992] and so lie above the ridge. This half of the search space may have a higher density of solutions to certain classes of problems (e.g. parity problems).

If we know or suspect that certain tree shapes have a higher chance of being solutions to a certain problem, size and depth limits offer a quick but crude way of guiding evolution. We could also initialise the population with shapes like those we expect the solutions to have. In addition, to "grow", "ramped" and "ramped half-and-half" [Koza, 1992] we could consider starting from a range of random shapes [Langdon, 2000], uniform sampling [Bohm and Geyer-Schulz, 1996, Iba, 1996] or even long thin trees [Soule, 1998]. ([Luke and Panait, 2001] finds no difference in performance when using a few initialisation techniques, but only considers three benchmarks where the solutions might have similar shapes). However, changes to our genetic operators may offer more sophisticated ways to improve GP. (Investigations of other crossover operators include [Ito *et al.*, 1999, O'Reilly, 1997]).

While the benchmarks used in Section 11.3 are deliberately simple, they show that, with tradition single-tree GP, subtree crossover is ineffective on large bloated programs. [Langdon and Nordin, 2000, Section 2] shows this can also be true of large non-bloated programs. Therefore, possibly if GP is to be used to evolve large complex programs, other representations or genetic operators may be needed. Multiple smaller trees, perhaps based on the ADF [Koza, 1994] approach might be a way forward.

Often GP analysis assumes that the population is made of full binary trees. This is obviously wrong but simple. Section 11.3.1 points out that GP populations tend to become like random trees. While apparently more complex than full binary trees, random trees do have nice mathematical properties that might make analysis of the evolution of populations of them tractable. In GP we often use function sets containing functions of more than one arity, which considerably complicates the analysis. Fortunately, it does not appear to dramatically change the properties of random trees or GP@İt may be possible to continue to approximate the behaviour of real GP populations by assuming only binary functions, or even take the apparently extreme position of assuming totally random arities. The mathematics of such random trees has also been analysed.

11.6 AntiBloat Techniques

In this section we briefly summarise techniques for counteracting bloat (see also [Zhang, 1997, Zhang, 2000]).

Size and depth limits are almost always used. Hard limits have the advantages of being easy to implement and being very effective. [Koza, 1992] introduced depth limits but, perhaps for efficiency reasons, size limits are now common, e.g. [Koza *et al.*, 1999]. Indeed, for efficiency reasons, some implementations (e.g. DGPC, GPQUICK and Discipulus) require a maximum size on each individual program to be specified.

As discussed in Section 11.5 (cf. also Section 11.4 and Chapter 10), GP populations will quickly be affected by even apparently generous limits. In general, whether limits help or hinder GP will depend both on the nature of the limit and the problem.

Size component to selection. Possibly the second most commonly used technique to combat bloat is to include some preference for smaller programs in the criterion used to select which programs will have children [Koza, 1992, Kinnear, Jr., 1993]. This goes under a number of names: parsimony pressure, Occam's[2] razor [Zhang and Mühlenbein, 1995, Zhang and Mühlenbein, 1996], Minimum Description Length [Iba *et al.*, 1994, Smith and Harries, 1998]. Common techniques involve creating a new fitness function which is a combination both of the program's score on the problem and its size. Often fitness is simply a weighted sum of the two. However, this immediately raises the question of how important the two components should be. ([Gathercole and Ross, 1996] suggests the size component be set so small that it only has an impact when comparing programs with identical scores).
Parsimony pressure creates a bias in favour of small programs. This is often what we want. (Small programs are generally regarded as more comprehensible than larger ones. It is suspected that they are less prone to overfitting, i.e. matching the training data but not generalising to new examples. Finally, they may consume less machine resources). However, as pointed out by [Rosca, 1997a] and [Soule, 1998], such a bias may be too big and result in the GP population converging towards small suboptimal programs, which the size bias makes it all but impossible for the population to escape.
In addition to including size reduction into a single objective fitness function, it may also be included as one of, possibly several, objectives which the GP population tries to match simultaneously [Langdon and Nordin, 2000, Ekart and Nemeth, 2001]. [Ryan, 1994] describes a different multi-objective approach that ties each objective with the breeding strategy.
Generally, parsimony is a fixed component of the fitness function, however it may be increased during a run. E.g. to increase the generality or the speed of an evolved solution found earlier in the GP run [Langdon, 1998c].

Genetic operators. Quite a range of genetic operators (crossover and mutation) have been proposed [Langdon, 1998c, Angeline, 1998b]. Some of these have a definite impact on program size. For example, Hoist mutation [Kinnear, Jr., 1994b] (similar to shrink mutation) creates programs from parent programs by removing code and thus they are certain to be smaller. Size fair crossover [Langdon, 2000] was designed to control bloat by balancing the size of the code fragments used in crossover. While one-point crossover automatically limits bloat since it cannot increase program depth above that of the deepest of the two parents.

---
[2] N51:17:53 W0:27:40

Poli also shows that we can reduce bloat by nullifying the evolutionary advantage of a high ineffective code ratio by using a constant mutation rate per *tree node* [Poli and Langdon, 1997b]. Programs containing a lot of ineffective code do not prosper, because the chance of damaging effective code is independent of the amount of ineffective code.

Code editing. These techniques attempt to automatically identify useless code from evolved programs and remove it [Blickle, 1996a, Hooper and Flann, 1996]. These techniques are more complex and so are used less often. Also, the rules to identify the code for removal are fixed, and there is a tendency for evolution to get the short-term advantages of bloat by creating useless code which rules cannot spot. [Brameier and Banzhaf, 2001] claims more success by considering just one class of introns in a linear genetic programming system. Edits are not written back to the population, so this is only used to speed up the GP. Code editing is often applied after a run to simplify evolved solution.

Modularity. One of the advantages claimed for automatically defined functions and other modularity techniques is their more powerful representation allows, and indeed encourages, GP to evolve smaller programs [Koza, 1994]. However the use of ADFs does not of itself always prevent code bloat [Langdon, 1995].

11.7 Conclusions

We have advanced the idea that bloat is a form of convergence in genetic programming. With populations of genotypes evolving along neutral networks towards the largest accessible part of the search space in an entropy-driven process like a random walk. In the case of binary function sets, we have mapped out the space and shown that populations evolve roughly along a near parabolic ridge at a subquadratic rate.

The average growth in program depth when using standard subtree crossover with standard random initial trees is almost linear. The rate is about one level per generation, but varies between problems. When combined with the known distribution of number of programs, this yields a prediction of subquadratic, rising to quadratic, growth in program size. There is considerable variation between runs, but on the average we observe subquadratic growth in size. In a continuous domain problem this rises apparently towards a quadratic power law limit, i.e.

$$\begin{array}{lll} \text{Discrete} & & \text{mean length} \leq O(\text{generations}^{2.0}) \\ \text{Continuous} & \lim_{g \to \infty} & \text{mean length} = O(\text{generations}^{2.0}) \end{array}$$

However in large discrete problems we observe a new type of genetic programming specific convergence: the whole population has the same fitness, even though its variety is 100%. This reduces the selection pressure on the population which, we suggest, is the reason why the growth remains subquadratic.

Most GP systems store each program separately and memory usage grows linearly with program size; i.e. O(generations$^{1.2-2}$). Run time is typically dominated by program execution time, which is proportional to program length [Langdon, 1998c, Appendix D.8], therefore run time is O(generations$^{2.2-3}$).

In other systems the whole population is stored in a directed acyclic graph (DAG) [Handley, 1994]. New links are created at a constant rate. However, memory usage may be less than O(generations), since in every generation programs are deleted. In the absence of side effects and with a fixed fitness function, it may be possible to avoid re-evaluating unchanged code by caching intermediate values. That is, only code from the crossover point to the root would be executed. Then run time should be proportional to the average height of trees. So run time O(generations^2).

Note we refer to standard subtree crossover; other genetic operators and/or representations have different bloat characteristics. For example, [Nordin and Banzhaf, 1995] suggest that program size increases exponentially with generations in their linear machine code representation and crossover operator. While new mutation [Langdon, 1998b] and crossover operators [Langdon, 2000] can reduce bloat in trees.

GP populations using standard subtree crossover (and no parsimony techniques) quickly reach the bounds on size or depth that are commonly used. When this will happen can be readily estimated. We suggest such bounds may have unanticipated (but problem-dependent) benefits.

To allow big programs (1,000,000 nodes) to evolve, we were restricted to simple problems which can be solved by small trees. However, our results do raise the question of how effective subtree crossover will be on complex discrete problems whose solutions are big programs. It may also be the case that subtree crossover will cease to be effective (i.e. explorative or disruptive) if the program is structured as big trees. GP may need to limit tree size (perhaps by evolving programs composed of many smaller trees) and/or alternative genetic operators may be required.

In Section 11.6 we outlined a few of the highly successful antibloat techniques and some of their potential disadvantages. So, while it might be claimed that bloat has been solved, it remains an interesting theoretical problem due to its widespread occurrence (many other phenomena are problem specific). Also, investigating bloat has lead to insights into GP dynamics and better understanding of GP biases. Which has lead to new operators with better biases.

This concluding chapter describes a conceptually simple and usable genetic programming theory, has tested it to new extremes in GP and shown, at least in part, that it works. In the case where its predictions were less than 100% correct we see an example of new GP behaviour: extreme phenotypic convergence.

12. Conclusions

The role of this concluding chapter is twofold: to summarise the major contributions but also to draw lessons.

The first two chapters introduced genetic programming, first as a search technique and secondly presenting the space it searches. I.e. the space of possible programs. Chapter 3 described early work on analysing populations of programs in terms of the components of the programs (specifically Price's Theorem and various component-based schema theorems). In Chapter 4 we introduced rigorous schema analysis into genetic programming, describing fixed size and shape schemata, and showing how genetic algorithm notions of schema order and defining length can be used in genetic programming. Since GP allows the representation to vary both in size and shape, we also introduced the notion of hyperspace. With these ideas we presented a family of crossover operators that are proper generalisations of the fixed representation crossover operators commonly used in GAs. Using GP one-point crossover and GP point mutation, we presented a classical schema theorem for GP, and showed that under these genetic operators, genetic programming will tend over time to behave like a fixed representation genetic algorithm.

The original schema theorem has been criticised because: (1) it only gives a lower bound and (2) it deals with the expected properties of the next generation. In Chapter 5 we presented the idea of a signal-to-noise ratio for schemata. This explicitly considers how the number of individuals in a finite genetic algorithm population, which match a schema, will vary about the average value. Note that this theory applies equally well to GAs and other population-based evolutionary computation techniques, even though originally developed for GP. Secondly, Chapter 5 presented a schema theorem which takes into account schema creation effects, and so yields an exact statement rather than a lower bound. Finally, Chapter 5 introduced an exact schema theorem for GP with Koza-style subtree crossover.

Chapter 6 started by describing various notions of effective fitness, and gave the equation for the exact effective fitness for GP with subtree crossover. It showed how effective fitness can yield insights into code bloat and code compression. Chapter 6 also described the various biases in GP crossover before turning to the important topics of building blocks, convergence, population sizing, problem difficulty and deception.

Chapters 7 and 8 took a step back. They are not primarily about the GP search process itself but about the space that GP (or indeed other techniques) searches. That is, the space of all possible computer programs. The space was quantified and shown to converge exponentially fast to a limiting distribution which can contain an exponentially large number of solutions. In some cases we were able to prove that the chance of randomly solving a problem falls exponentially with test-set size.

Chapters 9 and 10 used various analytical techniques from Chapters 2–8 to analyse two GP benchmark problems (the Santa Fe Ant and the MAX problems). They showed that GP hardness is often the result of various types of deception, and proposed and demonstrated improvements by reducing deception.

Chapter 11 returned to convergence and suggested that GA measures of convergence do not work well in genetic programming with subtree crossover, and so have tended to conceal the degree to which GP populations do actually converge. Further, we suggested that bloat is actually part of how GP converges. Chapter 11 also presented semi-quantitative program depth and size scaling laws.

The genetic programming search space is, at least in principle, infinite. This means it is is impossible to search it, or even sample it, without bias. We hope that we have provided a better understanding of the biases already present in genetic programming, and some suggestions for improving them. The existing subtree crossover appears to have some suitable biases but doubtless other, possibly problem-specific, improvements could be made to it and other genetic operators. The size and depth limits used in practical GP actually give GP very different biases. We have tentatively suggested how they might be exploited for specific applications. Similar suggestions can be made about the size and shape of programs in the initial population.

A common feature of the problems analysed in detail is they are deceptive, meaning that the fitness function and genetic operators conspire to drive the population away from solutions. That is, the effective fitnesses of partial solutions that lead in the wrong direction are too high. The fitness of programs that are not solutions is usually determined by the user. Therefore, more consideration about how much to reward partial solutions may yield dividends in practice. As part of this, GP users might like to consider how their GP populations are evolving. Is their population converging? If so, is it moving in the right direction? At present, many GP packages offer little in the way of tools to monitor populations.

We have mainly considered simple genetic programming, with one representation (typically one tree) and one (or two) genetic operators. In the last few years there have been many extensions to GP (e.g. ADFs, modular GP, grammars, strongly typed GP, dimensionally aware GP, linear GP, demes and other forms of structured population, multiobjective fitness functions, niching and sharing) which we have yet to tackle.

We hope that one of the nice results of this book will be a renaissance of interest in the theoretical foundations of genetic programming. Many of the results we have presented are generalisations for GP of genetic algorithm results. Thus, by working on the foundations of GP future researchers can, by using the appropriate specialisations, automatically apply their results to GAs. I.e. they get GA theory for free.

A. Genetic Programming Resources

The following list contains pointers to some additional sources of information about genetic programming:

1. Books
   - "Genetic Programming, An Introduction" [Banzhaf *et al.*, 1998a].
   - "Genetic Programming: On the Programming of Computers by Means of Natural Selection", along with follow-up volumes 2 and 3 [Koza, 1992, Koza, 1994, Koza *et al.*, 1999].
   - The three volumes of "Advances in Genetic Programming" [Kinnear, Jr., 1994a, Angeline and Kinnear, Jr., 1996, Spector *et al.*, 1999].
   - "Genetic Programming and Data Structures" [Langdon, 1998c].
2. Journal of "Genetic Programming and Evolvable Machines".
3. Conference Proceedings
   - "Genetic Programming 1996, 1997, 1998" [Koza *et al.*, 1996, Koza *et al.*, 1997, Koza *et al.*, 1998], and late-breaking papers collections [Koza, 1996, Koza, 1997, Koza, 1998].
   - "EuroGP" 1998, 1999, 2000, 2001" [Banzhaf *et al.*, 1998b, Poli *et al.*, 1999, Poli *et al.*, 2000, Miller *et al.*, 2001].
   - "Genetic and Evolutionary Computation COnference GECCO 1999, 2000, 2001" [Banzhaf *et al.*, 1999, Whitley *et al.*, 2000, Spector *et al.*, 2001], and late-breaking papers collections [Brave and Wu, 1999, Whitley, 2000, Goodman, 2001].
4. Many other conference and workshop proceedings have a few GP-related papers. Some examples are the International Conference on Genetic Algorithms (ICGA), Congress on Evolutionary Computation (CEC), Evolutionary Programming (EP93–98), Foundations of Genetic Algorithms (FOGA), Parallel Problem Solving from Nature (PPSN), International Conference on Evolutionary Computation (ICEC), Artificial Life (ALife) and European Conference on Artificial Life (ECAL).
   Other journals, notably the "IEEE Transactions on Evolutionary Computation" and "Evolutionary Computation", also have GP articles.
5. There is a lot of information about genetic programming and source code readily available via the Internet. For example, the GP bibliography [Langdon, 1996] is available at `http://www.cs.bham.ac.uk/~wbl/biblio/`. Unfortunately, Internet addresses (URLs) change very rapidly. However, a web search should be able to give the up to date location of particular resources (such as those documented in [Tufts, 1996]) or suggest alternatives.

Bibliography

[Alonso and Schott, 1995] Laurent Alonso and Rene Schott. *Random Generation of Trees.* Kluwer Academic Publishers, 1995.

[Altenberg, 1994a] Lee Altenberg. Emergent phenomena in genetic programming. In Anthony V. Sebald and Lawrence J. Fogel, editors, *Evolutionary Programming — Proceedings of the Third Annual Conference*, pages 233–241, San Diego, CA, USA, 24-26 February 1994. World Scientific Publishing.

[Altenberg, 1994b] Lee Altenberg. The evolution of evolvability in genetic programming. In Kenneth E. Kinnear, Jr., editor, *Advances in Genetic Programming*, chapter 3, pages 47–74. MIT Press, 1994.

[Altenberg, 1995] Lee Altenberg. The Schema Theorem and Price's Theorem. In L. Darrell Whitley and Michael D. Vose, editors, *Foundations of Genetic Algorithms 3*, pages 23–49, Estes Park, Colorado, USA, 31 July–2 August 1994 1995. Morgan Kaufmann.

[Angeline and Kinnear, Jr., 1996] Peter J. Angeline and Kenneth E. Kinnear, Jr., editors. *Advances in Genetic Programming 2.* MIT Press, 1996.

[Angeline, 1994] Peter J. Angeline. Genetic programming and emergent intelligence. In Kenneth E. Kinnear, Jr., editor, *Advances in Genetic Programming*, chapter 4, pages 75–98. MIT Press, 1994.

[Angeline, 1998a] Peter J. Angeline. Multiple interacting programs: A representation for evolving complex behaviors. *Cybernetics and Systems*, 29(8):779–806, November 1998.

[Angeline, 1998b] Peter J. Angeline. Subtree crossover causes bloat. In John R. Koza, Wolfgang Banzhaf, Kumar Chellapilla, Kalyanmoy Deb, Marco Dorigo, David B. Fogel, Max H. Garzon, David E. Goldberg, Hitoshi Iba, and Rick Riolo, editors, *Genetic Programming 1998: Proceedings of the Third Annual Conference*, pages 745–752, University of Wisconsin, Madison, Wisconsin, USA, 22-25 July 1998. Morgan Kaufmann.

[Bäck, 1996] Thomas Bäck. *Evolutionary Algorithms in Theory and Practice: Evolution Strategies, Evolutionary Programming, Genetic Algorithms.* Oxford University Press, New York, 1996.

[Banzhaf *et al.*, 1997] Wolfgang Banzhaf, Peter Nordin, and Markus Olmer. Generating adaptive behavior for a real robot using function regression within genetic programming. In John R. Koza, Kalyanmoy Deb, Marco Dorigo, David B. Fogel, Max H. Garzon, Hitoshi Iba, and Rick L. Riolo, editors, *Genetic Programming 1997: Proceedings of the Second Annual Conference*, pages 35–43, Stanford University, CA, USA, 13-16 July 1997. Morgan Kaufmann.

[Banzhaf *et al.*, 1998a] Wolfgang Banzhaf, Peter Nordin, Robert E. Keller, and Frank D. Francone. *Genetic Programming – An Introduction; On the Automatic Evolution of Computer Programs and its Applications.* Morgan Kaufmann, 1998.

[Banzhaf *et al.*, 1998b] Wolfgang Banzhaf, Riccardo Poli, Marc Schoenauer, and Terence C. Fogarty, editors. *Genetic Programming*, volume 1391 of *LNCS*, Paris, 14-15 April 1998. Springer-Verlag.

[Banzhaf *et al.*, 1999] Wolfgang Banzhaf, Jason Daida, Agoston E. Eiben, Max H. Garzon, Vasant Honavar, Mark Jakiela, and Robert E. Smith, editors. *GECCO-99: Proceedings of the Genetic and Evolutionary Computation Conference*, Orlando, Florida, USA, 13-17 July 1999. Morgan Kaufmann.

[Beyer, 2001] Hans-Georg Beyer. *The Theory of Evolution Strategies.* Natural Computing Series. Springer-Verlag, 2001.

[Blickle and Thiele, 1994] Tobias Blickle and Lothar Thiele. Genetic programming and redundancy. In J. Hopf, editor, *Genetic Algorithms within the Framework of Evolutionary Computation (Workshop at KI-94, Saarbrücken)*, pages 33–38, Im Stadtwald, Building 44, D-66123 Saarbrücken, Germany, 1994. Max-Planck-Institut für Informatik (MPI-I-94-241).

[Blickle and Thiele, 1995] Tobias Blickle and Lothar Thiele. A mathematical analysis of tournament selection. In Larry J. Eshelman, editor, *Genetic Algorithms: Proceedings of the Sixth International Conference*, pages 9–16, Pittsburgh, PA, USA, 15-19 July 1995. Morgan Kaufmann.

[Blickle, 1996a] Tobias Blickle. Evolving compact solutions in genetic programming: A case study. In Hans-Michael Voigt, Werner Ebeling, Ingo Rechenberg, and Hans-Paul Schwefel, editors, *Parallel Problem Solving From Nature IV. Proceedings of the International Conference on Evolutionary Computation*, volume 1141 of *LNCS*, pages 564–573, Berlin, Germany, 22-26 September 1996. Springer-Verlag.

[Blickle, 1996b] Tobias Blickle. *Theory of Evolutionary Algorithms and Application to System Synthesis.* PhD thesis, Swiss Federal Institute of Technology, Zurich, November 1996.

[Bohm and Geyer-Schulz, 1996] Walter Bohm and Andreas Geyer-Schulz. Exact uniform initialization for genetic programming. In Richard K. Belew and Michae D. Vose, editors, *Foundations of Genetic Algorithms IV*, pages 379–407, University of San Diego, CA, USA, 3–5 August 1996. Morgan Kaufmann.

[Brameier and Banzhaf, 2001] Markus Brameier and Wolfgang Banzhaf. A comparison of linear genetic programming and neural networks in medical data mining. *IEEE Transactions on Evolutionary Computation*, 5(1):17–26, February 2001.

[Brave and Wu, 1999] Scott Brave and Annie S. Wu, editors. *Late Breaking Papers at the 1999 Genetic and Evolutionary Computation Conference*, Orlando, Florida, USA, 13 July 1999.

[Chellapilla, 1997] Kumar Chellapilla. Evolutionary programming with tree mutations: Evolving computer programs without crossover. In John R. Koza, Kalyanmoy Deb, Marco Dorigo, David B. Fogel, Max H. Garzon, Hitoshi Iba, and Rick L. Riolo, editors, *Genetic Programming 1997: Proceedings of the Second Annual Conference*, pages 431–438, Stanford University, CA, USA, 13-16 July 1997. Morgan Kaufmann.

[Chung and Perez, 1994] S. W. Chung and R. A. Perez. The schema theorem considered insufficient. In *Proceedings of the Sixth IEEE International Conference on Tools with Artificial Intelligence*, pages 748–751, New Orleans, Nov 6–9 1994.

[Davis and Principe, 1993] Thomas E. Davis and Jose C. Principe. A Markov chain framework for the simple genetic algorithm. *Evolutionary Computation*, 1(3):269–288, 1993.

[Dawkins, 1976] R. Dawkins. *The Selfish Gene.* Oxford University Press, Oxford, 1976.

[De Jong *et al.*, 1995] Kenneth A. De Jong, William M. Spears, and Diana F. Gordon. Using Markov chains to analyze GAFOs. In L. Darrell Whitley and Michael D. Vose, editors, *Proceedings of the Third Workshop on Foundations of Genetic Algorithms*, pages 115–138, San Francisco, 31 July–2 August 1995. Morgan Kaufmann.

[D'haeseleer, 1994] Patrik D'haeseleer. Context preserving crossover in genetic programming. In *Proceedings of the 1994 IEEE World Congress on Computational Intelligence*, volume 1, pages 256–261, Orlando, Florida, USA, 27-29 June 1994. IEEE Press.

[Ekart and Nemeth, 2001] Aniko Ekart and S. Z. Nemeth. Selection based on the pareto nondomination criterion for controlling code growth in genetic programming. *Genetic Programming and Evolvable Machines*, 2(1):61–73, March 2001.

[Feller, 1970] William Feller. *An Introduction to Probability Theory and Its Applications*, volume 1. Wiley, 3rd edition, 1970.

[Flajolet and Oldyzko, 1982] Philippe Flajolet and Andrew Oldyzko. The average height of binary trees and other simple trees. *Journal of Computer and System Sciences*, 25:171–213, 1982.

[Fogel and Ghozeil, 1997] David B. Fogel and A. Ghozeil. Schema processing under proportional selection in the presence of random effects. *IEEE Transactions on Evolutionary Computation*, 1(4):290–293, 1997.

[Fogel and Ghozeil, 1998] David B. Fogel and A. Ghozeil. The schema theorem and the misallocation of trials in the presence of stochastic effects. In V. W. Porto, N. Saravanan, D. Waagen, and Agoston E. Eiben, editors, *Evolutionary Programming VII: Proc. of the 7th Ann. Conf. on Evolutionary Programming*, pages 313–321, Berlin, March 25–27 1998. Springer-Verlag.

[Gathercole and Ross, 1996] Chris Gathercole and Peter Ross. An adverse interaction between crossover and restricted tree depth in genetic programming. In John R. Koza, David E. Goldberg, David B. Fogel, and Rick L. Riolo, editors, *Genetic Programming 1996: Proceedings of the First Annual Conference*, pages 291–296, Stanford University, CA, USA, 28–31 July 1996. MIT Press.

[Gathercole, 1997] Chris Gathercole, 16 January 1997. Electronic communication.

[Gathercole, 1998] Chris Gathercole. *An Investigation of Supervised Learning in Genetic Programming*. PhD thesis, University of Edinburgh, 1998.

[Glover, 1989] Fred Glover. Tabu search – part I. *ORSA Journal on Computing*, 1(3):190–206, 1989.

[Goldberg, 1989a] David E. Goldberg. Genetic algorithms and Walsh functions: II. Deception and its analysis. *Complex Systems*, 3(2):153–171, April 1989.

[Goldberg, 1989b] David E. Goldberg. *Genetic Algorithms in Search, Optimization, and Machine Learning*. Addison-Wesley, Reading, Massachusetts, 1989.

[Goodman, 2001] Erik D. Goodman, editor. *Late Breaking Papers at the 2001 Genetic and Evolutionary Computation Conference*, San Francisco, California, USA, 7-11 July 2001.

[Greeff and Aldrich, 1997] D. J. Greeff and C. Aldrich. Evolution of empirical models for metallurgical process systems. In John R. Koza, Kalyanmoy Deb, Marco Dorigo, David B. Fogel, Max H. Garzon, Hitoshi Iba, and Rick L. Riolo, editors, *Genetic Programming 1997: Proceedings of the Second Annual Conference*, page 138, Stanford University, CA, USA, 13-16 July 1997. Morgan Kaufmann.

[Grefenstette, 1993] John J. Grefenstette. Deception considered harmful. In L. Darrell Whitley, editor, *Foundations of Genetic Algorithms 2*, San Mateo, CA, 1993. Morgan Kaufman.

[Gritz and Hahn, 1997] Larry Gritz and James K. Hahn. Genetic programming evolution of controllers for 3-D character animation. In John R. Koza, Kalyanmoy Deb, Marco Dorigo, David B. Fogel, Max H. Garzon, Hitoshi Iba, and

Rick L. Riolo, editors, *Genetic Programming 1997: Proceedings of the Second Annual Conference*, pages 139–146, Stanford University, CA, USA, 13-16 July 1997. Morgan Kaufmann.

[Gruau, 1994a] Frederic Gruau. *Neural Network Synthesis using Cellular Encoding and the Genetic Algorithm.* PhD thesis, Laboratoire de l'Informatique du Parallilisme, Ecole Normale Supirieure de Lyon, France, 1994.

[Gruau, 1994b] Frederic Gruau. Genetic micro programming of neural networks. In Kenneth E. Kinnear, Jr., editor, *Advances in Genetic Programming*, chapter 24, pages 495–518. MIT Press, 1994.

[Handley, 1994] Simon Handley. On the use of a directed acyclic graph to represent a population of computer programs. In *Proceedings of the 1994 IEEE World Congress on Computational Intelligence*, pages 154–159, Orlando, Florida, USA, 27-29 June 1994. IEEE Press.

[Handley, 1995] Simon Handley. Predicting whether or not a nucleic acid sequence is an E. coli promoter region using genetic programming. In *Proceedings of the First International Symposium on Intelligence in Neural and Biological Systems INBS-95*, pages 122–127, Herndon, Virginia, USA, 29-31 May 1995. IEEE Computer Society Press.

[Harries and Smith, 1997] Kim Harries and Peter Smith. Exploring alternative operators and search strategies in genetic programming. In John R. Koza, Kalyanmoy Deb, Marco Dorigo, David B. Fogel, Max H. Garzon, Hitoshi Iba, and Rick L. Riolo, editors, *Genetic Programming 1997: Proceedings of the Second Annual Conference*, pages 147–155, Stanford University, CA, USA, 13-16 July 1997. Morgan Kaufmann.

[Holland, 1973] John H. Holland. Genetic algorithms and the optimal allocation of trials. *SIAM Journal on Computation*, 2:88–105, 1973.

[Holland, 1975] John H. Holland. *Adaptation in Natural and Artificial Systems.* University of Michigan Press, Ann Arbor, USA, 1975.

[Hooper and Flann, 1996] Dale Hooper and Nicholas S. Flann. Improving the accuracy and robustness of genetic programming through expression simplification. In John R. Koza, David E. Goldberg, David B. Fogel, and Rick L. Riolo, editors, *Genetic Programming 1996: Proceedings of the First Annual Conference*, page 428, Stanford University, CA, USA, 28–31 July 1996. MIT Press.

[Iba *et al.*, 1994] Hitoshi Iba, Hugo de Garis, and Taisuke Sato. Genetic programming using a minimum description length principle. In Kenneth E. Kinnear, Jr., editor, *Advances in Genetic Programming*, chapter 12, pages 265–284. MIT Press, 1994.

[Iba, 1996] Hitoshi Iba. Random tree generation for genetic programming. In Hans-Michael Voigt, Werner Ebeling, Ingo Rechenberg, and Hans-Paul Schwefel, editors, *Parallel Problem Solving from Nature IV, Proceedings of the International Conference on Evolutionary Computation*, volume 1141 of *LNCS*, pages 144–153, Berlin, Germany, 22-26 September 1996. Springer-Verlag.

[Iba, 1999] Hitoshi Iba. Bagging, boosting, and bloating in genetic programming. In Wolfgang Banzhaf, Jason Daida, Agoston E. Eiben, Max H. Garzon, Vasant Honavar, Mark Jakiela, and Robert E. Smith, editors, *Proceedings of the Genetic and Evolutionary Computation Conference*, volume 2, pages 1053–1060, Orlando, Florida, USA, 13-17 July 1999. Morgan Kaufmann.

[Ito *et al.*, 1998] Takuya Ito, Hitoshi Iba, and Satoshi Sato. Non-destructive depth-dependent crossover for genetic programming. In Wolfgang Banzhaf, Riccardo Poli, Marc Schoenauer, and Terence C. Fogarty, editors, *Proceedings of the First European Workshop on Genetic Programming*, volume 1391 of *LNCS*, pages 71–82, Paris, 14-15 April 1998. Springer-Verlag.

[Ito *et al.*, 1999] Takuya Ito, Hitoshi Iba, and Satoshi Sato. A self-tuning mechanism for depth-dependent crossover. In Lee Spector, W. B. Langdon, Una-May O'Reilly, and Peter J. Angeline, editors, *Advances in Genetic Programming 3*, chapter 16, pages 377–399. MIT Press, 1999.

[Jacob, 1996] Christian Jacob. Evolving evolution programs: Genetic programming and L-systems. In John R. Koza, David E. Goldberg, David B. Fogel, and Rick L. Riolo, editors, *Genetic Programming 1996: Proceedings of the First Annual Conference*, pages 107–115, Stanford University, CA, USA, 28–31 July 1996. MIT Press.

[Janikow, 1996] Cezary Z. Janikow. A methodology for processing problem constraints in genetic programming. *Computers and Mathematics with Applications*, 32(8):97–113, 1996.

[Jefferson *et al.*, 1992] David Jefferson, Robert Collins, Claus Cooper, Michael Dyer, Margot Flowers, Richard Korf, Charles Taylor, and Alan Wang. Evolution as a theme in artificial life: The Genesys/Tracker system. In Christopher G. Langton, Charles Taylor, J. Doyne Farmer, and Steen Rasmussen, editors, *Artificial Life II*, volume X of *Santa Fe Institute Studies in the Sciences of Complexity*, pages 549–578. Addison-Wesley, February 1992.

[Juille and Pollack, 1996] Hugues Juille and Jordan B. Pollack. Massively parallel genetic programming. In Peter J. Angeline and Kenneth E. Kinnear, Jr., editors, *Advances in Genetic Programming 2*, chapter 17, pages 339–358. MIT Press, 1996.

[Kargupta and Goldberg, 1994] Hillol Kargupta and David E. Goldberg. Decision making in genetic algorithms: A signal-to-noise perspective. Technical Report IlliGAL Report No 94004, Illinois Genetic Algorithms Lab, 1994.

[Kargupta, 1995] Hillol Kargupta. Signal-to-noise, crosstalk, and long range problem difficulty in genetic algorithms. In Larry J. Eshelman, editor, *Proceedings of the 6th International Conference on Genetic Algorithms*, pages 193–200, Pittsburgh, PA, USA, 15–19 July 1995. Morgan Kaufmann.

[Keijzer, 1996] Maarten Keijzer. Efficiently representing populations in genetic programming. In Peter J. Angeline and Kenneth E. Kinnear, Jr., editors, *Advances in Genetic Programming 2*, chapter 13, pages 259–278. MIT Press, 1996.

[Kellel *et al.*, 2001] L. Kellel, B. Naudts, and A. Rogers, editors. *Theoretical Aspects of Evolutionary Computing*. Natural Computing Series. Springer-Verlag, 2001.

[Kennedy and Giraud-Carrier, 1999] Claire J. Kennedy and Christophe Giraud-Carrier. A depth controlling strategy for strongly typed evolutionary programming. In Wolfgang Banzhaf, Jason Daida, Agoston E. Eiben, Max H. Garzon, Vasant Honavar, Mark Jakiela, and Robert E. Smith, editors, *Proceedings of the Genetic and Evolutionary Computation Conference*, volume 1, pages 879–885, Orlando, Florida, USA, 13-17 July 1999. Morgan Kaufmann.

[Kinnear, Jr., 1993] Kenneth E. Kinnear, Jr. Evolving a sort: Lessons in genetic programming. In *Proceedings of the 1993 International Conference on Neural Networks*, volume 2, pages 881–888, San Francisco, USA, 28 March-1 April 1993. IEEE Press.

[Kinnear, Jr., 1994a] Kenneth E. Kinnear, Jr., editor. *Advances in Genetic Programming*. MIT Press, 1994.

[Kinnear, Jr., 1994b] Kenneth E. Kinnear, Jr. Alternatives in automatic function definition: A comparison of performance. In Kenneth E. Kinnear, Jr., editor, *Advances in Genetic Programming*, chapter 6, pages 119–141. MIT Press, 1994.

[Kirkpatrick *et al.*, 1983] S. Kirkpatrick, Jr. C. D. Gelatt, and M. P. Vecchi. Optimization by simulated annealing. *Science*, 220:671–680, 1983.

[Koza and Bennett III, 1999] John R. Koza and Forrest H Bennett III. Automatic synthesis, placement, and routing of electrical circuits by means of genetic programming. In Lee Spector, W. B. Langdon, Una-May O'Reilly, and Peter J. Angeline, editors, *Advances in Genetic Programming 3*, chapter 6, pages 105–134. MIT Press, 1999.

[Koza *et al.*, 1996] John R. Koza, David E. Goldberg, David B. Fogel, and Rick L. Riolo, editors. *Genetic Programming 1996: Proceedings of the First Annual Conference*, Stanford University, CA, USA, 28–31 July 1996. MIT Press.

[Koza *et al.*, 1997] John R. Koza, Kalyanmoy Deb, Marco Dorigo, David B. Fogel, Max H. Garzon, Hitoshi Iba, and Rick L. Riolo, editors. *Genetic Programming 1997: Proceedings of the Second Annual Conference*, Stanford University, CA, USA, 13-16 July 1997. Morgan Kaufmann.

[Koza *et al.*, 1998] John R. Koza, Wolfgang Banzhaf, Kumar Chellapilla, Kalyanmoy Deb, Marco Dorigo, David B. Fogel, Max H. Garzon, David E. Goldberg, Hitoshi Iba, and Rick Riolo, editors. *Genetic Programming 1998: Proceedings of the Third Annual Conference*, University of Wisconsin, Madison, WI, USA, 22-25 July 1998. Morgan Kaufmann.

[Koza *et al.*, 1999] John R. Koza, David Andre, Forrest H Bennett III, and Martin Keane. *Genetic Programming 3: Darwinian Invention and Problem Solving.* Morgan Kaufman, 1999.

[Koza, 1992] John R. Koza. *Genetic Programming: On the Programming of Computers by Means of Natural Selection.* MIT Press, 1992.

[Koza, 1994] John R. Koza. *Genetic Programming II: Automatic Discovery of Reusable Programs.* MIT Press, 1994.

[Koza, 1996] John R. Koza, editor. *Late Breaking Papers at the Genetic Programming 1996 Conference*, Stanford University, CA, USA, 28–31 July 1996. Stanford Bookstore.

[Koza, 1997] John R. Koza, editor. *Late Breaking Papers at the 1997 Genetic Programming Conference*, Stanford University, CA, USA, 13–16 July 1997. Stanford Bookstore.

[Koza, 1998] John R. Koza, editor. *Late Breaking Papers at the 1998 Genetic Programming Conference*, University of Wisconsin, Madison, WI, USA, 22-25 July 1998. Omni Press.

[Kuscu, 1998] Ibrahim Kuscu. Evolving a generalised behavior: Artificial ant problem revisited. In V. William Porto, N. Saravanan, D. Waagen, and Agoston E. Eiben, editors, *Seventh Annual Conference on Evolutionary Programming*, volume 1447 of *LNCS*, pages 799–808, Mission Valley Marriott, San Diego, California, USA, 25-27 March 1998. Springer-Verlag.

[Langdon and Banzhaf, 2000] W. B. Langdon and Wolfgang Banzhaf. Genetic programming bloat without semantics. In Marc Schoenauer, Kalyanmoy Deb, Günter Rudolph, Xin Yao, Evelyne Lutton, Juan Julian Merelo, and Hans-Paul Schwefel, editors, *Parallel Problem Solving from Nature - PPSN VI 6th International Conference*, volume 1917 of *LNCS*, pages 201–210, Paris, France, 16-20 September 2000. Springer-Verlag.

[Langdon and Nordin, 2000] W. B. Langdon and Peter Nordin. Seeding GP populations. In Riccardo Poli, Wolfgang Banzhaf, W. B. Langdon, Julian F. Miller, Peter Nordin, and Terence C. Fogarty, editors, *Genetic Programming, Proceedings of EuroGP'2000*, volume 1802 of *LNCS*, pages 304–315, Edinburgh, 15-16 April 2000. Springer-Verlag.

[Langdon and Poli, 1997a] W. B. Langdon and Riccardo Poli. An analysis of the MAX problem in genetic programming. In John R. Koza, Kalyanmoy Deb, Marco Dorigo, David B. Fogel, Max H. Garzon, Hitoshi Iba, and Rick L. Riolo, editors, *Genetic Programming 1997: Proceedings of the Second Annual Conference*, pages 222–230, Stanford University, CA, USA, 13-16 July 1997. Morgan Kaufmann.

[Langdon and Poli, 1997b] W. B. Langdon and Riccardo Poli. Fitness causes bloat. In P. K. Chawdhry, R. Roy, and R. K. Pant, editors, *Soft Computing in Engineering Design and Manufacturing*, pages 13–22. Springer-Verlag London, 23-27 June 1997.

[Langdon and Poli, 1998a] W. B. Langdon and Riccardo Poli. Fitness causes bloat: Mutation. In Wolfgang Banzhaf, Riccardo Poli, Marc Schoenauer, and Terence C. Fogarty, editors, *Proceedings of the First European Workshop on Genetic Programming*, volume 1391 of *LNCS*, pages 37–48, Paris, 14-15 April 1998. Springer-Verlag.

[Langdon and Poli, 1998b] W. B. Langdon and Riccardo Poli. Genetic programming bloat with dynamic fitness. In Wolfgang Banzhaf, Riccardo Poli, Marc Schoenauer, and Terence C. Fogarty, editors, *Proceedings of the First European Workshop on Genetic Programming*, volume 1391 of *LNCS*, pages 96–112, Paris, 14-15 April 1998. Springer-Verlag.

[Langdon and Poli, 1998c] W. B. Langdon and Riccardo Poli. Why ants are hard. In John R. Koza, Wolfgang Banzhaf, Kumar Chellapilla, Kalyanmoy Deb, Marco Dorigo, David B. Fogel, Max H. Garzon, David E. Goldberg, Hitoshi Iba, and Rick Riolo, editors, *Genetic Programming 1998: Proceedings of the Third Annual Conference*, pages 193–201, University of Wisconsin, Madison, Wisconsin, USA, 22-25 July 1998. Morgan Kaufmann.

[Langdon *et al.*, 1999] W. B. Langdon, Terence Soule, Riccardo Poli, and James A. Foster. The evolution of size and shape. In Lee Spector, W. B. Langdon, Una-May O'Reilly, and Peter J. Angeline, editors, *Advances in Genetic Programming 3*, chapter 8, pages 163–190. MIT Press, 1999.

[Langdon, 1995] W. B. Langdon. Evolving data structures using genetic programming. In Larry J. Eshelman, editor, *Genetic Algorithms: Proceedings of the Sixth International Conference (ICGA95)*, pages 295–302, Pittsburgh, PA, USA, 15-19 July 1995. Morgan Kaufmann.

[Langdon, 1996] W. B. Langdon. A bibliography for genetic programming. In Peter J. Angeline and Kenneth E. Kinnear, Jr., editors, *Advances in Genetic Programming 2*, chapter B, pages 507–532. MIT Press, 1996.

[Langdon, 1998a] W. B. Langdon. Better trained ants. In Riccardo Poli, W. B. Langdon, Marc Schoenauer, Terence C. Fogarty, and Wolfgang Banzhaf, editors, *Late Breaking Papers at EuroGP'98: the First European Workshop on Genetic Programming*, pages 11–13, Paris, France, 14-15 April 1998. CSRP-98-10, The University of Birmingham, UK.

[Langdon, 1998b] W. B. Langdon. The evolution of size in variable length representations. In *1998 IEEE International Conference on Evolutionary Computation*, pages 633–638, Anchorage, Alaska, USA, 5-9 May 1998. IEEE Press.

[Langdon, 1998c] W. B. Langdon. *Data Structures and Genetic Programming: Genetic Programming + Data Structures = Automatic Programming!*, volume 1 of *Genetic Programming*. Kluwer Academic Publishers, 1998.

[Langdon, 1999] W. B. Langdon. Size fair and homologous tree genetic programming crossovers. In Wolfgang Banzhaf, Jason Daida, Agoston E. Eiben, Max H. Garzon, Vasant Honavar, Mark Jakiela, and Robert E. Smith, editors, *Proceedings of the Genetic and Evolutionary Computation Conference*, volume 2, pages 1092–1097, Orlando, Florida, USA, 13-17 July 1999. Morgan Kaufmann.

[Langdon, 2000] W. B. Langdon. Size fair and homologous tree genetic programming crossovers. *Genetic Programming and Evolvable Machines*, 1(1/2):95–119, April 2000.

[Lee and Wong, 1995] Jack Y. B. Lee and P. C. Wong. The effect of function noise on GP efficiency. In X. Yao, editor, *Progress in Evolutionary Computation*, volume 956 of *Lecture Notes in Artificial Intelligence*, pages 1–16. Springer-Verlag, 1995.

[Luke and Panait, 2001] Sean Luke and Liviu Panait. A survey and comparison of tree generation algorithms. In Lee Spector, Erik D. Goodman, Annie Wu, W. B. Langdon, Hans-Michael Voigt, Mitsuo Gen, Sandip Sen, Marco Dorigo, Shahram Pezeshk, Max H. Garzon, and Edmund Burke, editors, *Proceedings of the Genetic and Evolutionary Computation Conference (GECCO-2001)*, pages 81–88, San Francisco, California, USA, 7-11 July 2001. Morgan Kaufmann.

[Luke and Spector, 1997] Sean Luke and Lee Spector. A comparison of crossover and mutation in genetic programming. In John R. Koza, Kalyanmoy Deb, Marco Dorigo, David B. Fogel, Max H. Garzon, Hitoshi Iba, and Rick L. Riolo, editors, *Genetic Programming 1997: Proceedings of the Second Annual Conference*, pages 240–248, Stanford University, CA, USA, 13-16 July 1997. Morgan Kaufmann.

[Luke, 2000a] Sean Luke. Code growth is not caused by introns. In L. Darrell Whitley, editor, *Late Breaking Papers at the 2000 Genetic and Evolutionary Computation Conference*, pages 228–235, Las Vegas, Nevada, USA, 8 July 2000.

[Luke, 2000b] Sean Luke. Two fast tree-creation algorithms for genetic programming. *IEEE Transactions on Evolutionary Computation*, 4(3):274–283, September 2000.

[Macready and Wolpert, 1996] William G. Macready and David H. Wolpert. On 2-armed Gaussian bandits and optimization. Sante Fe Institute Working Paper 96-05-009, March 1996.

[Maxwell, 1996] S. R. Maxwell. Why might some problems be difficult for genetic programming to find solutions? In John R. Koza, editor, *Late Breaking Papers at the Genetic Programming 1996 Conference*, pages 125–128, Stanford University, CA, USA, 28–31 July 1996. Stanford Bookstore.

[McKay *et al.*, 1995] Ben McKay, Mark J. Willis, and Geoffrey W. Barton. Using a tree structured genetic algorithm to perform symbolic regression. In A. M. S. Zalzala, editor, *First International Conference on Genetic Algorithms in Engineering Systems: Innovations and Applications, GALESIA*, volume 414, pages 487–492, Sheffield, UK, 12-14 September 1995. IEE.

[McPhee and Miller, 1995] Nicholas Freitag McPhee and Justin Darwin Miller. Accurate replication in genetic programming. In Larry J. Eshelman, editor, *Genetic Algorithms: Proceedings of the Sixth International Conference (ICGA95)*, pages 303–309, Pittsburgh, PA, USA, 15-19 July 1995. Morgan Kaufmann.

[McPhee and Poli, 2001] Nicholas Freitag McPhee and Riccardo Poli. A schema theory analysis of the evolution of size in genetic programming with linear representations. In Julian F. Miller, Marco Tomassini, Pier Luca Lanzi, Conor Ryan, Andrea G. B. Tettamanzi, and W. B. Langdon, editors, *Genetic Programming, Proceedings of EuroGP'2001*, volume 2038 of *LNCS*, pages 108–125, Lake Como, Italy, 18-20 April 2001. Springer-Verlag.

[Menke, 1997] R. Menke. A revision of the schema theorem. Technical Report CI 14/97, Collaborative Research Center SFB 531, University of Dortmund, 1997.

[Michalewicz, 1994] Zbigniew Michalewicz. *Genetic Algorithms + Data Structures = Evolution Programs*. Springer-Verlag, 2 edition, 1994.

[Miller *et al.*, 2001] Julian F. Miller, Marco Tomassini, Pier Luca Lanzi, Conor Ryan, Andrea G. B. Tettamanzi, and W. B. Langdon, editors. *Genetic Programming, Proceedings of EuroGP'2001*, volume 2038 of *LNCS*, Lake Como, Italy, 18-20 April 2001. Springer-Verlag.

[Montana, 1995] David J. Montana. Strongly typed genetic programming. *Evolutionary Computation*, 3(2):199–230, 1995.

[Nix and Vose, 1992] Allen E. Nix and Michael D. Vose. Modeling genetic algorithms with Markov chains. *Annals of Mathematics and Artificial Intelligence*, 5:79–88, 1992.

[Nordin and Banzhaf, 1995] Peter Nordin and Wolfgang Banzhaf. Complexity compression and evolution. In Larry J. Eshelman, editor, *Genetic Algorithms: Proceedings of the Sixth International Conference (ICGA95)*, pages 310–317, Pittsburgh, PA, USA, 15-19 July 1995. Morgan Kaufmann.

[Nordin *et al.*, 1995] Peter Nordin, Frank D. Francone, and Wolfgang Banzhaf. Explicitly defined introns and destructive crossover in genetic programming. In Justinian P. Rosca, editor, *Proceedings of the Workshop on Genetic Programming: From Theory to Real-World Applications*, pages 6–22, Tahoe City, California, USA, 9 July 1995.

[Nordin *et al.*, 1996] Peter Nordin, Frank D. Francone, and Wolfgang Banzhaf. Explicitly defined introns and destructive crossover in genetic programming. In Peter J. Angeline and Kenneth E. Kinnear, Jr., editors, *Advances in Genetic Programming 2*, chapter 6, pages 111–134. MIT Press, 1996.

[Nordin *et al.*, 1997] Peter Nordin, Wolfgang Banzhaf, and Frank D. Francone. Introns in nature and in simulated structure evolution. In Dan Lundh, Bjorn Olsson, and Ajit Narayanan, editors, *Bio-Computation and Emergent Computation*, Skovde, Sweden, 1-2 September 1997. World Scientific Publishing.

[Nordin, 1997] Peter Nordin. *Evolutionary Program Induction of Binary Machine Code and its Applications.* PhD thesis, der Universitat Dortmund am Fachereich Informatik, 1997.

[O'Neill and Ryan, 1999] Michael O'Neill and Conor Ryan. Evolving multi-line compilable C programs. In Riccardo Poli, Peter Nordin, W. B. Langdon, and Terence C. Fogarty, editors, *Genetic Programming, Proceedings of EuroGP'99*, volume 1598 of *LNCS*, pages 83–92, Goteborg, Sweden, 26-27 May 1999. Springer-Verlag.

[O'Reilly and Oppacher, 1995] Una-May O'Reilly and Franz Oppacher. The troubling aspects of a building block hypothesis for genetic programming. In L. Darrell Whitley and Michael D. Vose, editors, *Foundations of Genetic Algorithms 3*, pages 73–88, Estes Park, Colorado, USA, 31 July–2 August 1994 1995. Morgan Kaufmann.

[O'Reilly, 1995] Una-May O'Reilly. *An Analysis of Genetic Programming.* PhD thesis, Carleton University, Ottawa-Carleton Institute for Computer Science, Ottawa, Ontario, Canada, 22 September 1995.

[O'Reilly, 1997] Una-May O'Reilly. Using a distance metric on genetic programs to understand genetic operators. In *IEEE International Conference on Systems, Man, and Cybernetics, Computational Cybernetics and Simulation*, volume 5, pages 4092–4097, Orlando, Florida, USA, 12-15 October 1997.

[Page *et al.*, 1999] J. Page, Riccardo Poli, and W. B. Langdon. Smooth uniform crossover with smooth point mutation in genetic programming: A preliminary study. In Riccardo Poli, Peter Nordin, W. B. Langdon, and Terence C. Fogarty, editors, *Genetic Programming, Proceedings of EuroGP'99*, volume 1598 of *LNCS*, pages 39–49, Goteborg, Sweden, 26-27 May 1999. Springer-Verlag.

[Perkis, 1994] Tim Perkis. Stack-based genetic programming. In *Proceedings of the 1994 IEEE World Congress on Computational Intelligence*, volume 1, pages 148–153, Orlando, Florida, USA, 27-29 June 1994. IEEE Press.

[Pinto-Ferreira and Mamede, 1995] C. Pinto-Ferreira and N. J. Mamede, editors. *Progress in Artificial Intelligence (Proceedings 7th Portuguese Conference on*

Artificial Intelligence, EPIA'95), number 990 in Lecture Notes in Artificial Intelligence. Springer-Verlag, 1995.

[Poli and Langdon, 1997a] Riccardo Poli and W. B. Langdon. An experimental analysis of schema creation, propagation and disruption in genetic programming. In Thomas Bäck, editor, *Genetic Algorithms: Proceedings of the Seventh International Conference*, pages 18–25, Michigan State University, East Lansing, MI, USA, 19-23 July 1997. Morgan Kaufmann.

[Poli and Langdon, 1997b] Riccardo Poli and W. B. Langdon. Genetic programming with one-point crossover. In P. K. Chawdhry, R. Roy, and R. K. Pant, editors, *Soft Computing in Engineering Design and Manufacturing*, pages 180–189. Springer-Verlag London, 23-27 June 1997.

[Poli and Langdon, 1997c] Riccardo Poli and W. B. Langdon. A new schema theory for genetic programming with one-point crossover and point mutation. In John R. Koza, Kalyanmoy Deb, Marco Dorigo, David B. Fogel, Max H. Garzon, Hitoshi Iba, and Rick L. Riolo, editors, *Genetic Programming 1997: Proceedings of the Second Annual Conference*, pages 278–285, Stanford University, CA, USA, 13-16 July 1997. Morgan Kaufmann.

[Poli and Langdon, 1998a] Riccardo Poli and W. B. Langdon. On the search properties of different crossover operators in genetic programming. In John R. Koza, Wolfgang Banzhaf, Kumar Chellapilla, Kalyanmoy Deb, Marco Dorigo, David B. Fogel, Max H. Garzon, David E. Goldberg, Hitoshi Iba, and Rick Riolo, editors, *Genetic Programming 1998: Proceedings of the Third Annual Conference*, pages 293–301, University of Wisconsin, Madison, Wisconsin, USA, 22-25 July 1998. Morgan Kaufmann.

[Poli and Langdon, 1998b] Riccardo Poli and W. B. Langdon. Schema theory for genetic programming with one-point crossover and point mutation. *Evolutionary Computation*, 6(3):231–252, 1998.

[Poli and Langdon, 1999] Riccardo Poli and W. B. Langdon. Sub-machine-code genetic programming. In Lee Spector, W. B. Langdon, Una-May O'Reilly, and Peter J. Angeline, editors, *Advances in Genetic Programming 3*, chapter 13, pages 301–323. MIT Press, 1999.

[Poli and Logan, 1996] Riccardo Poli and Brian Logan. The evolutionary computation cookbook: Recipes for designing new algorithms. In *Proceedings of the Second Online Workshop on Evolutionary Computation*, Nagoya, Japan, March 1996.

[Poli and McPhee, 2001a] Riccardo Poli and Nicholas F. McPhee. Exact GP schema theory for headless chicken crossover and subtree mutation. In *Proceedings of the 2001 Congress on Evolutionary Computation CEC2001*, pages 1062–1069, COEX, World Trade Center, 159 Samseong-dong, Gangnam-gu, Seoul, Korea, 27-30 May 2001. IEEE Press.

[Poli and McPhee, 2001b] Riccardo Poli and Nicholas F. McPhee. Exact schema theorems for GP with one-point and standard crossover operating on linear structures and their application to the study of the evolution of size. In *Genetic Programming, Proceedings of EuroGP 2001*, LNCS, Milan, 18-20 April 2001. Springer-Verlag.

[Poli and McPhee, 2001c] Riccardo Poli and Nicholas F. McPhee. Exact schema theory for GP and variable-length GAs with homologous crossover. In *Proceedings of the Genetic and Evolutionary Computation Conference (GECCO-2001)*, San Francisco, California, USA, 7-11 July 2001. Morgan Kaufmann.

[Poli and Page, 2000] Riccardo Poli and Jonathan Page. Solving high-order boolean parity problems with smooth uniform crossover, sub-machine code GP and demes. *Genetic Programming and Evolvable Machines*, 1(1/2):37–56, April 2000.

[Poli *et al.*, 1998] Riccardo Poli, W. B. Langdon, and Una-May O'Reilly. Analysis of schema variance and short term extinction likelihoods. In John R. Koza, Wolfgang Banzhaf, Kumar Chellapilla, Kalyanmoy Deb, Marco Dorigo, David B. Fogel, Max H. Garzon, David E. Goldberg, Hitoshi Iba, and Rick Riolo, editors, *Genetic Programming 1998: Proceedings of the Third Annual Conference*, pages 284–292, University of Wisconsin, Madison, Wisconsin, USA, 22-25 July 1998. Morgan Kaufmann.

[Poli *et al.*, 1999] Riccardo Poli, Peter Nordin, W. B. Langdon, and Terence C. Fogarty, editors. *Genetic Programming, Proceedings of EuroGP'99*, volume 1598 of *LNCS*, Goteborg, Sweden, 26-27 May 1999. Springer-Verlag.

[Poli *et al.*, 2000] Riccardo Poli, Wolfgang Banzhaf, W. B. Langdon, Julian F. Miller, Peter Nordin, and Terence C. Fogarty, editors. *Genetic Programming, Proceedings of EuroGP'2000*, volume 1802 of *LNCS*, Edinburgh, 15-16 April 2000. Springer-Verlag.

[Poli *et al.*, 2001] Riccardo Poli, Jon E. Rowe, and Nicholas F. McPhee. Markov chain models for GP and variable-length GAs with homologous crossover. In *Proceedings of the Genetic and Evolutionary Computation Conference (GECCO-2001)*, San Francisco, California, USA, 7-11 July 2001. Morgan Kaufmann.

[Poli, 1997] Riccardo Poli. Evolution of graph-like programs with parallel distributed genetic programming. In Thomas Bäck, editor, *Genetic Algorithms: Proceedings of the Seventh International Conference*, pages 346–353, Michigan State University, East Lansing, MI, USA, 19-23 July 1997. Morgan Kaufmann.

[Poli, 1999a] Riccardo Poli. Parallel distributed genetic programming. In David Corne, Marco Dorigo, and Fred Glover, editors, *New Ideas in Optimization*, chapter 27. McGraw-Hill, 1999.

[Poli, 1999b] Riccardo Poli. Schema theorems without expectations. In Wolfgang Banzhaf, Jason Daida, Agoston E. Eiben, Max H. Garzon, Vasant Honavar, Mark Jakiela, and Robert E. Smith, editors, *Proceedings of the Genetic and Evolutionary Computation Conference*, volume 1, page 806, Orlando, Florida, USA, 13-17 July 1999. Morgan Kaufmann.

[Poli, 2000a] Riccardo Poli. Hyperschema theory for GP with one-point crossover, building blocks, and some new results in GA theory. In Riccardo Poli, Wolfgang Banzhaf, W. B. Langdon, Julian F. Miller, Peter Nordin, and Terence C. Fogarty, editors, *Genetic Programming, Proceedings of EuroGP'2000*, volume 1802 of *LNCS*, pages 163–180, Edinburgh, 15-16 April 2000. Springer-Verlag.

[Poli, 2000b] Riccardo Poli. Exact schema theorem and effective fitness for GP with one-point crossover. In L. Darrell Whitley, David E. Goldberg, Erick Cantu-Paz, Lee Spector, Ian Parmee, and Hans-Georg Beyer, editors, *Proceedings of the Genetic and Evolutionary Computation Conference (GECCO-2000)*, pages 469–476, Las Vegas, Nevada, USA, 10-12 July 2000. Morgan Kaufmann.

[Poli, 2000c] Riccardo Poli. Recursive conditional schema theorem, convergence and population sizing in genetic algorithms. In William M. Spears and Worthy Martin, editors, *Proceedings of the Foundations of Genetic Algorithms Workshop (FOGA 6)*, Charlottesville, VA, USA, 21–23 July 2000. In press.

[Poli, 2000d] Riccardo Poli. Why the schema theorem is correct also in the presence of stochastic effects. In *Proceedings of the Congress on Evolutionary Computation (CEC 2000)*, pages 487–492, San Diego, USA, 6-9 July 2000.

[Poli, 2001a] Riccardo Poli. Exact schema theory for genetic programming and variable-length genetic algorithms with one-point crossover. *Genetic Programming and Evolvable Machines*, 2(2):123–163, June 2001.

[Poli, 2001b] Riccardo Poli. General schema theory for genetic programming with subtree-swapping crossover. In Julian F. Miller, Marco Tomassini, Pier Luca

Lanzi, Conor Ryan, Andrea G. B. Tettamanzi, and W. B. Langdon, editors, *Genetic Programming, Proceedings of EuroGP'2001*, volume 2038 of *LNCS*, pages 143–159, Lake Como, Italy, 18-20 April 2001. Springer-Verlag.

[Price, 1970] George R. Price. Selection and covariance. *Nature*, 227, August 1:520–521, 1970.

[Radcliffe, 1991] Nicholas J. Radcliffe. Forma analysis and random respectful recombination. In *Proceedings of the Fourth International Conference on Genetic Algorithms*, pages 222–229. Morgan Kaufmann, 1991.

[Radcliffe, 1994] Nicholas J. Radcliffe. The algebra of genetic algorithms. *Annals of Maths and Artificial Intelligence*, 10:339–384, 1994.

[Radcliffe, 1997] Nicholas J. Radcliffe. Schema processing. In Thomas Bäck, David B. Fogel, and Zbigniew Michalewicz, editors, *Handbook of Evolutionary Computation*, pages B2.5–1–10. Oxford University Press, 1997.

[Ratle and Sebag, 2000] Alain Ratle and Michele Sebag. Genetic programming and domain knowledge: Beyond the limitations of grammar-guided machine discovery. In Marc Schoenauer, Kalyanmoy Deb, Günter Rudolph, Xin Yao, Evelyne Lutton, Juan Julian Merelo, and Hans-Paul Schwefel, editors, *Parallel Problem Solving from Nature - PPSN VI 6th International Conference*, pages 211–220, Paris, France, 16-20 September 2000. Springer-Verlag. LNCS 1917.

[Ridley, 1993] Matt Ridley. *The Red Queen, Sex and the Evolution of Human Nature.* Penquin, 1993.

[Rosca and Ballard, 1996] Justinian P. Rosca and Dana H. Ballard. Discovery of subroutines in genetic programming. In Peter J. Angeline and Kenneth E. Kinnear, Jr., editors, *Advances in Genetic Programming 2*, chapter 9, pages 177–202. MIT Press, 1996.

[Rosca, 1995] Justinian P. Rosca. Entropy-driven adaptive representation. In Justinian P. Rosca, editor, *Proceedings of the Workshop on Genetic Programming: From Theory to Real-World Applications*, pages 23–32, Tahoe City, California, USA, 9 July 1995.

[Rosca, 1997a] Justinian P. Rosca. Analysis of complexity drift in genetic programming. In John R. Koza, Kalyanmoy Deb, Marco Dorigo, David B. Fogel, Max H. Garzon, Hitoshi Iba, and Rick L. Riolo, editors, *Genetic Programming 1997: Proceedings of the Second Annual Conference*, pages 286–294, Stanford University, CA, USA, 13-16 July 1997. Morgan Kaufmann.

[Rosca, 1997b] Justinian P. Rosca. Fitness-size interplay in genetic search. Position paper at the Workshop on Evolutionary Computation with Variable Size Representation at ICGA-97, 20 July 1997.

[Rosca, 1997c] Justinian P. Rosca. *Hierarchical Learning with Procedural Abstraction Mechanisms.* PhD thesis, Department of Computer Science, The College of Arts and Sciences, University of Rochester, Rochester, NY 14627, USA, February 1997.

[Rudolph, 1994] Günter Rudolph. Convergence analysis of canonical genetic algorithm. *IEEE Transactions on Neural Networks*, 5(1):96–101, 1994.

[Rudolph, 1997a] Günter Rudolph. Genetic algorithms. In Thomas Bäck, David B. Fogel, and Zbigniew Michalewicz, editors, *Handbook of Evolutionary Computation*, pages B2.4–20–27. Oxford University Press, 1997.

[Rudolph, 1997b] Günter Rudolph. Models of stochastic convergence. In Thomas Bäck, David B. Fogel, and Zbigniew Michalewicz, editors, *Handbook of Evolutionary Computation*, pages B2.3–1–3. Oxford University Press, 1997.

[Rudolph, 1997c] Günter Rudolph. Stochastic processes. In Thomas Bäck, David B. Fogel, and Zbigniew Michalewicz, editors, *Handbook of Evolutionary Computation*, pages B2.2–1–8. Oxford University Press, 1997.

[Russell and Norvig, 1995] S. J. Russell and P. Norvig. *Artificial Intelligence: A Modern Approach.* Prendice Hall, Englewood Cliffs, New Jersey, 1995.

[Ryan, 1994] Conor Ryan. Pygmies and civil servants. In Kenneth E. Kinnear, Jr., editor, *Advances in Genetic Programming*, chapter 11, pages 243–263. MIT Press, 1994.

[Sedgewick and Flajolet, 1996] Robert Sedgewick and Philippe Flajolet. *An Introduction to the Analysis of Algorithms.* Addison-Wesley, 1996.

[Smith and Harries, 1998] Peter W. H. Smith and Kim Harries. Code growth, explicitly defined introns, and alternative selection schemes. *Evolutionary Computation*, 6(4):339–360, Winter 1998.

[Soule and Foster, 1997] Terence Soule and James A. Foster. Code size and depth flows in genetic programming. In John R. Koza, Kalyanmoy Deb, Marco Dorigo, David B. Fogel, Max H. Garzon, Hitoshi Iba, and Rick L. Riolo, editors, *Genetic Programming 1997: Proceedings of the Second Annual Conference*, pages 313–320, Stanford University, CA, USA, 13-16 July 1997. Morgan Kaufmann.

[Soule and Foster, 1998a] Terence Soule and James A. Foster. Effects of code growth and parsimony pressure on populations in genetic programming. *Evolutionary Computation*, 6(4):293–309, Winter 1998.

[Soule and Foster, 1998b] Terence Soule and James A. Foster. Removal bias: a new cause of code growth in tree based evolutionary programming. In *1998 IEEE International Conference on Evolutionary Computation*, pages 781–186, Anchorage, Alaska, USA, 5-9 May 1998. IEEE Press.

[Soule *et al.*, 1996] Terence Soule, James A. Foster, and John Dickinson. Code growth in genetic programming. In John R. Koza, David E. Goldberg, David B. Fogel, and Rick L. Riolo, editors, *Genetic Programming 1996: Proceedings of the First Annual Conference*, pages 215–223, Stanford University, CA, USA, 28–31 July 1996. MIT Press.

[Soule, 1998] Terence Soule. *Code Growth in Genetic Programming.* PhD thesis, University of Idaho, Moscow, Idaho, USA, 15 May 1998.

[Spears, 2000] William M. Spears. *Evolutionary Algorithms: The role of Mutation and Recombination.* Natural Computing Series. Springer-Verlag, 2000.

[Spector and Alpern, 1994] Lee Spector and Adam Alpern. Criticism, culture, and the automatic generation of artworks. In *Proceedings of Twelfth National Conference on Artificial Intelligence*, pages 3–8, Seattle, Washington, USA, 1994. AAAI Press/MIT Press.

[Spector *et al.*, 1999] Lee Spector, W. B. Langdon, Una-May O'Reilly, and Peter J. Angeline, editors. *Advances in Genetic Programming 3.* MIT Press, 1999.

[Spector *et al.*, 2001] Lee Spector, Erik D. Goodman, Annie Wu, W. B. Langdon, Hans-Michael Voigt, Mitsuo Gen, Sandip Sen, Marco Dorigo, Shahram Pezeshk, Max H. Garzon, and Edmund Burke, editors. *Proceedings of the Genetic and Evolutionary Computation Conference, GECCO-2001*, San Francisco, California, USA, 7-11 July 2001. Morgan Kaufmann.

[Spiegel, 1975] Murray R. Spiegel. *Probability and Statistics.* McGraw-Hill, New York, 1975.

[Stephens and Vargas, 2000] C. R. Stephens and J. Mora Vargas. Effective fitness as an alternative paradigm for evolutionary computation I: General formalism. *Genetic Programming and Evolvable Machines*, 1(4):363–378, October 2000.

[Stephens and Vargas, 2001] C. R. Stephens and J. Mora Vargas. Effective fitness as an alternative paradigm for evolutionary computation II: Examples and applications. *Genetic Programming and Evolvable Machines*, 2(1):7–32, March 2001.

[Stephens and Waelbroeck, 1997] C. R. Stephens and H. Waelbroeck. Effective degrees of freedom in genetic algorithms and the block hypothesis. In Thomas

Bäck, editor, *Proceedings of the Seventh International Conference on Genetic Algorithms (ICGA97)*, pages 34–40, East Lansing, 1997. Morgan Kaufmann.

[Stephens and Waelbroeck, 1999] C. R. Stephens and H. Waelbroeck. Schemata evolution and building blocks. *Evolutionary Computation*, 7(2):109–124, 1999.

[Stephens, 1999] C. R. Stephens. "effective" fitness landscapes for evolutionary systems. In Peter J. Angeline, Zbyszek Michalewicz, Marc Schoenauer, Xin Yao, and Ali Zalzala, editors, *Proceedings of the Congress on Evolutionary Computation*, volume 1, pages 703–714, Mayflower Hotel, Washington D.C., USA, 6-9 July 1999. IEEE Press.

[Syswerda, 1989] Gilbert Syswerda. Uniform crossover in genetic algorithms. In J. David Schaffer, editor, *Proceedings of the third international conference on Genetic Algorithms*, pages 2–9, George Mason University, 4-7 June 1989. Morgan Kaufmann.

[Syswerda, 1991] Gilbert Syswerda. A study of reproduction in generational and steady state genetic algorithms. In Gregory J. E. Rawlings, editor, *Foundations of genetic algorithms*, pages 94–101. Indiana University, 15-18 July 1990, Morgan Kaufmann, 1991.

[Tackett, 1993] Walter Alden Tackett. Genetic programming for feature discovery and image discrimination. In Stephanie Forrest, editor, *Proceedings of the 5th International Conference on Genetic Algorithms, ICGA-93*, pages 303–309, University of Illinois at Urbana-Champaign, 17-21 July 1993. Morgan Kaufmann.

[Tackett, 1994] Walter Alden Tackett. *Recombination, Selection, and the Genetic Construction of Computer Programs.* PhD thesis, University of Southern California, Department of Electrical Engineering Systems, USA, 1994.

[Teller and Veloso, 1996] Astro Teller and Manuela Veloso. PADO: A new learning architecture for object recognition. In Katsushi Ikeuchi and Manuela Veloso, editors, *Symbolic Visual Learning*, pages 81–116. Oxford University Press, 1996.

[Teller, 1998] Astro Teller. *Algorithm Evolution with Internal Reinforcement for Signal Understanding.* PhD thesis, School of Computer Science, Carnegie Mellon University, Pittsburgh, USA, 5 December 1998.

[Tufts, 1996] Patrick Tufts. Genetic programming resources on the world-wide web. In Peter J. Angeline and Kenneth E. Kinnear, Jr., editors, *Advances in Genetic Programming 2*, chapter A, pages 499–506. MIT Press, 1996.

[Vose, 1999] Michael D. Vose. *The simple genetic algorithm: Foundations and theory.* MIT Press, 1999.

[Whigham and Crapper, 1999] Peter A. Whigham and Peter F. Crapper. Time series modelling using genetic programming: An application to rainfall-runoff models. In Lee Spector, W. B. Langdon, Una-May O'Reilly, and Peter J. Angeline, editors, *Advances in Genetic Programming 3*, chapter 5, pages 89–104. MIT Press, 1999.

[Whigham, 1995] Peter A. Whigham. A schema theorem for context-free grammars. In *1995 IEEE Conference on Evolutionary Computation*, volume 1, pages 178–181, Perth, Australia, 29 November - 1 December 1995. IEEE Press.

[Whigham, 1996a] Peter A. Whigham. Search bias, language bias, and genetic programming. In John R. Koza, David E. Goldberg, David B. Fogel, and Rick L. Riolo, editors, *Genetic Programming 1996: Proceedings of the First Annual Conference*, pages 230–237, Stanford University, CA, USA, 28–31 July 1996. MIT Press.

[Whigham, 1996b] Peter A. Whigham. *Grammatical Bias for Evolutionary Learning.* PhD thesis, School of Computer Science, University College, University of New South Wales, Australian Defence Force Academy, Canberra, Australia, 14 October 1996.

[Whitley *et al.*, 2000] L. Darrell Whitley, David E. Goldberg, Erick Cantu-Paz, Lee Spector, Ian Parmee, and Hans-Georg Beyer, editors. *Proceedings of the Genetic and Evolutionary Computation Conference (GECCO-2000)*, Las Vegas, Nevada, USA, 10-12 July 2000. Morgan Kaufmann.

[Whitley, 1993] L. Darrell Whitley. A genetic algorithm tutorial. Technical Report CS-93-103, Department of Computer Science, Colorado State University, August 1993.

[Whitley, 1994] L. Darrell Whitley. A Genetic Algorithm Tutorial. *Statistics and Computing*, 4:65–85, 1994.

[Whitley, 2000] L. Darrell Whitley, editor. *Late Breaking Papers at the 2000 Genetic and Evolutionary Computation Conference*, Las Vegas, Nevada, USA, 8 July 2000.

[Wolpert and Macready, 1997] David H. Wolpert and William G. Macready. No free lunch theorems for optimization. *IEEE Transactions on Evolutionary Computation*, 1(1):67–82, April 1997.

[Wong and Leung, 2000] Man Leung Wong and Kwong Sak Leung. *Data Mining Using Grammar Based Genetic Programming and Applications*, volume 3 of *Genetic Programming*. Kluwer Academic Publishers, 2000.

[Wright, 1931] Sewall Wright. Evolution in mendelian populations. *Genetics*, 16:97–159, March 1931.

[Wright, 1932] Sewall Wright. The roles of mutation, inbreeding, crossbreeding and selection in evolution. In D. F. Jones, editor, *Proceedings of the Sixth International Congress on Genetics*, volume 1, pages 356–366, 1932.

[Yu and Clack, 1998] Tina Yu and Chris Clack. Recursion, lambda abstractions and genetic programming. In John R. Koza, Wolfgang Banzhaf, Kumar Chellapilla, Kalyanmoy Deb, Marco Dorigo, David B. Fogel, Max H. Garzon, David E. Goldberg, Hitoshi Iba, and Rick Riolo, editors, *Genetic Programming 1998: Proceedings of the Third Annual Conference*, pages 422–431, University of Wisconsin, Madison, Wisconsin, USA, 22-25 July 1998. Morgan Kaufmann.

[Zhang and Mühlenbein, 1995] Byoung-Tak Zhang and Heinz Mühlenbein. Balancing accuracy and parsimony in genetic programming. *Evolutionary Computation*, 3(1):17–38, 1995.

[Zhang and Mühlenbein, 1996] Byoung-Tak Zhang and Heinz Mühlenbein. Adaptive fitness functions for dynamic growing/pruning of program trees. In Peter J. Angeline and Kenneth E. Kinnear, Jr., editors, *Advances in Genetic Programming 2*, chapter 12, pages 241–256. MIT Press, 1996.

[Zhang, 1997] Byoung-Tak Zhang. A taxonomy of control schemes for genetic code growth. Position paper at the Workshop on Evolutionary Computation with Variable Size Representation at ICGA-97, 20 July 1997.

[Zhang, 2000] Byoung-Tak Zhang. Bayesian methods for efficient genetic programming. *Genetic Programming and Evolvable Machines*, 1(3):217–242, July 2000.

List of Special Symbols

$\#$ — In a schema # is a "don't care" symbol. In bit string genetic algorithms # stands for exactly one bit (0 or 1).
In genetic programming trees # stands for any valid subtree located at this particular point in the tree.
In variable arity hyperschema the meaning of # depends upon whether it is used as an internal or external (leaf) code in the schema. If it is a leaf then # is any subtree. While if it is internal, # means exactly one function, whose arity is not smaller than the number of subtrees connected to it in the schema.

$!$ — Factorial. $N! = 1 \times 2 \times 3 \times \cdots \times N$.

$\binom{M}{k}$ — The number of ways of choosing k events from M.
$\binom{M}{k} = \frac{M!}{k!(M-k)!}$. Sometimes represented as C_k^M.

$\lceil \, \rceil$ — Ceiling function, which returns the smallest integer not smaller than its input.

$\lfloor \, \rfloor$ — Floor function, which returns the largest integer not bigger than its input.

$\sqrt{\ }$ — Square root.

$*$ — The "don't care" symbol for fixed length genetic algorithm schema. In bit string GAs * matches both `0` and `1` at exactly one location in the string (chromosome).
(In a few figures, * is used to represent multiplication as it is often used in computer programs).

$=$ — "Don't care" symbol representing a single function or terminal in fixed-size-and-shape genetic programming schema.

$\emptyset$ — The null or empty set. That is the set which contains no members.

$\alpha(H, t)$ — The chance of a child matching schema H at generation $t + 1$.

$B(H)$ — The event that the one-point crossover point lies between the defining (i.e. the non-$=$) nodes of schema H.

$C(h, \hat{h})$ — The set of crossover points in the common region between programs h and $\hat{h}$.

$\delta(x)$ — $\delta(x)$ is a function that returns 1 if x is true and 0 otherwise.

D0, ..., D_{n-1} — n leaves of program tree. Each corresponds to an input to the program.

$D_c(H)$ — The event "schema H is disrupted when a program h matching H is crossed over". That is, the child of h does not match H.

$D_{c_1}(H)$ — The first way that $D_c(H)$ can occur. That is, the two parents have different shapes (as well as schema H being disrupted).

$D_{c_2}(H)$ — The second way that $D_c(H)$ can occur. That is, the two parents have the same shape (however schema H is still disrupted).

ϵ — Some very small positive number. Used to indicate limiting behaviour.

$E[\,]$ — The expectation operator. $E[x]$ is the limit of the mean of x over a large number of independent trials. For example, suppose x is the number of children an individual has. Clearly $x = 0, 1, 2, \ldots$, so x is an integer but $E[x]$ need not be. For a particularly fit individual, $E[x]$ might be 1.3.

$\bar{f}(t)$ — The mean fitness of individuals in the population at time t.

$f(H,t)$ — The mean fitness of the individuals matching schema H in the population at time t.

H — A schema.

h — A program.

$\hat{h}$ — Another program (possibly different from h).

$i(H,t)$ — The number of times schema H matches individuals within the population at generation t. Note in general that H may match (be instantiated in) an individual multiple times. With bit string GAs and bit string schemata, only zero or one match between schema and individual is possible, so $i(H,t) = m(H,t)$. However, in general, $i(H,t) \geq m(H,t)$.

$I(M,i,z)$ — Total number of individuals that must be processed to yield a desired result by generation i with a probability z using a population of size M.

iff — If and only if.

l — Length of a program. In linear programs, the number of instructions executed. In tree programs, this equals the total size of the tree, i.e. the sum of the number of internal node (functions) and external nodes (leaves). $l \equiv N$.

$\mathcal{L}(H)$ — The defining length of schema H. In bit string genetic algorithms, the maximum distance between two defining (i.e. 0 or 1) symbols in schema H.

In a fixed size and shape tree GP schema, $\mathcal{L}(H)$ is the number of links in the minimum tree fragment including all the defining (i.e. non-$=$) symbols within schema H.

$L(H,i)$ The left or lower part of schema H. For a fixed length genetic algorithm, $L(H,i)$ is the schema obtained by replacing with "don't care" symbols ($*$) all the elements of H from position $i+1$ to the end of the string.
In trees, the hyperschema representing the lower part of schema H. $L(H,i)$ is made by replacing all the nodes on the path between position i and the root node with = nodes, and all the subtrees connected to those nodes with # nodes. Note that $L(H,i)$ is typically a smaller tree than H. See Figure 5.2 page 76.

$L(H,i,j)$ "Lower building blocks". $L(H,i,j)$ are defined on page 91 in the context of variable-arity hyperschemata. See also Figure 5.6 on page 94.

$l(H,i)$ The lower part of tree schema H. $l(H,i)$ is the schema obtained by replacing with "don't care" symbols (=) all the nodes of H above crossover point i. See Figure 5.1 page 75.

l.h.s. Left hand side.

M The population size.

$m(H,t)$ The number of individuals (e.g. programs) within the population at generation t that match the schema H.

N Length of a program or bit string chromosome.

$N(H)$ Length of schema H, i.e. the total number of nodes in H.

NB Nota Bene, note well.

$\mathbf{NC}(h,\hat{h})$ Number of nodes in the tree common region of program h and program $\hat{h}$

O() Big O notation. Used to indicate limiting behaviour is not worse than some bound. For example, $\mathrm{O}(x^2)$ means in the limit of large x it does not grow with increasing x bigger than some constant times x^2.

$\mathcal{O}(H)$ The order of schema H. For example, for a bit string GA, the number of 0s and 1s (i.e. non-# symbols) that the schema H contains. Similarly, for a GP schema, the number of non-don't care symbols in the schema.

$p_{\mathrm{diff}}(t)$ The probability that the child produced by one-point crossover between two programs h and $\hat{h}$ does not match schema H, given that h matches H and $\hat{h}$ is a different shape to H (and therefore does not match it). Symbolically, $p_{\mathrm{diff}}(t) = \Pr\{D_c(H)|\hat{h} \notin G(H)\}$.

$p(H, t)$ The probability of selection of the schema H. In fitness proportionate selection $p(H, t) = m(H, t)f(H, t)/(M\bar{f}(t))$.

$P_d(H, t)$ The expected proportion of children created by crossover alone (i.e. ignoring mutation) which do not match schema H produced from parents in the population at time t which did match schema H. This is know as *disruption* of schema H.

p_m The probability that a location in a chromosome (a locus) will be mutated (i.e. changed) when a child is created.
There is scope for confusion here as in some genetic programming literature p_m is defined as the fraction of children created by mutation and so is analogous to p_{xo}. In contrast, here (and in the genetic algorithm literature) p_m is the chance of mutation per location. To avoid confusion we use p'_m to indicate the fraction of children created by mutation.
Note that if p_m is constant, the chance of a change being made somewhere rises as the genetic material gets bigger (actually it is $1 - (1 - p_m)^N$).

p'_m The probability of applying mutation. Note that $p'_m \neq p_m$, since p_m is defined as the probability per location in the chromosome, while p'_m, like p_{xo}, is defined per child.

$\Pr\{x\}$ The probability of event x.

p_{xo} The probability of using crossover to create a new member of the population.

$R(H, i)$ The right part of schema H. For a fixed-length genetic algorithm, $R(H, i)$ is the schema obtained by replacing with "don't care" symbols (*) all the elements of H from position 1 to position i.

r.h.s. Right hand side.

T Number of individuals in a tournament selection group from which the fittest will be chosen to be a parent of a child in the next generation.

t Generation number.
Mostly we deal with evolution proceeding via distinct generations, each of which is numbered (i.e. $t = 0, 1, 2, \ldots$). However more continuous models in which populations overlap (often called "steady state") are also popular.

$U(H, i)$ The hyperschema representing the upper part of schema H. $U(H, i)$ is formed by replacing the subtree rooted at i with a # node. See Figure 5.2 page 76.
In fact "upper building blocks", in the context of variable-arity hyperschemata as defined on page 91 are the same as ordinary hyperschemata $U(H, i)$. See also Figure 5.5 on page 93.

$u(H, i)$	The upper part of tree schema H. $u(H, i)$ is the schema obtained by replacing with "don't care" symbols (=) all the nodes of H below crossover point i. See Figure 5.1 page 75.
vs.	Versus. Against.
w.r.t.	With respect to.
XOR	Binary Boolean function which performs exclusive-or. I.e. result is true iff its inputs are different from each other

Glossary

Arity Number of arguments that a function has. For example, + usually adds two numbers together. The numbers are its arguments. That is, + is a binary function and so has an arity of 2.

Binary trees In binary trees, each function (internal node) has exactly two inputs. So the number of functions = $\lfloor l/2 \rfloor$ and number of terminals = $\lceil l/2 \rceil$. Since l is odd, number of terminals = number of functions + 1.

Binomial distribution The probability of exactly k independent events each with a probability p in M trials is $\binom{M}{k} p^k (1-p)^{M-k}$, were $\binom{M}{k} = M!/(k!(M-k)!)$. The mean of the Binomial distribution is pM and its variance is $p(1-p)M$.

Boolean Dealing only with the two logical values: true (1) and false (0). Named after the mathematician George Boole.

Building block hypothesis [Goldberg, 1989b, page 41] talking of binary bit string genetic algorithms says "Short, low order, and highly fit schemata are sampled, recombined" (crossed over), "and resampled to form strings of potentially higher fitness. In a way, by working with these particular schemata (the building blocks), we have reduced the complexity of our problem; instead of building high-performance strings by trying every conceivable combination, we construct better and better strings from the best partial solutions of past samplings." "Just as a child creates magnificent fortresses through the arrangement of simple blocks of wood" (building blocks), "so does a genetic algorithm seek near optimal performance through the juxtaposition of short, low-order, high-performance schemata, or building blocks." Note [Goldberg, 1989b] suggests that building blocks are highly fit schemata with only a few defined bits (low order), and that these are close together (short).

Convergence Precisely every individual in the population is identical. Such convergence is seldom seen in genetic programming using Koza's subtree swapping crossover. However, populations often stabilise after a time, in the sense that the best programs all have a common ancestor and their behaviour is very similar (or identical) both to each other and to that of high fitness programs from the previous (and future?) generations. Often the term "convergence" is loosely used.

Deception Deception is where the gradient in the fitness landscape leads away from locations with the best possible fitness. This is because the genetic algorithm, genetic programming or other search technique, is "deceived" into being attracted towards local peaks rather than the global best-fitness location.

Defining Length $\mathcal{L}(H)$ The maximum distance between two defining (i.e. non-#) symbols in schema H. In tree GP schemata, $\mathcal{L}(H)$ is the number of links in the minimum tree fragment including all the non-= symbols within a schema H

Diploid A species is diploid if its genetic chromosomes are paired. (In general, each cell has multiple pairs of chromosomes). A species is haploid if its chromosomes are not paired. In diploid species, the chromosome pairs separate during sexual reproduction. When they come together again in the child, one chromosome in each pair comes from the mother and the other from the father.

Disruption If the child of an individual that matches schema H does not itself match H, the schema is said to have been disrupted.

Diversity Variation between individuals in the population. Typically diversity refers to genetic variation. In bit string genetic algorithms, diversity may be measured by Hamming distance (i.e. counting the number of bits that are different) between bit strings. [Koza, 1992] defines "variety" in the population by the number of unique programs it contains, but this measure takes no notice of the fact that the behaviour of genetically different programs can be very similar or even identical.

Dominance When one gene overrides the effect of another. Typically this refers to genes at the same locus of a pair of chromosomes (see diploid). A potential "use" is one gene acts as a backup for the other, taking on its function should the first one be damaged (e.g. by a mutation).

Effective fitness The effective fitness of a schema is its "fitness" to take into account crossover and mutation. This can be thought of as the fitness that the schema would need to have to grow/shrink as it does with crossover and mutation if they were not there. $f_{\text{eff}}(H,t) = \frac{\alpha(H,t)}{\alpha(H,t-1)} \times \bar{f}(t)$

Eigenvector An eigenvector v of a square matrix M has the property that multiplying M by it yields another vector which is parallel to v. That is, $vM = \lambda v$ where λ is a (possibly) complex number known as the eigenvalue of v.

Enumeration A search in which all possible solutions are tried in sequence.

Epistasis Non-linear interaction between genes.

Genetic drift Genes are inherently digital, in that an individual has an integer number of copies of each gene, and its children inherit an integer number of them. That is, an individual cannot have half a gene. While fitness may play a part in how many children inherit how many copies of a gene, in each individual child there is a large element of chance. Genetic drift is where random (i.e. not related to fitness) fluctuations in the genetic material in the population occur and lead to macroscopic changes to the population. Naturally genetic drift is more important in smaller rather than larger populations.

Genetic drift As a crude example, consider a rare gene, say it occurs in 1% of the population but has no effect on fitness. So we expect it to occur in 1% of the next generation, and 1% of the next and so on. However, what happens if there are only 100 individuals in the population? In the current population there is one individual with the gene. In the next population (for simplicity also of size 100) we expect on average there to be one individual with the gene. However, it is quite likely there will be two or more, or that there will be none. Obviously if there are none, then there can be none in the next generation. While if there are two then on average the third population will also have twice as many as in the initial population. So, even though this gene is unrelated to fitness, its concentration is liable to drift with time. The number of generations for significant change to occur randomly is $O(|\text{population}|^2)$.

Genotype The complete list of genes contained within an individual. In general, the position of the genes, as well as which ones are present, may be important.

Haploid Unlike, diploid species, the chromosomes are not paired (although each cell may have more than one chromosome).

Hits A program scores a "hit" when its output is sufficiently close or equal to the target value for a particular test input. In some problems a program's fitness may be given by the number of hits it collects when run on the entire test set.

Instantiate An instance of. For example, an individual program may match a similarity template, such as a schema, one or more times. Each match is an instantiation (instance) of the schema.

Leaf Outermost part of program tree. In contrast to functions or internal nodes, leaves have no arguments. Also called a terminal. In many cases in GP leaves are the inputs to the program.

Length of a schema The total number of nodes in the schema is called the *length* $N(H)$ of a schema H. $N(H)$ is also equal to the number of nodes in the programs matching H.

Linear tree A program composed of a variable number of unary functions and a single terminal. Note linear tree GP differs from bit string genetic algorithms since a population may contain programs of different lengths and there may be more than two types functions or more than two types of terminals.

Local peaks These are locations in the fitness landscape where the fitness is below the best possible, and where every movement away from them leads to other locations all of whom have lower fitness.

Loci Plural of locus.

Locus A specific point of the chromosome. In bit string genetic algorithms it is a particular bit. More generally, it is a specific location in the genetic material that can take on one of a number of values, known as alleles. Commonly in genetic algorithms a locus controls a specific parameter, but many-to-many mappings between gene loci and parameters are possible (and believed to be prevalent in real life).

Markov process A discrete random process where the chance of moving to another state depends only on the current state of the process and not on any earlier history.

Mating pool While we mainly treat the population as a whole, an equivalent approach is to separate from it and place into a mating pool those individuals that will have children. Fitness selection is performed separately before genetic operations. Children are created from parents chosen only from the mating pool.

Monte Carlo In a Monte Carlo search, points in the search space are sampled randomly. If sufficient independent points are sampled, a reasonable estimate of the whole search space can be made.

Mutation In Nature, a mutation is viewed as a mistake when DNA is copied. E.g. when a cell divides to create two cells. Such an error introduces a random change to the DNA. Such changes are usually considered to be harmful, and Nature has elaborate

error-detection techniques which considerably reduce the rate at which errors are introduced. Note if the cell with the error survives then all the cells it produces will also have the error. If the cell is a germ cell, i.e. used to create new individuals, then the children produced using it will inherit the error.
In artificial evolution, e.g. GAs and GP, mutation is used to mean any inherited random change to the genes. However, mutation is not used to describe crossover, i.e. the process whereby children are created from a combination of genes from two (or more) parents in which genes are copied correctly. In GAs it is common to use mutation (i.e. random changes) in combination with crossover. In traditional genetic algorithms mutation means choosing at random bits in the chromosome and flipping them. Each bit is considered independently and a decision is made with low probability p_m if it is to be changed or not. Note the number of bits changed is variable, and lies between 0 and N (where N is the number of bits in the chromosome) but is on average $p_m \times N$. It is commonly recommended to set p_m to about $\frac{1}{N}$ [Bäck, 1996].
In evolutionary programming, evolutionary strategies and real-valued genetic algorithms, it is common to apply mutation to every gene by adding a small random value to the gene. In genetic programming mutation is becoming increasingly often used. However, there are many different types of random changes that can be made to programs (see [Langdon, 1998c, pages 34–36]). Also, some authors recommend using a number of different types of mutations in combination with each other [Angeline, 1998a]. One unfortunate source of confusion is p_m. In GAs it means the chance of changing each gene per generation. In some cases in GP, p_m is used to mean the chance that a child will be produced using mutation, rather than crossover. That is, treating p_m as analogous to p_{xo}. Here we use p'_m to denote this second meaning.

Order $\mathcal{O}(H)$ The number of defining symbols in schema H.

Parity problems Benchmark problems widely used in GP but inherited from the artificial neural network community. Parity is calculated by summing all the binary inputs and reporting if the sum is odd or even. This is considered difficult because: (1) a very simple artificial neural network cannot solve it and (2) all inputs need to be considered and a change to any one of them changes the answer.

Polyploid A species with more than two chromosomes grouped together. See diploid and haploid.

Premature convergence This loosely means that something has gone wrong. More precisely, that the population has converged (every individual in the population is identical, see convergence) to a suboptimal solution. Often the term "premature convergence" is loosely used.

Propagation The inheritance of characteristics of one generation by the next. For example, a schema is propagated if individuals in the current generation match it and so do those in the next generation. Those in the next generation may be (but don't have to be) children of parents who matched it.

Recombination This is creating children by combining the genetic material from two (or more) parents. Effectively it is another name for crossover.

Roulette-wheel selection The archetypal selection scheme for genetic algorithms. We imagine a biased roulette wheel where each individual in the population has its own slot and the width of its slot is proportional to its fitness. That is, above-average individuals have wider-than-average slots. Individuals are selected to be parents of children in the next generation independently, one at a time, by spinning the wheel and dropping a ball into it. The chance of drawing a particular individual is proportional to the width of its slot, i.e. proportional to its fitness. However, others selection techniques, such as stochastic universal sampling [Bäck, 1996, page 120] or tournament selection, are often used in practice. This is because they have less stochastic noise, or are fast, easy to implement and have a constant selection pressure [Blickle, 1996b].

Schema A set of programs or bit strings that have some genotypic similarity. Usually the set is specified by defining a similarity template which members of the set must match. The template specifies the fixed part, which the programs must match, and the variable part. Don't care symbols are used to define the variable part. In tree schemata, both the content and the shape of the tree must be considered.

Schemata Plural of schema.

Stochastic matrix A matrix whose elements lie between 0 and 1 and where the sum of all the elements in each row is 1. A doubly stochastic matrix has the additional property that its transpose (i.e. the matrix whose i, j element is equal to the j, i element of the original matrix) is also a stochastic matrix.

Terminal Another name for leaf.

Unary function A function that takes one argument.

Wild-card symbol This is also known as a "don't care" symbol. These are =, * and #. They indicate how a schema can match an actual program (or bit string) in the population.

Index

Printing (Computer to Film): Saladruck Berlin
Binding: Stürtz AG, Würzburg